*Aims and Scope*
Optimization has been expanding in all directions at an astonishing rate during the last few decades. New algorithmic and theoretical techniques have been developed, the diffusion into other disciplines has proceeded at a rapid pace, and our knowledge of all aspects of the field has grown even more profound. At the same time, one of the most striking trends in optimization is the constantly increasing emphasis on the interdisciplinary nature of the field. Optimization has been a basic tool in all areas of applied mathematics, engineering, medicine, economics, and other sciences.

The series *Springer Optimization and Its Applications* publishes undergraduate and graduate textbooks, monographs and state-of-the-art expository work that focus on algorithms for solving optimization problems and also study applications involving such problems. Some of the topics covered include nonlinear optimization (convex and nonconvex), network flow problems, stochastic optimization, optimal control, discrete optimization, multi-objective programming, description of software packages, approximation techniques and heuristic approaches.

More information about this series at http://www.springer.com/series/7393

Ivan V. Sergienko • Mikhail Mikhalevich
Ludmilla Koshlai

# Optimization Models in a Transition Economy

Springer

Ivan V. Sergienko
Mikhail Mikhalevich
Ludmilla Koshlai

V.M. Glushkov Institute
of Cybernetics
National Academy of
Sciences of Ukraine
Kiev, Ukraine

ISSN 1931-6828 ISSN 1931-6836 (electronic)
ISBN 978-1-4899-7888-2 ISBN 978-1-4899-7544-7 (eBook)
DOI 10.1007/978-1-4899-7544-7
Springer New York Heidelberg Dordrecht London


Softcover reprint of the hardcover 1st edition 2014


Printed on acid-free paper

Springer is part of Springer Science+Business Media (www.springer.com)

# Preface

Every day before leaving the comfort of our home we wonder what the weather is like. The weather forecast helps people feel more comfortable and prepared when heading outside. Numerous media sources report a wide range of weather characteristics such as precipitation, wind strength, direction, speed, temperature, humidity, atmospheric pressure, cloudiness, visibility, atmospheric phenomena that include fog, blizzards, storms, and other meteorological occurrences.

The weather undergoes continuous changes that can be very noticeable from one day to the next and sometimes even from one minute to the next. Depending on the forecast a person chooses protective clothing and items such as raincoats, umbrellas, and glasses.

For transportation on land, water, and in the air meteorological factors such as precipitation, ice, wind strength, and direction are crucial. In aviation for instance, it would be important to note a sharp increase in wind and turbulence.

The weather has always existed, and modern man still has almost no control over the weather. Unlike the weather, economy was created by man. The word "economy" itself was introduced by a great scholar of ancient times, Aristotle. It was formed by combining two words: "oikos"—"household" and "nomos," from "nemein"—"management." Modern scholars have given the following interpretation of this concept: the economy is a life support system that is consciously built and utilized by people in all spheres of human activities.

Many people take part in daily economic activities, live in a shared economic environment and sometimes use economic terms such as "money," "prices," "inflation," "wages," "taxes," "income," "costs," "volumes of production," "exchange rate," "unemployment," or "the market."

Economic information reveals the pros and cons of different economic systems and public choice options, the difficult choice between equality and efficiency, electoral process, activities of firms and industrial organization, marketing and management, labor market, capital and money markets, family economics, consumer behavior, and national and global economy.

The fall of the Berlin Wall, the collapse of the communist empire in Eastern Europe in the early 1990s led to significant changes in the economies of the newly formed countries. It also created new interactions and greater influence between the "open market" and the "transition" economies.

This book is written on the basis of the experience of the authors and their colleagues in mathematical modeling of many transition economies' characteristics. The main conclusion reached by this book's authors is that the concept of a transition economy opened many avenues for further studies of both open and transition economies. Mathematical modeling of weather phenomena merely allows us to forecast certain essential weather parameters without any possibility of changing them. In contrast, modeling of transition economies gives us the freedom to not only predict changes in important indexes of all economy types but also influence them more effectively in the desired direction. In other words, any economy including a transitional one can be controlled by humans.

This book is useful to anyone who wants to increase profits within their business and improve the quality of their family life and the economic area they live in. Professional economists will greatly benefit from new interdependencies between the well-known micro- and macroeconomic parameter behaviors. Moreover, they will appreciate the new possibilities in steering and changing both transitional and open market economies. Mathematicians will be introduced to new mathematical models as well as efficient methods of finding either feasible or optimal solutions for those models. Employees of state planning and statistical organizations will be able to use the proposed models and solutions for improving existing information technologies and operational abilities of their departments. This book is also beneficial for undergraduate and graduate students specializing in the fields of Economic Informatics, Economic Cybernetics, Applied Mathematics, and Large Information Systems.

The roots of this book arose from pioneering publications, many organizational efforts, and the critical view of our colleague and former director of Glushkov Institute of Cybernetics, Vladimir S. Mikhalevich. Many of our colleagues worldwide have read, discussed, and commented on some of the book's ideas as well as the book in its entirety. We express our deepest gratitude to Yu.M. Ermol'ev and N.Z. Shor for laying the foundation that allowed the final version of this book to be fully realized.

The financial support from the Glushkov Institute of Cybernetics, National Academy of Sciences of Ukraine is greatly appreciated.

Kiev, Ukraine Ivan Sergienko
May 2014

# Contents

# Chapter 1
# Financial Stabilization Models

## 1.1 Financial Stabilization: The Prime Goal of Market Reforms

Financial crisis in the form of malfunction in money turnover had become one of the first serious problems faced by the post-communist countries in their transition from centralized planned economy to an open-market economy. The most dangerous and destructive forms of this malfunction at that time were the inflation and non-payment crisis. The banking sector had been getting huge rents from inflation, mainly because of the negative real interest rates paid to depositors. The financial sector, for example in Russia, received about 8 % of the Gross Domestic Product (GDP) in 1992 through arbitrage exploiting the negative real interest rates paid to depositors. Moreover, with the so-called stabilization policies provided to banks there exist other forms of profit opportunities via the GKO-OFZs (in Russian: **G**osudarstvennoye **K**ratkosrochnoye **O**byazatyelstvo-**O**bligatsyi **F**ederal'novo **Z**aima), i.e., Government Short-Term Commitments-Federal Loan Obligations, which are government bonds issued by the state of Russia. Taking into account that access to auctions was restricted to preselected dealers preventing form participation the foreign investors and ordinary Russians, the GKO-OFZs made up for the loss of rents from reduced inflation and gave the banks a stake in fighting inflation to keep the GKO-OFZ returns high (see Ronald [30]).

In early 1990s of the twentieth century the inflation was presented by a significant raise of prices and accompanied by the decrease of money purchasing power which has been typical almost for all states of Central and Eastern Europe under reforms according to the "big bang" approach. However, would be incorrect to consider that this inflation was provoked just by implementation of the "shock therapy." Intensive inflation was experienced also by several countries in Asia treated by means of the gradual approach to reforms, for instance, in Vietnam (see Dang Thi Hieu Li [8]). Several periods of price instability (although not so intense as in the



I.V. Sergienko et al., *Optimization Models in a Transition Economy*,
Springer Optimization and Its Applications 101, DOI 10.1007/978-1-4899-7544-7_1

post-communist European countries) took place in China in 1980s–1990s of the twentieth century. Any analysis of financial sector conditions prior to reforms may lead to the conclusion about inevitability of the inflation increase at the first stage of reforms' implementation.

Several hardly connected monetary contours existed in planned economy. They include cash circulation, which serves for the consumer goods market with fixed, centrally defined prices, still commodity turnover also depended on demand, which was not completely controlled by the state. Also there was cashless turnover, used by the state enterprises. In this contour were planned not only prices, but also the volumes and structure of commodities turnovers. This way prices and flows of money played secondary role comparing with commodity flows (those were primary target of planning), becoming an instrument of value measurement of differentiated goods with the goal of their accounting. Special cashless money (image of hard currency) was used in the sphere of foreign trade. Considering the state monopoly on any foreign trade neither private persons nor state enterprises inside the country could use them. Payments among the USSR and European socialist countries were made using clearing pseudo money which could not be converted into any hard currency. There were also cash surrogates of hard currency ("Vneshposiltorg" checks, bones for sailors and military men serving abroad, etc.) in turnover.

Considering the weak connections among the above-mentioned contours, monetary policy of planned economy was held independently in each of them. This policy did not influence the prices (which were set on basis of centralized decisions and, often, spontaneously) and weakly influenced volumes of commodity turnover (regarding cashless money such influence was absent at all). As a result, changes in money supply did not play any regulating role; they have ensured earlier planned volumes of production and sales and were strictly tied to current fixed prices. Transition to "free" (market) prices unavoidably had lead to deep disproportions between mass of goods and amount of money. Therefore, disorganization of money turnover (including inflation) was inevitable at the early stage of market reforms.

It should be noted that the significant inflation potential was accumulated in former socialist states, especially in the USSR, during the last decade prior to reforms. Abrupt raise of government expenses (first of all for military purposes), attempts to slow down technological lag by significant increase of investments in R/D projects and scientific researches (usually without any rationalization of their structure) and a number of other factors contributed to this pre-chaotic non-controlled state. Especially strong impact on the USSR finances was made by the drop of world prices on oil in the middle of 1980s and related to this shortage of state budget incomes. The problems, appeared with money supply, usually were solved by a monetary emission. This approach especially often was used in the sphere of cashless turnover. In the planned economy, as was mentioned above, the growth of nominal "internal" cashless money supply could influence neither prices nor volume of the commodity turnover. However with the transition to market economy, the unification of cash and cashless money turnover is necessary, as a result, an earlier formed inflation potential immediately was opened up.

Of course, the above-mentioned aspects were taken into consideration when the market reforms were planned. Still it was widely considered opinion that the pre-reforms disproportions will be quickly removed by the market self-regulatory mechanisms, the inflation will not be lasting and intensive. This viewpoint was based on a belief that these disproportions are the only reason of inflation and partially appealed to financial stabilization experience in the countries with developed market economy (see, for instance, [39]).

In reality, almost in all post-communist countries of Europe and the former USSR inflation was not only more intensive and more lasting than expected; it has also been more destructive, while its forms and consequences differed from those which be leaded by the prior reasons. In all these countries there was more or less expressive stagflation—combination of inflation with a drop in production and decline of employment. These processes could not be explained only as results of previously made errors in monetary policy.

Economic theory carries out two forms of inflation—demand inflation, caused by increase of consumers' incomes, and cost inflation, caused by raise of prime cost of the produced goods. In the well-known AD/AS model primary reason for the inflation of the first type is the up-and-right shift of the demand curve AD, and down-and-right shift of the supply curve AS is the reason for the second form of inflation (Fig. 1.1).

Here $Q$ is the commodity turnover, $p$ with different super indices are prices, $(p^0, Q^0)$ is the initial point of the price and turnover, $q^0$—the initial primary cost of production, $AD$, $AS$—initial position of demand and supply curves, $AD'$—position of demand curve when demand inflation of is started, $AS'$—position of supply curve when costs inflation arises, $q^1$—new primary prices of production in this case, $(p^1, Q^1)$, $(p^2, Q^2)$—new equilibrium states for the cases of demand and cost inflation.

As you can see in Fig. 1.1, the demand inflation leads to an increase of commodity turnover as its short-time consequence, while cost inflation immediately leads to a decrease of the turnover and to a recession.

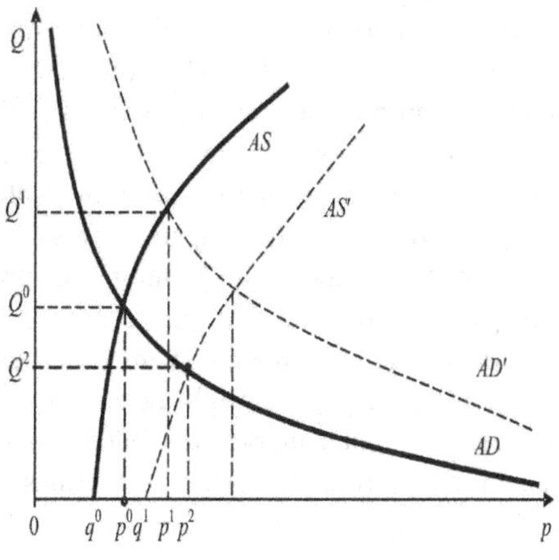


**Fig. 1.1** AS/AD model

It should be noted that this classification of inflation is rather symbolic. For example, wage growth in the real sector causes both raise of the customers' demand (as result of their incomes increase) and raise of the primary production cost (as a result of labor expenses increase). Nevertheless, the conclusion that the above-mentioned disproportions in the currency turnover should cause first of all the demand inflation is very persuasive. In the reality, the inflation processes, especially at the late stages, were more similar to the cost inflation. Therefore, the reasons of inflation in the transition economy within the former communist states located in parts of Europe and Asia, were not connected only with the errors in monetary policy; they go far beyond the sphere of currency turnover.

Now let us highlight some factors that differentiate the transition economy from the market economy with the purpose to impact on the cost inflation process weightily. This includes first of all a high level of monopolization and spreading of other forms of imperfect competition such as oligopolies and double-sided monopolistic competition [33]. The excessive concentration and production specialization were encouraged in the centralized economy as factors, which simplified planning. This was provoking monopolies that continued to exist even after diminishing of large branch structures, such as industrial Ministries in the former Soviet Union. All sizes of diminishing were unable to solve this problem completely. First of all, there are technology cycle limits to diminishing. The huge chemical and metallurgical plants could not be divided into large number of small enterprises. Second, new owners, usually coming from the previous economic elite, are tied by old corporate relations, depending on the same financial institutions, common infrastructures, etc. They used these ties and infrastructures for extra profit receiving and for pressure at "out-of-system" competitors. Therefore, the monopolism in the post-communist states transition economy is much more spread, deep and diverse than in a stable market economy.

It is well known that the inflation processes under the conditions of imperfect competition may be principally different comparing to the competitive market. In particular, sudden raise of prices might be a reaction of monopolist on either increase or decrease of demand (example of this phenomena is presented in Chap. 2). Consequently, inflation in a highly monopolistic economy may not be followed by an adjusted increase of nominal money supply.

The imperfection of production technologies, their high energy, materials and labor consumption per capita is another factor that influences on the inflation in transition economy. The paternalistic relations between the state and former state enterprises, which turn into form of "soft budget constrains" for the enterprises (see [19]), reinforces the impact of this factor. In situation, when losses of the non-state (but the former state!) enterprises are covered by the state budget, their top management does not experience a need to change the existing technologies and to improve the used management. The huge state expenses, related to bringing to life such a policy, also might be an important reason of inflation, because having undeveloped stock market the budget deficit in many cases was financed in the transition countries through money emission.

Some more important factors of inflation are the old style behavior of the state officials and enterprise managers, the absence of their market experience, and usage of non-rational strategies conducting business. Especially weighty these factors were at the early stage of market reforms.

In the development of inflation processes in the transition economy states an important role belongs to some external factors, such as the discrepancy of local prices to the prices on the world market, instability of national currency (last factor was especially significant for the countries, which earned their independence at the beginning of 1990s and did not have enough reserves in the foreign currencies), monopolism in relations among countries and others.

Hence, the inflation in countries with the transition economy is a complicated multiaspected process that goes beyond the monetary turnover. Its analysis requires an implementation of various methods and approaches; the economic-mathematical modeling is only one of them. Starting from the middle of the twentieth century, a number of fundamental investigations of inflation processes using quantitative methods were conducted. Among those should be named works by Cagan [4], Gandolfo [17]. The models of enterprises' behavior, build on a base of adaptive and rational expectations theories, were created and analyzed in work of Sargent and Wallance [31]. The experience of an anti inflation policy implementation was generalized in publications of Straffa [32], Hahn [11], and Frenkel [10]. Nonetheless, due to the mentioned features of inflation in the transition economy, the received results for the market economy are required serious analysis and adjustment to be done.

Currently a number of mathematical models on this subject are developed; some of those will be described in following sections of this chapter. Our special attention was paid, besides authors' investigations conducted in Cybernetics Institute of NAS of Ukraine, to the results of other two scientific schools. The first of those schools was shaped on the base of Central Economic-Mathematical Institute (CEMI) Russian Academy of Sciences (RAS), the second one—on the base of Systems Research Institute of Polish Academy of Sciences. Our comparison of the obtained results allows to conclude regardless difference in approaches and specifics of the countries as follows: the results are similar and in many points complementing each other. The same conclusion applies to the model developed in Bulgaria and presented in Sect. 1.3.

The models of demand inflation that describes inflation processes at the early stage of reforms are considered in Sect. 1.2. The key goal of these models is to evaluate consequences of price liberalization under conditions of misbalance between supply and demand on commodity goods market. We assume that this misbalance is caused by monetary disproportions of the prior to reforms period. In the same section there are models analyzing the influence of monopolism on prices with an existence of demand inflation and choice of actions for the state in the sphere of credit-bank and fiscal policies, which could reduce such an influence.

The models of costs inflation taking into account the imperfect competition factors are presented in Sect. 1.3. Their analysis allows to define the mechanisms of this process which lay both inside and outside of the monetary policy sphere.

The analysis of impact of inflationary expectations in the transition economy is done in Sect. 1.4. In this Section the models of another negative process in the sphere of money turnover, called as payments crisis, are viewed. The roots of this phenomenon also tracking back to the pre-reform period. Under the conditions of hegemony of the state property the payment discipline was not viewed as an unavoidable condition of cooperation between enterprises and, especially, between enterprises and the state itself. Every 5–10 years large-scale write off of all debts of agricultural companies took place starting from middle of the 1950s in the former USSR, up to one third of the industrial enterprises were considered unprofitable and received subsidies from the government then. Naturally, within outlined conditions, top managers were not interested in economy of money resources of their companies. The real inflation and inflation expectations together with development of "shadow" economy have intensified all kinds of non-payments. The models, examined in Sect. 1.4, allow to formulate our recommendations on monetary, banking, fiscal and industrial policy in order to reduce negative consequences within the payment crisis.

## 1.2 Two Macromodels Under Inflation Conditions

We studied two macromodels: a dynamic model of supply-and-demand disbalance and a model of monopolist–producer behavior under conditions of inflation. Both models are based on assumptions of a pricing mechanism in which the rate of price change is proportional to the difference between supply and demand. As has been noted in [27], this assumption is a fairly good approximation of the market-pricing mechanism.

We shall examine a dynamic macromodel with continuous time that describes an economy in terms of the most aggregated indices: the GDP, national income, price index (price level), savings, etc. We shall start with the assumption that the price level corresponds to the discrepancy between the actual solvent demand and supply. Thus, for the given model, the demand and prices are not exogenous parameters, unlike in the model from Petrov et al. [28], while the relationships that reflect the dynamics of GDP and money supply are similar to those in that model.

Let $x(t)$ be a gross product at time $t$ measured in stable (comparable) prices; then $x(t) = ax(t) + y(t), 0 < a < 1$, where $a$ is the share of intermediate consumption in the value of the gross product and $y(t)$ is the net product. It is assumed that the last goes to consumption $R(t)$ and savings $I(t)$. Let $w$ be a percentage of savings in the net product; then $y$, $R$, and $I$ are related by the following equalities

$$y(t) = R(t) + I(t), \quad I(t) = wy(t), \quad I(t) = b_1 \frac{dx}{dt}, \quad 0 \le w \le 1.$$

Hence, we obtain the equations: $R(t) = (1 - w)y(t)$ for the supply of goods; $dx\,l\,dt = (1 - a)wx(t)/b_1, \ 0 \le b_1 \le 1$, where $b_1$ is the stock capacity (the reciprocal of the effectiveness of capital expenditures, for the increase in gross product; and $dy/dt = w(l - a)y(t)/b_1$ for the growth of net product.

Let the change in the money supply $D(t)$ corresponds to the difference between the consumers' incomes and their expenditures. It is assumed that incomes are proportional (with a coefficient $q$) to newly produced product, i.e., the net product measuring in current prices. Then we obtain

$$dD/dt = qp(t)y(t) - p(t)\min\left\{S(t), R(t)\right\}, \quad 0 \le q \le 1,$$

where $p(t)$ is the price level (the ratio of current prices to a stable level assumed as a unit) and $S(t)$ is the consumers' demand at the time instant $t$. The latter is determined by the formula $S(t) = D(t)/p(t)$.

The price level varies with the discrepancy between supply and demand as follows:

$$\frac{dp}{dt} = m\left[S(t) - R(t)\right], \quad 0 \le m \le 1, \tag{1.1}$$

where $m$ is a parameter for price-level adjustment. The formula (1.1) is an analogue of Samuelson's equation (see [27]). Thus, the model has the form

$$\begin{aligned} p' &= m\left[S - R\right], \\ S' &= \frac{qR}{1-w} - \min\left\{S, R\right\} - m\frac{S(S-R)}{p}, \\ R' &= \frac{wR}{b}, \quad b = \frac{b_1}{1-a}, \end{aligned} \tag{1.2}$$

$$p(t) \ge 0, \quad S(t) \ge 0, \quad R(t) \ge 0, \quad 0 \le t \le T, \quad 0 \le a, \quad q,\ m,\ w \le 1.$$

The system of Eq. (1.2) is nonlinear with a nonsmooth right side of the second equation. To analyze this system, we turn to its phase portrait. It is presented at the plane $SOR$ in Fig. 1.2. We draw on the phase plane a line $S = R$ and call it $\Gamma$, which divides the phase plane into two domains. Then the qualitative behavior of both sides of the system will be analyzed. The qualitative study begins with the setting of fixed (stationary) points, which are determined by the equalities

$$\begin{gathered} m\left[S - R\right] = 0, \\ \frac{qR}{1-w} - \min\left\{S, R\right\} - m\frac{S(S-R)}{p} = 0, \\ \frac{wR}{b} = 0. \end{gathered} \tag{1.3}$$

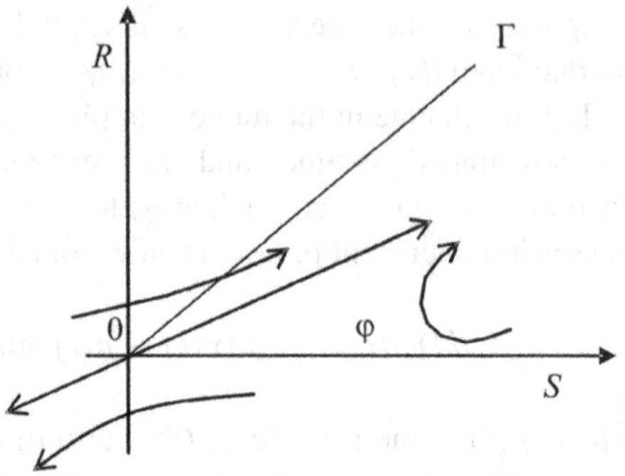


**Fig. 1.2** The phase portrait of the system (1.2)

For the case of $w \neq 0$, the values $S = 0$, $R = 0$, and $p \neq 0$ serve as a solution of system (1.3). We investigate the stability of the stationary state in a linear approximation in each of the regions.

1. $S < R$. The matrix $A = \left[\partial f_i / \partial x_j\right]$, $i, j = \overline{1, 3}$ of the linearized system has the form

$$A = \begin{bmatrix} 0 & m & -m \\ mS(S-R) & 1 - \dfrac{2mS - mR}{\overline{p}} & \dfrac{q}{1-w} + \dfrac{mS}{\overline{p}} \\ 0 & 0 & w/b \end{bmatrix}_{(p,S,R)=(\overline{p},0,0)}$$

$$= \begin{bmatrix} 0 & m & -m \\ 0 & -1 & \dfrac{q}{1-w} \\ 0 & 0 & w/b \end{bmatrix}.$$

Matrix $A$ has a rank of 2, and its structure allows the equation for $p$ to be separated from the equations for $S$ and $R$. We examine projections of the trajectories of the linearized system onto the plane $p = \text{const}$:

$$A' = \begin{pmatrix} -1 q/(1-w) \\ 0 w/b \end{pmatrix}.$$

The eigenvalues of matrix $A'$ are $u = -1$ and $u = w/b$. The stationary (saddle) point is unstable. The phase portrait is shown in Fig. 1.2. The slope of line $L$ is determined by formula

$$tg\varphi = k = \frac{(w / b)k}{-1 + \dfrac{kq}{1-w}}; \quad k = 0,$$

$$k = \frac{1-w}{q}\left(1 - \frac{w}{b}\right)\begin{cases} < 1, \\ = 1, \\ > 1, \end{cases} \left(\frac{q}{1-w} - 1\right)\begin{cases} < w / b, \\ = w / b, \\ > w / b. \end{cases}$$

2. $R < S$. The matrix of the linearized system has the form:

$$A = \begin{bmatrix} 0 & m & -m \\ \dfrac{mS(S-R)}{\overline{p}^2} & \dfrac{-2mS + mR}{\overline{p}} & \dfrac{q}{1-w} - 1 + \dfrac{mS}{\overline{p}} \\ 0 & 0 & w/b \end{bmatrix}_{(p,S,R)=(\overline{p},0,0)}$$

$$= \begin{bmatrix} 0 \; m & -m \\ 0 \; 0 & \dfrac{q}{1-w} - 1 \\ 0 \; 0 & w/b \end{bmatrix}.$$

The phase portrait is provided in Fig. 1.3, where $tg\,\varphi = \dfrac{w}{b}\dfrac{1-w}{q+w-1}$.

Thus, the stationary point is unstable in both domains, which is readily explained – the absence of production and. accordingly the consumption is not a normal state of the system. Now let us consider the phase portrait of the initial nonlinear system. Since it is difficult to represent its solution in the form of explicit formulas, we turn to the isocline method. We examine system (1.2) and note the following characteristics:

(a) there is only one fixed point – the coordinate origin; and
(b) the isocline $R' = 0$ coincides with the $S$ axis and for $S \neq 0$ on the isocline the relation $S' = -mS^2 / p < 0$ is satisfied.

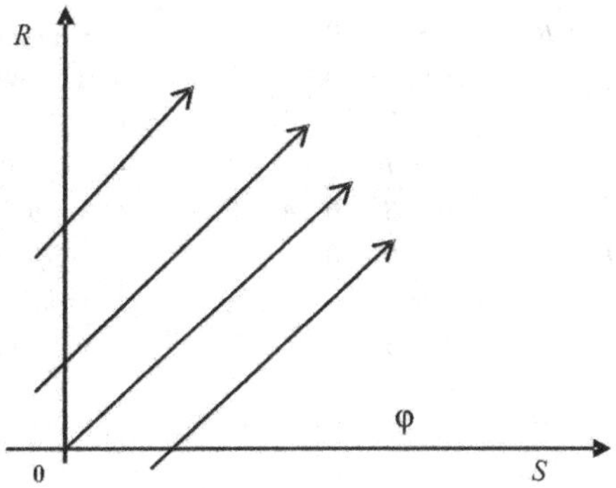


**Fig. 1.3** The phase portrait of the linearized system (1.3)

Therefore, there exists a trajectory that coincides with the positive semi-axis of $S$ and is directed toward the coordinate origin. For $S < R$, the isocline equation corresponding to slope $C$ has the form

$$R = \frac{S^2 + \frac{p}{m} \cdot S}{\frac{p}{m}\left(\frac{q}{1-w} - \frac{w}{b}C\right) + S} = S - (A - B) + \frac{A(A-B)}{S+A},$$

$$A = \left(\frac{q}{1-w} - \frac{w}{b}\right)C, \qquad B = \frac{p}{m}.$$

For $R < S$, the equation of an isocline with slope $C$ has the form

$$R = \frac{S^2}{\frac{p}{m}\left(\frac{q}{1-w} - 1 - \frac{w}{b}C\right) + S} = S - (A - B) + \frac{(A-B)^2}{S + (A-B)}.$$

Depending on the different values of $A$ and $C$ (see Table 1.1), Fig. 1.4 represents the curve of the isocline.

The corresponding phase portraits for $p = \text{const}$ (projections at the plane $R0S$) will have qualitatively different forms, depending on the relationships of parameters $q$, $w$, and $b$:

1. $\frac{q}{(1-w)} - 1 < \frac{q}{(1-w)} < \frac{w}{b}$ (Fig. 1.5a).
2. $\frac{q}{(1-w)} - 1 < \frac{w}{b} < \frac{q}{(1-w)}$ (Fig. 1.5b).
3. $\frac{w}{b} < \frac{q}{(1-w)} - 1 < \frac{q}{(1-w)}$ (Fig. 1.5c),

$$B = p/m, \qquad D = \frac{p}{m}\left(1 - \frac{q}{1-w} + \frac{w}{b}\right).$$

**Table 1.1** Isocline curves

| Region | Parameter relationships | | Curve no. |
|---|---|---|---|
| $S < R$ | $A - B > 0,\ A > 0$ | $C < [q/(1-w) - 1](b/w)$ | 1 |
| | $A - B > 0,\ A < 0$ | $[q/(1-w) - 1](b/w) < C < qb/[(1-w)w]$ | 2 |
| | $A - B < 0,\ A = 0$ | $C = [q/(1-w) - 1](b/w)$ | 3 |
| | $A = B$ | $C = [q/(1-w) - 1](b/w)$ | 4 |
| | $A - B < 0,\ A < 0$ | $C < qb/[(1-w)w]$ | 5 |
| $R < S$ | $A - B < 0$ | $C > [q/(1-w) - 1](b/w)$ | 6 |
| | $A = B$ | $C = [q/(1-w) - 1](b/w)$ | 7 |
| | $A - B > 0$ | $C < [q/(1-w) - 1](b/w)$ | 8 |

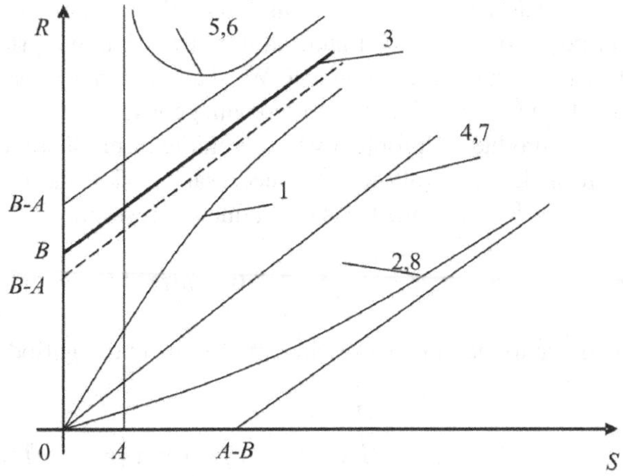


**Fig. 1.4** The curve of the isoclines depending on different values A and C

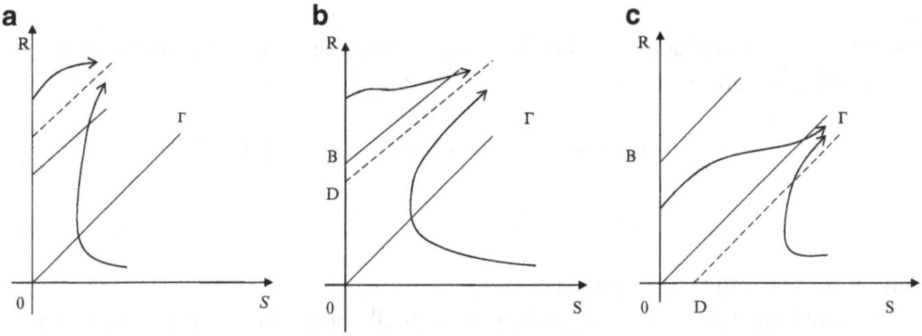


**Fig. 1.5** $a, b, c$ The phase portraits for $p = \text{const}$ at the ROS plane

Having analyzed this model, we draw the following conclusions.

1. If a pricing mechanism is not used for the control, then the stability can be achieved only by establishing the correspondences among the parameters $g$, $w$, and $b$. But this is problematic, since these parameters are governed by various mechanisms in a real economy.
2. The presence of a market-pricing mechanism does not in itself ensure stability. As can be seen from the phase portraits, there is an overproduction crisis when $g/(1-w) - 1 < q/(1-w) < w/b$ and a demand crisis when $w/b < g/(1-w) - 1 < q/(1-w)$. The relationship $g/(1-w) - 1 < w/b < q/(1-w)$ must be maintained to ensure relative stability of supply and demand $\left(|S - R| < p/m\right)$.

Further we shall examine the producer's model behavior under conditions of the transition to a market economy when prices are not fixed at a particular level but their change is depending on the relationship between the supply and demand. With rapidly rising prices, it is profitable for the producer to delay the production time (offering in the form of supply) of goods until prices rise up to a corresponding level.

We shall consider, on one hand, the effect of price dynamics on the decision to put goods on the market and, on the other side, the price change as a function of decisions made by the producer. We shall attempt to answer the following question: how long a production delay is advantageous?

The producer's problem whose volume of production affects the price level can be formulated as follows. It is necessary to find the production volumes $u(t)$ for times $t \in [0, T]$ on the basis of production restrictions

$$0 \le u(t) \le \overline{u} \tag{1.4}$$

such that to maximize gross income for the entire period

$$\int_0^T \exp(-\nu t)\, p(t) \min\left\{u(t), S(t)\right\} dt, \tag{1.5}$$

here $\nu$ is the discount factor, $S(t)$ is the demand function, and $p(t)$ is the price level, which varies as follows

$$p' = m(S - u), \quad p(t) \ge 0, \quad 0 \le m \le 1, \tag{1.6}$$

$$p(0) = p_0 \tag{1.7}$$

Here $m$ is the coefficient of price elasticity.

The problem (1.4)–(1.7) is a problem of optimal control in which $u$ is the control parameter (further referred to as control) and $p$ is the phase coordinate. Let the demand vary by the following exponent law

$$S(t) = \exp(\mu t),$$

with $\mu$ is the rate of demand growth.

If we assume that $\overline{u} > \exp(\mu t) \ge S(t)$, then the constraint (1.4) is not significant and it is required just for limitation of the set of controls.

The following assertion is valid.

**Theorem 1.1.** *Let the equation*

$$\exp(-\nu\tau)\left(\frac{m}{\mu}\exp(\mu\tau) + p_0 - \frac{m}{\mu}\right) = \frac{\exp\left((\mu-\nu)T\right) - \exp\left((\mu-\nu)\tau\right)}{\mu - \nu} m \tag{1.8}$$

*have a solution $\tau \in [0; T]$; then an optimal strategy for problem* (1.4)–(1.7) *consists of two parts: $u(t) = 0,\ t \in [0; \tau]$ and $u(t) = S(t),\ t \in [\tau; t]$.*

*Proof.* Let us map the Eq. (1.6) onto the multiplier $\psi(t)$ and write the Hamiltonian

$$H(t, p, \psi) = \exp(-\nu t)\, p\, \min\{u, S\} + \psi\, m\, (S - u).$$

Let $h = \max_u H(t, p, \psi)$. According to the necessary conditions of optimality theorem [8], if curve $p(t)$ is a solution to the problem (1.4)–(1.7), then there is a curve $\psi(t)$ such that

$$\begin{bmatrix} -d\psi \,/\, dt \\ dp \,/\, dt \end{bmatrix} \in \partial h(t, p, \psi) \tag{1.9}$$

almost everywhere.

Thus, we should study the solution of this differential inclusion under the boundary conditions $p(0) = p_0$ and $\psi(T) = 0$ by means of computing the following function

$$h : h = \max_u \left\{ \exp(-\nu t)\, p\, \min\{u, S\} + \psi m (S - u) \right\}.$$

Since this function is a piecewise-linear function with respect to $u$, its maximum is attained at either the extreme points of set of permissible $u$ values or the following points of nondifferentiability: $u = 0$, $u = S$, and $u = \overline{u}$. Thus

$$h = \max_u \left\{ m\psi S,\ \exp(-\nu t)\, pS,\ \exp(-\nu t)\, pS + \psi m\, (S - \overline{u}) \right\}.$$

Let us denote by $\alpha = m\psi$, $\beta = \exp(-\nu t)\, p$. Note that $\beta \geq 0$ ($p \geq 0$). Three cases are possible.

**A.** $\alpha < 0,\ \beta > 0.$ Thus

$$h = \max_u \left\{ \beta S,\ \beta S + \alpha(S - \overline{u}) \right\} = \beta S + \alpha(S - \overline{u}) = \exp(-\nu t) pS + \psi m(S - \overline{u}),$$

$$\partial h = \begin{bmatrix} \exp(-\nu t) S \\ m(S - \overline{u}) \end{bmatrix}. \tag{1.10}$$

Hence

$$\psi(t) = \frac{-\exp\left((\mu - \nu)t\right)}{\mu - \nu} + C_1, \quad p(t) = \frac{m}{\mu} \exp(\mu t) - \overline{u} m t + C_2. \tag{1.11}$$

**B.** $0 \le \alpha \le \beta$. Hence

$$h = \max_u \left\{ \alpha S,\ \beta S,\ \beta S + \alpha(S - \overline{u}) \right\} = \beta S = \exp(-\nu t)\, pS,$$

$$\partial h = \begin{bmatrix} \exp(-\nu t) S \\ 0 \end{bmatrix}, \tag{1.12}$$

$\psi$ varies in accordance with Eq. (1.11), and $p(t) = \text{const.}$

**C.** $\alpha > \beta > 0$. Thus

$$h = m\psi S, \quad \partial h = \begin{bmatrix} 0 \\ mS \end{bmatrix}.$$

Hence

$$p(t) = \frac{m}{\mu} \exp(\mu t) + C_2, \quad \psi(t) = \text{const.} \tag{1.13}$$

Now we are ready to analyze the behavior of the following system of differential inclusions Eq. (1.9).

1. Neither condition $A$ nor $C$ are satisfied for $(\psi, p)$ at time $t = T$. In fact, $\psi(t) = 0$ means that $m\psi(t) = 0 > exp(-\nu T)\, p(t)$ for $B$. Considering that $\psi$ is a constant in case $C$, this conflicts to the condition $p(t) \ge 0$.

   Therefore, $(\psi, p)$ belongs to $A$ or $B$ at the time instant $t = T$. Hence, $\psi(t)$ is determined by formula (1.11), where $C_1$ is specified by the condition $\psi(t) = 0$:

$$C_1 = \frac{\exp\Big((\mu - \nu)T\Big)}{\mu - \nu}.$$

   Now for the interval $(\tau, T]$ $\psi(t)$ has the following form

$$\psi(t) = \frac{\exp\Big((\mu - \nu)T\Big) - \exp\Big((\mu - \nu)\tau\Big)}{\mu - \nu}. \tag{1.14}$$

   Let us examine the sign of $\psi(t)$ in Eq. (1.14) for various values of parameters $\mu$ and $\nu$:

   (a) $\mu - \nu > 0$; then $\exp\Big((\mu = \nu)t\Big) > \exp\Big((\mu - \nu)t\Big)$ and $\psi(t) > 0$;

   (b) $\mu - \nu < 0$; then $\exp\Big((\mu - \nu)T\Big) < \exp\Big((\mu - \nu)t\Big)$ and $\psi(t) > 0$;

   (c) $\mu = \nu$; then $d\psi \,/\, dt = -1$, $\psi(t) = T - t\ \psi(t) > 0$.

Thus, $\psi(t)$ takes only positive values, i.e., $(\psi, p)$ cannot fall into case $A$, where $m\psi < 0\ (m > 0)$.

2. The curve $(\psi, p)$, having fallen into case $B$, remains there. The condition under which $(\psi, p)$ finds itself in case $B$ will be the validity of inequality $\exp(-\nu t)p(t) > m\psi(t)$. For case $B$ $(\psi, p)$ varies in accordance with the inclusion (1.12). Hence, it follows that with the unchanged left side the right side of this inequality is decreased:

$$p(t) > m\exp(\nu t)\psi(t) = m\exp(\nu t)\frac{\exp\Big((\mu-\nu)T\Big) - \exp\Big((\mu-\nu)t\Big)}{\mu-\nu},$$

$$p(t) > \frac{\exp(\mu T) - \exp(\mu t)}{\mu - \nu},$$

i.e., the inequality is preserved for any $t$.

Therefore, an optimal production mode can be characterized as a policy consisting of two phases with switching at the time instant $\tau$. Now it is necessary to set $u = 0$ at the interval $[0, \tau]$ and $u = S = \exp(\mu t)$ in the interval $(\tau, T]$. The switching time instant $t$ will be determined by the following boundary conditions $p(0) = p_0$ and $\psi(T) = 0$, Eqs. (1.10) and (1.11), and the condition $\exp(-\nu t)p(\tau) = m\psi(\tau)$.

Here, $p$ will be extracted from Eqs. (1.13) and (1.7) as follows:

$$p = \frac{m}{\mu}\exp(\mu t) + p_0 - \frac{m}{\mu}.$$

To summarize, the switching time instant is determined by the equation

$$\exp(-\nu t)\left(\frac{m}{\mu}\exp(\mu\tau) + p_0 - \frac{m}{\mu}\right) = \frac{\exp\Big((\mu-\nu)T\Big) - \exp\Big((\mu-\nu)\tau\Big)}{\mu-\nu}m.$$

The theorem is proved.

Let us consider the Eq. (1.8) for $\nu = 0$ (the absence of discounting). In this case we have that

$$\exp(\mu t) = \frac{1}{2}\left(1 - p_0\frac{m}{\mu}\right) + \frac{1}{2}\exp(\mu T)$$

and

$$\tau = \frac{1}{2}\ln\frac{1}{2}\left(1 - p_0\frac{\mu}{m} + \exp(\mu T)\right). \tag{1.15}$$

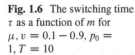

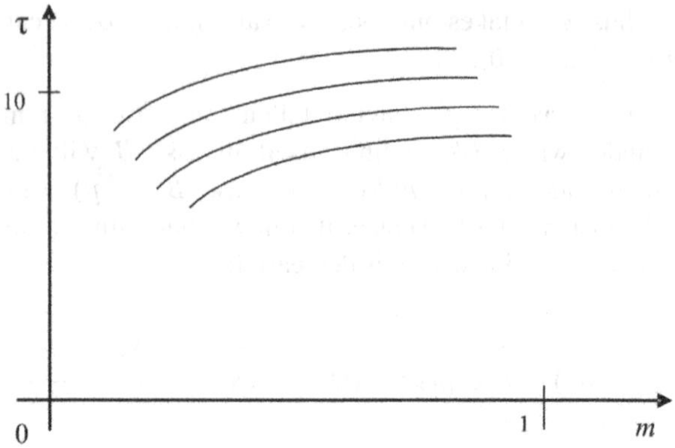


**Fig. 1.6** The switching time $\tau$ as a function of $m$ for $\mu, v = 0.1 - 0.9$, $p_0 = 1, T = 10$

For the existence of the logarithm, it is necessary that

$$m > \frac{p_0 \mu}{1 + \exp(\mu T)}. \tag{1.16}$$

This condition cannot be satisfied only if $p_0 > \left(1 + \exp(\mu t)\right) / \mu$ (for $m > 1$), which conflicts with the condition (1.6). Otherwise, $\tau$ in Eq. (1.15) is a function of $m$: as follows $\tau = \tau(m)$ increases until $\tau = t$ by an increment of $m$ if equality takes place in Eq. (1.16). The increment rate of $\tau$ with an increase in $m$ is reduced.

For the case of $\nu \neq 0$, the Statgraf package was used to plot graphs (Fig. 1.6) of the switching time $\tau$ as a function of $m$ for $\mu = 0.1$–$0.9$, $\nu = 0.1$–$0.9$, $p_0 = 1$, and $T = 10$.

As is apparent, the function $\tau = \tau(m)$ increases with an increase in the coefficient of price elasticity $m$. The effect of a change is, in turn, a function of the relationship between $\mu$ and $\nu$: this effect becomes more substantial with an increase in $(\mu - \nu)$. In addition, $\tau$ is to a considerable extent a function of the coefficient $\mu$ (rate of growth in demand). For any $\nu$, the increase in $\tau$ is considerable when $\mu$ increases from 0.1 to 0.9. Figure 1.6 shows that a rise in $\nu$ for small $\mu$, results in a decrease in $\tau$.

Thus, the production delay $\tau$ is a function of the coefficient of price elasticity $m$, the discount factor $\nu$, and the rate of demand growth $\mu$. The delay increases to a considerable extent with an increase in demand. This trend can be restrained by increasing the discount factor. An increase in the discount factor enhances the effect of price "tempering" for the duration of the delay; the greater the price "freedom" implies the greater the delay in the production process.

Thus, a delay in placing goods on the market can be combated in two ways. The first way is by means of increasing the charges for money held in banks (this increases the discount factor) or in the second way by means of anti-inflationary measures that do not interfere with market-pricing mechanisms – for example, by increasing the currency stability (this reduces the coefficient of price elasticity).

The deficit of a state budget is usually considered among main reasons of the demand inflation. Several models use this indicator for a description of the price instability. The following macromodel which is proposed by Bruno [3], is one of them. This model first was developed and calibrated for Israel, when the inflation of 1980s had created serious problems; later on it was modified for transition economy.

Our analysis for the case of a transition economy like in Ukraine extracts the steady-state relationship between deficits and inflation which might be justified within the classical monetary model of price determination in the medium term. Our research line will be based on the framework used by Bruno [3] to describe the inflationary periods in Israel, and will be adjusted to the Ukrainian reality. This analysis has to be modified in a number of ways in order to take into account the institutional differences between the Israeli and the Ukrainian economies.

Let $D$ be that portion of the total consolidated public sector deficit to be financed by the monetary emissions (residual financing of the state deficit, plus subsidized Central Bank credit). In a steady-state that deficit must be equal to the inflation tax plus any continuing non-inflationary increases in the monetary base of the inflation tax. Let the monetary base of the inflation tax be $MB$:

$$D' = \left(\frac{\dot{P}}{P} + g\right)M, \tag{1.17}$$

where $g$ is the rate of growth of real GDP. (Note that the steady-state growth of MB at the rate $g$ is non-inflationary.) By dividing the both sides by nominal GDP, $y$, and dividing through by $MB/y$, and rearranging terms implies

$$\frac{\dot{P}}{P} = \left(\frac{D'}{Y}\right)\left(\frac{Y}{MB}\right) - g. \tag{1.18}$$

In words, the steady-state tendency of inflation is the total deficit to be financed by emissions (as a percentage of nominal GDP) times the velocity of the monetary base of the inflation tax, $y/MB$, minus the rate of growth of real GDP.

In order to understand the nature of this simplification, a simple model of deposits can be employed:

$$SD1 = \frac{dSD1}{dt} = NS(t) - DS(t), \tag{1.19}$$

where $SD(t)$ is the derivative of the level of deposits and $NS(t)$ and $DS(t)$ are, respectively, the inflow and outflow, both positive, i.e., flowing one-way. The variable NS can be interpreted as the flow of savings while $SD$ as the flow of dissavings.

An increase of $S$ often causes the increase of $SD1(t)$ ($SD1(t) > 0$), yet $S$ is not a sole cause of changes in $SD1(t)$. It is obvious that the increase of $SD1(t)$ can be an effect of the following behavior of both $S$ and $DS$:

(i) $NS$ increasing and $DS$ decreasing or steady (obvious);
(ii) both $S$ and $DS$ increasing ($DS$ increasing slower);
(iii) $NS$ decreasing or steady and $DS$ decreasing ($DS$ decreasing faster).

Point (iii) shows the danger of a simplified reasoning, for it reveals the contradiction between the reasoning based on the net changes of the stock variable ($SD1(t)$), and that based on the behavior of flows ($NS$ and $DS$). Thus, the increase of $SD1(t)$ can be a result of the qualitatively different processes (ii) and (iii). Such an analysis can be applied also to the other stock variables analyzed in the banking sector.

In such circumstances should the central bank in the same way interpret and react to the different processes: (i)–(iii), which generate the same result in the level of a given stock?

In practice, the attempts to identify the processes (i) through (iii) are seldom made. The currently employed method of collecting data does not foster such an analysis, therefore it is not easy to overcome this. On the other hand, the models are being adapted to the available data so the loop is closed. This problem is reflected in the way the banking sector is modeled in most econometric models, for example such as NiDEM from UK or Czerwinski et al. [7] and Welfe et al. [37] from Poland.

The aim of our analysis is to present an approach that might help to solve some of the above mentioned modeling problems. Our approach is based on the concept that contrary to most other markets. In a money market the money does not change owners but it is almost always deposited/lent for a time period agreed upon. After some time period, not necessarily equal to the one agreed upon, money is returned to the depositor/bank.

In each time period the depositors make two decisions: one concerning the amount to be deposited and the other concerning the expected duration of such a deposit. The money deposited for a given time period leaves the deposits not all at once—the resultant outflow lasts for some time. This is so because the money deposited on a given day is then withdrawn in parts after different periods. It may be assumed that the duration of a deposit can be described by some statistical distribution (uniform, normal, etc.).

This hypothesis is justified by the available data. An example of such a justification is presented in Fig. 1.7. However, the shown structure represents not the very structure of the real average duration of deposits but the intended one –

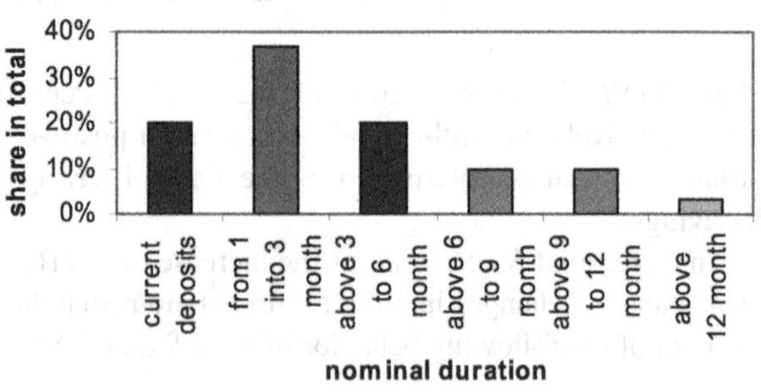


**Fig. 1.7** The statistical distribution for the duration time of a deposit

a structure, chosen by the bank customers to distribute their money among time deposits with different duration. This means that, for example, money in the time deposits with duration from 3 to 6 months could in average stay there much longer (or shorter) than 6 months. The discrepancy between the intended and effective structures cannot be eliminated because there is no other reliable source of information.

At this stage an analysis of the variables that co-determine the flow of the savings and dissavings is necessary.

One can assume that the flow of savings is determined in the significant part by the consumption function, e.g., the Keynesian one (see Keynes [18]). The savings in the Keynesian model should be interpreted as net savings made in the economy as a whole. This makes it possible for the different rates of savings and dissavings: savings in the economy are being made by the individuals even in times when net savings are negative indicating that the dissavings are greater than savings.

At this point, however, an important distinction has to be made. Namely, in the Keynesian consumption function the savings (or dissavings) denote the net changes of the stock of savings while in the considered our model, the savings are a non-consumed part of the flow income. One part of the flow of savings consists of the involuntary savings related to the periodical (in the short-run) character of income and the transactional motive of saving. This part of savings is in turn partly transformed into the short-term deposits (current, checking). A certain part of the income flows through the deposits. The intensity of that flow is determined by many factors, of which the development of the banking sector is a crucial one. Another part of the flow of savings can be associated with the voluntary savings of a longer duration.

Agents who save take into account not only the duration of new deposits but also the duration of the deposits created in earlier periods. In effect there emerges a divergence between the nominal and effective duration of deposits. This is so because the agents put money into different time deposits according to the past preferences revealed at the time of making deposits while the effective time distribution of deposits reflects the decisions made under the influence of the recent situation.

The core of the model of deposits is based on Eq. (1.19) describing the stock of deposits and the following equation determining the flow of the dissavings:

$$DS(t) = \int_0^\infty w(t,\tau) NS(t-\tau) d\tau, \tag{1.20}$$

where $w(t, \tau) > 0$ is the probability density function of the duration of a deposit at time $t$ and

$$\int_0^\infty w(t,\tau)\, d\tau = 1, \quad w(t,\tau) \geq 0. \tag{1.21}$$

One can note in (1.21) that the distribution of duration of a deposit can vary. The average duration of a deposit at time $t$ is denoted by $TD(t)$ and might be represented as the following function:

$$TD(t) = \int_0^\infty \tau w(t,\tau)\, d\tau. \tag{1.22}$$

The above expression can be, in a sense, misleading as it expresses the formal definition of the average duration of deposits. In the causal relationship the implication can be the opposite: depositors' decisions concerning the duration of their deposits result in the changes in the average duration of a deposit, and in consequence, the distribution of the average duration of deposits changes.

When developing such a model it is also necessary to analyze whether the changes in the deposit interest rate affect the flow of savings, and whether the propensity to save is influenced by the interest rate on deposits. However, at the start of such an analysis one does not need to assume such a relationship because other factors determining savings may be of the higher order. The other factors mentioned could include the following: the interest rates other than the deposit interest rate, the so-called sentiment, the demographic processes affecting the functioning of pension funds, the technological changes in the circulation of money (e.g., plastic money), etc.

Another question is related to a possible dependency between the average duration of deposits and the rate of savings. Do the deponents leave the distribution of the deposit duration and the average duration of deposits unchanged or, just the opposite, change either of these? Common wisdom suggests that the decrease of savings might be matched by the shrinking average duration of deposits. And the opposite: one can expect that the increase of savings is accompanied by the extension of the average duration of deposits. Such hypotheses concerning the interaction between the rate of savings and the average duration of deposits caused by the impact of the changes in the deposit interest rate have to be adopted in the further stages of modeling.

## 1.3 Modeling of Costs Inflation

Variant calculations were made on the basis of macromodel (1.2) and its modifications using the information about the economy of Ukraine in 1992–1993. Variants differ by suppositions about the values of model parameters, in particular, coefficient of price elasticity $m$. The results of calculations in comparison with actual dynamics of prices are presented in Table 1.2. According to this information, a model exactly enough described the dynamics of prices until the autumn of 1992, especially with taking into account the fact that the first sharp jump of prices, forecasted on

**Table 1.2** Dynamics of prices index on produced goods and services (to 1990)

| Year | Actually | On the basis of modeling | |
|---|---|---|---|
| | | Variant 1 | Variant 2 |
| 1990 | 1 | 1 | 1 |
| 1991 | 2.21 | 6.03[a] | 6.03[a] |
| 1992 | 25.9 | 16.87 | 16.87 |
| 1993 (5 months) | 206.1 | 23.71 | 24.3 |

[a] Taking into account the forecasted jump of prices in November–December 1991

November–December, 1991, actually happened in January–February, 1992. This delay was resulted by political factors, which were not directly connected with economic processes, such as disintegration of USSR and race for power in Ukraine.

However, afterwards the actual rates of inflation were substantially greater in the comparison with forecasted.

Discussing this fact in spring and summer of 1993 with the research group on transitional economy in the Glushkov's Cybernetics Institute (The National Academy of Sciences of Ukraine), Academician Mikhalevich V. S. paid special attention to necessity for search of reason of inflation acceleration in the specific of strongly monopolized real sector of transition economy. The relations between the groups of monopolists, for example, within the former industrial branches, could create the basis for such processes. It is quite difficult to analyze such effects using the aggregated macromodels and hence the development of more detailed models is required.

One of such models was considered in our publication [21]. The Samuelson's "equalizations of motivation" given in [27] creates the basis of our model, the same reasons were used for the macromodel (1.2). Our model differs from (1.2) by its detailing in the respect of indicators for separate industries. This allows us to analyze some structural effects. We consider the opened model; export and import values are taken into account in our open model. We assume that the real sector of economy consists of $N$ industries (branches), among them $N_1$ industries ($N_1 < N$) produce investment goods. Similar to the model (1.2), our model is dynamic, the time is changing continuously.

To summarize our model is based on the following assumptions.

1. The general volume of the exported and imported goods consists of two parts, namely:

   – critical import, which is necessary for maintenance of the supply of consuming goods for each industry at minimum possible level (in the model this level assuming to be equal to zero); the volume of critical import is determined irrespective to world market prices for these goods;
   – the commercial export–import, which volume is assumed to be proportional to the expected profits, i.e., to the differences in the world and internal prices.

2. The amount of investment goods directed to the expansion of production and replacement of drop-out equipment are distributed between the industries proportionally to the expected income in each branch.
3. The consumer demand is proportional to users' paying capacity; the components of the demand (i.e., the demand on the products of each branch) are proportional between themselves.
4. The price on products of each branch changes proportionally to unsatisfacted demand, but cannot be low than the costs of products.
5. $N+2$ groups of consumers are considered in the model, namely:

   – the persons, who are employed in $N$ industries and who receive income proportionally to the products made them (group numbers $1-N$);
   – the persons employed in the social sphere, who receive pay for labor proportionally to the size of the real national income (group number $N+1$);
   – the persons, who receive an income from export–import operations (group number $N+2$).

6. The volumes of domestic and imported goods of each branch realized at the internal market are assumed to be proportional to total production and import volumes.
7. The model is realized on the time interval $[0;T]$; the effects related to the presence of the delay in mastering of investments for expansion of the production are not taken into account. It is assumed that all parameters of model are continuous functions of the time $t$.

**The following decision variables and parameters are used in our model**:

$x_j, j=\overline{1,N}$, is the value of the gross product of $j$-th industry;
$y_j, j=\overline{1,N}$, is the value of the final product of $j$-th industry;
$I_j, j=\overline{1,N}$, is the value of the critical import of products of $j$-th industry;
$U_{ij}, j=\overline{1,N}$, $i=\overline{1,N_i}$, is the volume of products of $i$-th industry, which is used for expansion of production and replacement of equipment in $j$-th industry;
$R_j, j=\overline{1,N}$, is the supply of products of the $j$-th industry proposed for final consumption;
$\widehat{R}_{lj}$, $l=\overline{1,N+2}$, $j=\overline{1,N}$, is the share of products of $j$-th industry in the total consumption of the $l$-th consumer group;
$\overline{e}_j, j=\overline{1,N}$, is the ratio of net import (the difference between import and export) of the $j$-th industry to its final product;
$\widehat{V}_j, j=\overline{1,N}$, is the unit costs of the production of $j$-th industry instead of wages;
$V_j, j=\overline{1,N}$, is the expected profit from realization of the unit production of $j$-th industry;

$\lambda_j, j = \overline{1, N}$, is the share of expenditures directed to the expansion of production and to replacement of drop-outs equipment in $j$-th industry, in the general volume of such expenditures;

$\widehat{S}_{l_j}$, $l = \overline{1, N+2}$, $j = \overline{1, N}$, is the final demand of the $l$-th consumer group on the products of the $j$-th industry;

$S_j, j = \overline{1, N}$, is the total demand on the products of the $j$-th industry;

$\delta_j, j = \overline{1, N}$, is the price for production of $j$-th industry;

$D_l$, $l = \overline{1, N+2}$, are the total money accumulations of the consumers from $l$-th group;

$\widehat{D}_j, j = \overline{1, N}$, are the total money accumulations of the enterprises of $j$-th industry;

$z_j, j = \overline{1, N}$, is the volume of realization of the production of $j$-th industry.

**The input data of our model are denoted as**:

$a_{ij}$, $i = \overline{1, N}$, $j = \overline{1, N}$, are the coefficients of the direct expenditures of the products of $i$-th industry on the unit production in $j$-th industry;

$b_{ij}$, $i = \overline{1, N_1}$, $j = \overline{1, N}$; are the direct expenditures of the investment products of $i$-th industry on the unit growth of production in $j$-th industry;

$d_{ij}^0$, $i = \overline{1, N_1}$, $j = \overline{1, N}$; are the minimal level of expenditures of products of the $i$-th industry necessary for maintenance the existed level of production in $j$-th industry (it is the level of the expenditures for replacement of drop-outs equipment);

$W_i, i = \overline{1, N_1}$, is the share of expenses of the $i$-th industry products directed to replacement and investments, in the general final production of this industry;

$\overline{W}_j, j = \overline{1, N}$, is the share of the final production of $j$-th industry, which is consumed out of the mentioned consumer groups (includes consumption for defense, security, health care, etc);

$q_j, j = \overline{1, N}$, is the share of incomes of consumers from $l$-th group ($l = j$) in the final product of the $j$-th industry;

$q_{N+1}$ is the share of incomes of consumers of $N+1$-th group in the real national income;

$\overline{q}_j, j = \overline{1, N}$, is the share of profits of $j$-th industry remaining in its disposal;

$\alpha_j^l, j = \overline{1, N}$, is the share of the demand on the products of $j$-th industry in total demand on the $l$-th consumer group;

$n_j$, $\widehat{n}_j$, $m_j$, $j = \overline{1, N}$, are the coefficients for proportion in equations;

$\gamma$ is the value of interest rates (which assumed to be equal for all industries and consumer groups);

| | |
|---|---|
| $\widehat{p}_j, j = \overline{1, N},$ | is the price for the products of $j$-th industry at the world market; |
| $\widehat{q}_j, j = \overline{1, N},$ | is the share of value added in the costs of products be made by $j$-th industry. (All costs indexes, except the money accumulation of industries and consumers, cost value and expected income, are expressed in comparable (stable) prices.) |

In the following we outline the block structure of our model features. It allows us to develop and modify our model without any substantial additional expenditures.

Our model consists of five groups of differential and functional equations, namely:

Block PRODUCTION.
Block MOTIVATION.
Block DEMAND AND PRICES.
Block FINANCES OF CONSUMERS.
Block FINANCES OF INDUSTRIES.

Let us now consider these blocks in the sequentially indicated order as follows.

**The block PRODUCTION** includes the following equations.

The balance of gross and final products in industries is expressed by a formula:

$$y_i = x_i - \sum_{j=1}^{n} a_{ij} x_j + I_i, \quad I = \overline{1, N}.$$

The size of critical import concerns so:

$$I_i = \max\Big(0, \sum_{j=1}^{N} a_{ij} x_j - x_i\Big), \quad i = \overline{1, N},$$

and expenditures on replacement and expansion of funds

$$U_{ij} = \frac{\lambda_i}{\sum\limits_{k=1}^{N} \lambda_k} (1 + \overline{e}_i) W_i y_i, \quad i = \overline{1, N_i}, \quad j = \overline{1, N}.$$

The dynamics of gross product changes in industries is described by equation

$$x_j' = \min_{i=\overline{1, N_i}} \Big(b_{ij}^{-1} \left(U_{ij} - d_{ij}^0\right)\Big), \quad j = \overline{1, N}.$$

The supply of each industry is equal to:

$$R_j = \begin{cases} (1+\overline{e}_j)(1 - W_j - \overline{W}_j)\, y_j, \text{ if} & 1 \le j \le N_i, \\ (1+\overline{e}_j)(1 - \overline{W}_j)\, y_j, & \text{if } N_i + 1 \le j \le N. \end{cases}$$

The supply distribution in consumer groups is carried out proportionally to demand:

$$\hat{R}_{lj} = \frac{\hat{S}_{lj}}{\sum\limits_{k=1}^{N+2} \hat{S}_{kj}} R_j, \quad l = \overline{1, N+2}, \quad j = \overline{1, N}.$$

**Block MOTIVATION** consists of the following equations and inequalities.

The change of the ratio of net import to final production is determined by the equation:

$$\overline{e}'_j = n_j \left(p_j - \hat{p}_j\right), \quad j = \overline{1, N}.$$

The export cannot exceed the produced final goods:

$$\overline{e}_j \ge -1, \quad j = \overline{1, N}.$$

The unit costs of the production in current prices is estimated as follows:

$$\hat{v}_j = \sum_{k=1}^{N} a_{kj}\, p_k + \frac{1}{x_j}\Big(\sum_{i=1}^{N_i} U_{ij}\, p_i + \hat{p}_j I_j\Big), \quad j = \overline{1, N}.$$

The expected income equals to $\nu_j = p_j - \hat{v}_j - q_j p_j,\ j = \overline{1, N}$.

The change of distributing of expenses on expansion in industries concerns by equation

$$\lambda'_j = \hat{n}_j \left(\nu_j \overline{q}_j\right), \quad \lambda_j \ge 0, \quad j = \overline{1, N}.$$

**Block DEMAND AND PRICES** consists of the following equations and inequalities.

The change of demand is equal to the change of consumers' pay ability:

$$\hat{S}'_{lj} = \alpha^l_j \left(\frac{D_l}{p_j}\right)', \quad l = \overline{1, N+2}, \quad j = \overline{1, N}, \quad \hat{S}_{lj} \ge 0.$$

The total demand is determined as follows:

$$S_j = \sum_{l=1}^{N+2} \hat{S}_{lj}, \quad j = \overline{1, N}.$$

The prices are determined by Samuelson equations:

$$p_j' = m_j(S_j - R_j), \quad j = \overline{1, N}.$$

The price cannot be below the costs (taking into account an admissible level of the profit):

$$p_j \geq \hat{v}_j + p_j q_j + p_j \hat{q}_j, \quad j = \overline{1, N},$$

from which implies that

$$p_j \geq \frac{1}{1 - q_j - \hat{q}_j - a_{jj}} \left( \sum_{\substack{k=1 \\ k \neq j}}^{N} a_{kj} p_k + \frac{1}{x_j} \Big( \sum_{i=1}^{N_i} U_{ij} p_i + \hat{p}_j I_j \Big) \right), \quad j = \overline{1, N}.$$

**Block FINANCES OF CONSUMERS** includes the following differential and functional equations.

The balance of incomes and expenditures for employed in material production is described by equations

$$D_l' = \gamma D_l + (1 + \gamma) \left( q_l p_l x_l - \sum_{j=1}^{N} p_j \min \left( \hat{S}_{lj}, \hat{R}_{lj} \right) \right), \quad l = \overline{1, N}.$$

Balance of incomes and expenditures for employed in social sphere looks like

$$D_{N+1}' = \gamma D_{N+1} + (1 + \gamma) \left( q_{N+1} \sum_{j=1}^{N} p_j y_j - \sum_{j=1}^{N} p_j \min \left( \hat{S}_{N+1j}, \hat{R}_{N+1j} \right) \right).$$

The volumes of production realization for industries are the following:

$$Z_j = \begin{cases} \sum_{k=1}^{N} a_{jk} x_k + \sum_{k=1}^{N} U_{jk} + \min(S_j, R_j), & \text{if} \quad 1 \leq j \leq N_i, \\ \sum_{k=1}^{N} a_{jk} x_k + \min(S_j, R_j), & \text{if} \quad N_i + 1 \leq j \leq N. \end{cases}$$

The balance of incomes and expenditures from commercial export and import has the form

$$D'_{N+2} = \gamma D_{N+2} + (1+\gamma)\left(\sum_{j=1}^{N}\left(\max\left(\frac{\max(0,\overline{e}_j y_j)}{x_j + \max(0,\overline{e}_j y_j)} p_j Z_j,\ -\hat{p}_j \overline{e}_j y_j\right)\right.\right.$$
$$\left.\left. - \max\left(\hat{p}_j \overline{e}_j y_j, -p_j \overline{e}_j y_j\right)\right) - \sum_{j=1}^{N} p_j \min\left(\hat{S}_{N+2j}, \hat{R}_{N+2j}\right)\right).$$

**Block is FINANCES OF INDUSTRIES** includes the equations of balance of the incomes and expenditures for industries.

$$\hat{D}'_j = \gamma \hat{D}_j + (1+\gamma)\left(\frac{x_j \overline{q}_j}{x_j + \max\left(0, \overline{e}_j y_j\right)} p_j Z_j - \hat{v}_j x_j - q_j p_j x_j\right), \quad j = \overline{1, N}.$$

In addition, the initial (in the moment of time $t = 0$) values of the following variables are assumed to be known:

$$x_j,\ j = \overline{1, N};\quad \overline{e}_j,\ j = \overline{1, N};\quad \lambda_j,\ j = \overline{1, N};$$

$$\hat{S}_{lj},\ l = \overline{1, N+2},\ j = \overline{1, N};\quad \delta_j,\ j = \overline{1, N};\quad D_l,\ l = \overline{1, N+2}\quad \hat{D}_j,\ j = \overline{1, N}.$$

A complicated structure and large dimension of the model did not allow us to apply the methods of the quality analysis. Therefore, basic attention was spared to the variants calculations on its basis. Such calculations were conducted repeatedly because of inaccuracy of initial data, taking into account different assumptions about the values of inaccurate parameters. Thus our main attention was spared to the search of conformities to the law, which were common for all or for majority of variants.

In particular, the calculations made using the dynamic detailed model of transition economy (with different model information) demonstrated the possibility of development of the crisis scenario getting the name structural inflationary crisis.

We will mark the features of such crisis development. The situation of supply–demand misbalance toward greater demand is its starting point. Therefore, the initial stage of crisis development looks like the demand crisis: the prices and nominal money accumulations increase for industries and related consumer groups. However due to very fast growth of the prices the supply begins to exceed the demand (first for some, then for all types of products). The price growth, nevertheless, is not slowed, and vice versa, accelerated. As the result first the enterprises of the considered $N$ branches and then consumers employed at them become insolvent. The decrease of investment activity (diminishing of $\lambda_j$) to the minimum level stopped up in our model. It is accompanied by the increase of the ratio of size of export to the volume of produced goods and by falling of the size of real national income. The misbalance between supply and demand grows; as the result, all consumer groups (except for the persons engaged in export–import operations) become insolvent. The change in prices is not conditioned by the final demand and supply, and hence, is the main reason of such type of crisis.

A high degree of branch monopolism in combination with the practice of costs pricing based on averaged expenditure level in industries have created the preconditions for development of the costs inflation. A certain role was herein played by the external factors (e.g., the prices for basic energy resources growth in Ukraine in 1993 in dollar equivalent up to 15–20 times) [34]. The considered below model of the costs pricing also allows exposing the internal factors which in turn are able to cause the permanent growth of prices.

Let us consider the system consisting of $n$ branches (industries). We denote by $a_{ij}$ the direct expenditures of the production of $i$-th branch for the unit production in $j$-th branch in the comparable prices, $q_j$ is the share of value added in the costs of products of $j$-th branch, $H_j$ are other components of the price including "shadow" incomes. Here the term "shadow" is not related to the so-called shadow prices, for example, in Linear Programming and means, in fact, "illegal" income from many points view except our modeling purposes. The costs of products of $j$-th industry at the time moment $t + \Delta t$ is denoted as $p_j(t + \Delta t)$ and is determined by known prices $p_j(t)$ of products in other industries as

$$p_j(t + \Delta t) = \sum_{i=1}^{n} a_{ij} p_i(t) + q_j p_j(t + \Delta t) + H_j, \quad j = \overline{1, n}. \tag{1.23}$$

It should be noted that in (1.23) the values of $q_j$, $H_j$ can also depend on $t$. From here

$$p_j(t + \Delta t) = \frac{1}{1 - q_j} \sum_{i=1}^{n} a_{ij} p_i(t) + \frac{H_j}{1 - q_j}, \quad j = \overline{1, n},$$

or in a vector form

$$p(t + \Delta t) = \overline{A} p(t) + \overline{u}, \tag{1.24}$$

where $p$ is the vector of prices $(p_1, \ldots, p_n)$; $\overline{A}$ is a matrix $\left\{\overline{a}_{ij}\right\}$,

$$\overline{a}_{ij} = \frac{a_{ij}}{1 - q_j}, \quad \overline{u} = (\overline{u}_i, \ldots, \overline{u}_n), \quad \overline{u}_j = \frac{H_j}{1 - q_j}, \quad i, j = \overline{1, n}.$$

We will consider the equation

$$p = \overline{A} p + \overline{u}. \tag{1.25}$$

Formula (1.24) can be considered as the procedure of the solution search for the last equation by the simple iteration method, presented, for instance, in [20]. The productivity of matrix , i.e., implementation of inequalities:

$$\sum_{j=1}^{n} \overline{a}_{ij} < 1 \quad \text{for} \quad i = \overline{1,n} \qquad \text{or} \qquad \sum_{i=1}^{n} \overline{a}_{ij} < 1 \quad \text{for} \quad j = \overline{1,n},$$

is the sufficient condition for the solvability of Eq. (1.25) and for the convergence of the procedure (1.24) to its solution. If this condition is not fulfilled then the unlimited growth is possible for some components of the vector-function $p(t)$. In particular, if

$$\min_{j=\overline{1,n}} \left( \sum_{i=1}^{n} \overline{a}_{ij} \right) = \mu > 1,$$

then from (1.24) follows the estimate:

$$\min_{j=\overline{1,n}} \Big( p_j(t + \Delta t) \Big) \geq \mu \min_{j=\overline{1,n}} \Big( p_j(t) \Big)$$

and the prices will increase with the permanent rates.

Such an inflationary scenario is considered in [21] and it was called as "the structural hyperinflation crisis." The structural disproportions, presence of the industries with the high level of expenditures (for them $\sum_{i=1}^{n} \overline{a}_{ij} > 1$ hold) and with the high production demand on their processing (for them $\sum_{j=1}^{n} \overline{a}_{ij} > 1$ hold) are presenting our scenario origins.

The industries with the high level of expenditures are aiming to improve their financial position, by the increase of their production prices. However, this price growth accelerated by industries with the high production demand results in passing ahead the growth of costs in industries initiating the avalanche shaped rise in prices, and push them to the new price growth.

Our analysis of multibranch balances for 1990–1993 made in [22] has indicated the presence of pre-conditions for such inflationary scenario in Ukraine at the beginning of 90th. So, the high level of expenditures was presented in coal and food industries. Ferrous metallurgy, chemical and light industries also entered in the number of such industries in 1993 as the result of monopolistic increase of averaged profitability. The high production demand existed in electro energy supply, oil and gas industry and in machinery. Thus, almost all industries of material production were involved in the described scenario. The value of multiplicator $\mu$ estimated for supposition of quarter prices increase, was equal to 3.52.

It should be noted that the procedure (1.24) is similar to the dynamic price multibranch model, which is applied to the analysis of price shocks in [9]. These shocks (both external and internal) are rather typical for early stages of transition with its high monopolism.

Let us define two vectors of the price indices as $P_t$, and $P\nu_t$, using the prices from the previous period as a base; for $t = 0$ they are set up to elements which are all equal to 1.

Let us now introduce a shock into the price system by substituting $P\nu_1$, for $P\nu_0$, thus obtaining, with $K_i = \nu P\nu_i$,

$$P_i = AP_0 + K_i. \tag{1.26}$$

The vector $P_t$ indicates how much the sectoral prices should be modified if firms, while maintaining a constant mark-up, were to pass on to the variations of prices in current costs only. In fact, the firms perceive the current costs variations as the price variations are necessary to keep the profit margins at a constant level. So the firms pass on the costs variations to prices, keeping the mark-up fixed, and then start the trading at the level of new prices.

As soon as the trade takes off again at the new level of prices, the firms have to submit to further costs variations on account of the fact that all other firms have behaved in a similar way and they have simultaneously included the previous costs variations in their prices. Thus the firms have to modify their prices yet again and if the adoption of these new prices should modify the costs/prices equilibrium, this equilibrium will give rise to another round of price variations to which the historical margins would be added. This process arrives at a stable position of equilibrium only if the costs of the period $n - 1$ are completely covered by the prices in the period $n$, as soon as these prices are adopted simultaneously by all firms and do not provoke any further modifications in the production costs with the firms charging their desired profit margins.

Therefore (1.26) can still express a situation of disequilibrium. If $P_i$ express the price variations needed to cover the costs variations, we should adjust them as follows:

$$P_2 = AP_1 + K_1. \tag{1.27}$$

If $P_2 \neq P_1$ then this means that the sectoral indices are still is an expression of the disequilibrium between the unit costs and prices such that to cover the new variations in costs. Moreover, the prices ought now should be modified with respect to the previous period to the extent indicated by the vector $P_2$. In general, with the $t$th iteration, we have

$$P_t = AP_{t-1} + K_1. \tag{1.28}$$

We now calculate the vector $\Delta_t$, given by the difference in modulus between $P_t$ and $P_0$, i.e.,

$$\Delta_t = \left|P_t - P_0\right|.$$

It is obvious that we can calculate the cumulative variation in sectoral prices at the period $t$, with respect to the initial period by the sum of $\Delta_1, \Delta_2, \ldots, \Delta_t$.

Let us define the indices vector $P_t^0$ as

$$P_t^0 = P_0 + \sum_{j=1}^{t} \Delta_j = P_0 + \sum_{j=1}^{t} \left| P_j - P_0 \right|. \tag{1.29}$$

Now, in each period we have two different vectors $P_t$ and $P_t^0$ where $P_t^0$ represents the variations of the sectoral price indices with respect to the equilibrium prices of the base period.

Naturally, at this point the recursive algorithm represented by (1.28) should be processed under the following conditions: for each $\varepsilon > 0$ there exists an $n_0$ such that for each $n > n_0$, $\Delta_n^* = |P_n - P_{n-1}| < \varepsilon$.

By substituting recursively in (1.28) for $P_3, \ldots, P_n$, $n \to \infty$, we obtain

$$P_n = A^{n-1} P_1 + \left( A^{n-2} + A^{n-3} + \ldots + A + I \right) K_1$$

and if $n$ is the last term of the series, then $A^{n-1}$ converges to the null matrix and the series within brackets converges to the limit $(I - A)^{-1}$, so that we obtain

$$P_n = (I - A)^{-1} K_1 \tag{1.30}$$

which proves the convergence to the stationary solution (see Waugh [36]).

Returning to (1.29) and substituting it recursively for $P_t$ as far as $t = n$ that satisfies (1.30), we obtain

$$P_n^0 = P_0 + \Delta_i + \Delta_2 + \ldots + \Delta_n = P_0 + \sum_{t=1}^{n} \Delta_t \tag{1.31}$$

which represents the *cumulative variation of sectoral prices* at the final period $n$, with respect to the initial period.

The state economic policy has large influence on the development of inflationary processes in a transitional economy. It is necessary to select two substantial aspects analyzing such policy impact on prices.

First, this is the direct price control, or, more frequently, unsuccessful attempts of such control using different sorts of subsidies, another payments from the state budget, "freezing" of prices, out-of-market distribution of products, etc. Pricing unavoidable becomes one of sick places of economy during market reforms. Even in an ideal case the transition must be accompanied by radical changes in the price functions and, consequently, by the sharp oscillations of their values. The deep pre-markets disproportions, absence of necessary infrastructure, domination of "black market," and other factors aggravate the negative consequences of such changes. Therefore, in spite of the criticism from the side of economic theory, the direct state price control is still considered by practical politicians as a method to "lower the costs of reforms," soften social tension and to restrict the destructive influence of inflation. The fact that the methods of the direct price management, lets even in the limited scales, are used in the developed countries also serves on behalf of such a policy.

Second, the state in transitional economy has strong unrealized influence on prices through its economic policy, first through taxation, state credits and investments, monetary policy and other macroeconomic aspects which are more or less connected with budgetary policy.

The budgetary deficit and related monetary and credit emissions are traditionally considered among main inflationary factors. In the transition to an open market economy their influence increases by the considerable part of the national income redistributed through the state budget (e.g., in Ukraine in 1992–1993 it was 0.65–0.7, by using of greater part of budget incomes for production purposes, by chronic lack of money under the conditions of high inflation, and by a number of other factors. The correlation between the monetary emissions and inflationary processes requires a special study for the transitional economy; such a study will be done in the next subsection.

The mentioned aspects are closely connected. For instance, subsidies for price control together with credits for unprofitable enterprises in public sector are one of the main reasons for large budgetary deficit. If the state tries to decrease the budgetary deficit then the state just increasing the tax rates that results in the decline of production. As a consequence, the real budget incomes are not increased, but diminished.

The complex system study of all aspects of state influence on prices is therefore necessary. The mathematical modeling can be a basis of such analysis. Some of the mathematical models will be designed and considered below.

It should be noted that models, similar to the considered above the dynamic detailed model were developed in 90th for several countries of East and Central Europe. The dynamical macromodel of Polish economy presented in [14] is one of them.

The model of Polish economy consists of five production sectors, the ones producing, respectively, the agricultural intermediaries $A$, non-agricultural intermediaries $M$, food $F$, industrial consumption goods $C$, and investment goods $I$, the banking sector $B$, the household sector $H$, the public sector (government plus public services) $G$, as well as the labor market. Flows of goods, services and labor between sectors and the outside world are presented in Table 1.2, while money flows are presented in Table 1.3.

These tables indicate only the structure of links between sectors (Table 1.4). Intensities of flows are determined by interaction between these sectors and the principle of minimum of supply and demand. The flows from $M$ and $A$ in Table 1.3 represent the intermediary inputs of, respectively, industrial and agricultural raw materials, while inflows to $H$ represent consumption of households.

Now let us consider how every sector is described in this model. We start from the production sectors. Each production sector uses intermediaries $M$ (sectors $A$ and $F$

**Table 1.3** Flows of goods, services, and labor between sectors

| | | To | | | | | | | |
|---|---|---|---|---|---|---|---|---|---|
| | C | Sectors | A | M | I | F | C | H | G |
| | o | A | x | | | x | | | |
| | u | M | x | x | x | x | x | | x |
| F | n | I | x | x | x | x | x | | x |
| r | t | F | | | | | | x | |
| o | r | C | | | | | | x | |
| m | y | H[a] | x | x | x | x | x | | x |
| | World | | x | x | x | x | x | x | |

[a] Labor supply

**Table 1.4** Money flows between sectors

| | | To | | | | | | | | | |
|---|---|---|---|---|---|---|---|---|---|---|---|
| | | Sectors | A | M | I | F | C | H | G | B | World |
| | C | A | X | x | x | | | x | x | x | x |
| | o | M | | x | x | | | x | x | x | x |
| F | u | I | | x | x | | | x | x | x | x |
| r | n | F | X | x | x | | | x | x | x | X |
| o | t | C | | x | x | | | x | x | x | x |
| m | r | H | | | | x | x | | x | x | x |
| | y | G | X | x | x | | | x | | x | x |
| | | B | x | x | x | x | x | x | x | x | x |
| | World | | x | x | x | x | x | x | x | x | |

including the intermediaries $A$ purchased from the domestic and foreign suppliers and delivers its output both to the domestic market and to export. A production sector determines its optimum production rate by maximizing the expected short-run profit:

$$\max_{L} \Pi^* = \max_{L} \left\{ P^* Q(L) - C\big[Q(L)\big] \right\}, \quad K(t) = \overline{K}, \tag{1.32}$$

where:

- $P^*$ is the index of expected sales price based on the freely determined price defined further;
- $C\big[Q(L)\big]$ is the costs of producing $Q(L)$ at full production capacity:

$$C\Big[Q(L)\Big] = p_K a K(t) + wL + \Big(p_A q_A + p_M q_M\Big)(1+r)Q(L) + D, \tag{1.33}$$

where $a$ is the depreciation rate; $w$ is gross wage rate; $L$ is the number of employees; $P_K$ – the price index of fixed assets; $P_A$ – the price index of supplies of product $A$ to a given sector; $P_M$ – the price index of supplies of product $M$ to a given sector; $q_A$ – the intensity of input $A$ in a given sector; $q_M$ – the intensity of input $M$ in a given sector; $r$ – the interest rate; $D$ – the repayment of interest on long-term loans.

The production capacity of a given production sector is described by the following production capacity function (Gadomski [14]):

$$Q = F(K, L) = LP_L = LP_L^* \left[ \frac{U}{U^*} \exp\left(1 - \frac{U}{U^*}\right) \right]^{\beta}, \tag{1.34}$$

where

$P_L^*$ – the parameter defining the maximum value of the average labor productivity;
$U$ – the capital to labor ratio: $U = K/L$;
$U^*$ – the value of capital to labor ratio for which $P_L(U^*) = P_L^*$;
$\beta$ – the substitution coefficient ($\beta > 1$).

This production capacity function is derived from the concept that wrong proportions of the production factors are a potential source of the production inefficiency. An isoquant of the production capacity function (1.34) is presented in Fig. 1.8. It can be noticed that the production is efficient whenever value of the capital-to-labor ratio $U$, $U = K/L$, satisfies the condition $U^{**} < U < U^*$, i.e., belongs to the substitution range. An attempt to produce using the value of $U$ from outside the substitution range results in inefficiency because the same output can be produced using less input of at least one production factor.

Having defined the costs and production functions, a relationship between "purely" technical and economic aspects of the production has been established. Taking into account that:

$MC$ – the marginal costs, $MC = \dfrac{dC}{dQ} = w\dfrac{1}{\dfrac{dQ}{dL}} + \left(p_A q_A + p_M q_M\right)(1 + r)$,

$AC$ – the average costs, $AC = C(K, L)/Q(L)$,

$AVC$ – the average variable costs, $AVC = wL + \left(p_A q_A + p_M q_M\right)(1 + r)$

one can notice that there is a relationship between the minimal points of $AVC$ and $MC$ as functions of the production volume and $L^*$ and $L_p$, corresponding respectively to the maximum of the productivity of labor and the inflection point of the short-term production function.

In Gadomski and Woroniecka [15] it was shown that the production capacity function (1.34) provides the correspondence between the short-term production costs and the employment for full utilization of the production capacity, Fig. 1.8. The average productivity of labor reaches the maximum at $L^*$ if the break-even point of perfect competition, i.e., when capital-to-labor ratio is equal to $U^*$. Further the increase of employment over the level of $L^*$ causes an increase of the productivity of capital and production but decrease of the average productivity of labor. The maximum productivity of the capital and the maximum production

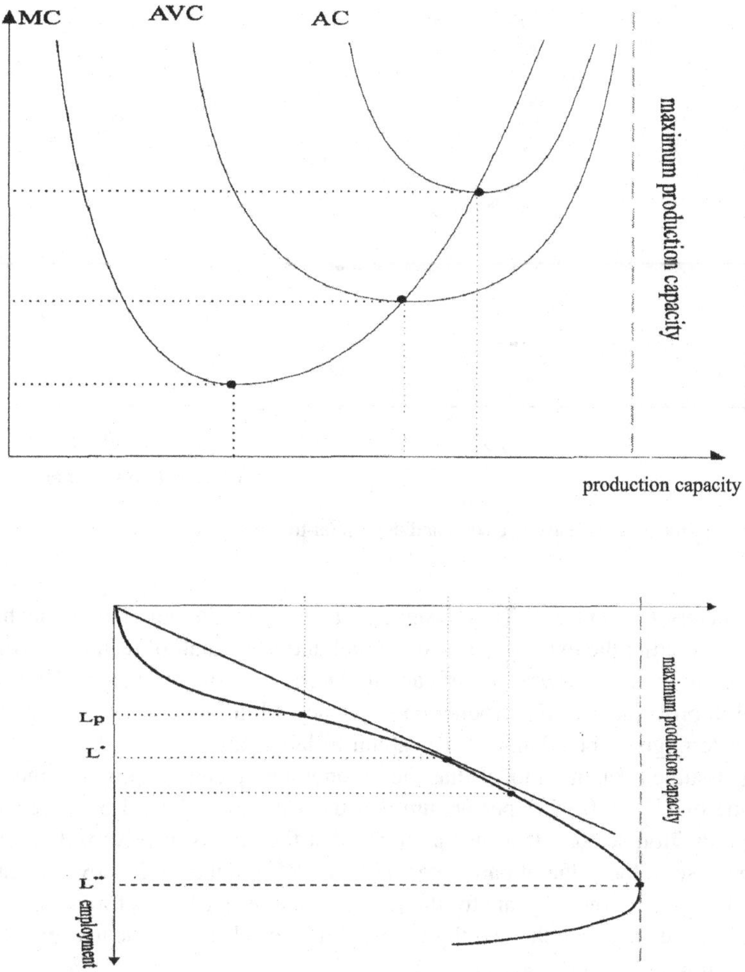


**Fig. 1.8** Correspondence between employment and production costs

is attained at $L^{**}$, i.e., when the capital-to-labor ratio is equal to $U^{**}$, $U^{**} = U^*(\beta - 1)/\beta$. The maximum marginal productivity is attained at $L_p$ (a root of the equation $\partial^2 Q/\partial L^2 = 0$) or, equivalently, at $U_p$. An optimum production and optimum employment are determined by the equality of the marginal costs and the marginal revenues.

In the intermediate-term analysis, i.e., when the inputs of both production factors change while the production technique remains constant, the average production costs under full utilization of production capacities are the function of the capital-to-labor ratio and are independent of the scale of production (see Fig. 1.9).

In the long-term the quantities of production factors as well as the production technique undergo changes. The technical progress is expressed by the changes

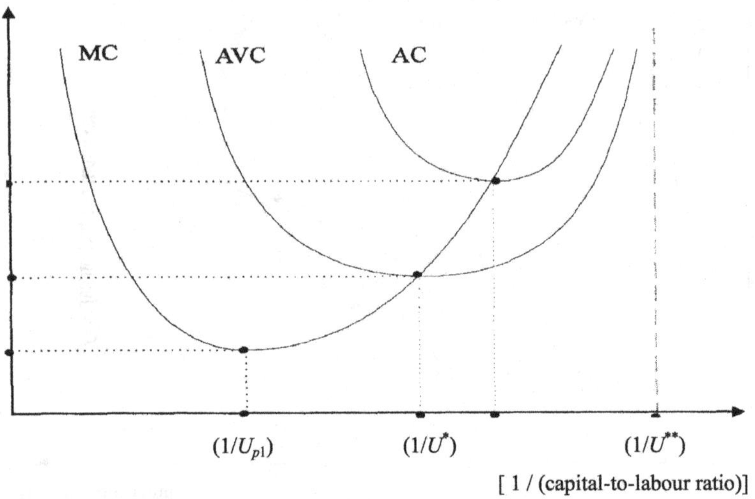


**Fig. 1.9** Intermediate-term average costs and the capital-to-labor ratio

of parameters $U^*$ and $P_L^*$. They change as a result of the autonomous technical progress affecting the existing stock of capital and as a result of technical progress embodied in the new generations of capital. An increase of parameter $P_L^*$ reflects a general improvement of the production efficiency while increase of $U^*$ represents the long-term growth tendency of the capital-to-labor ratio.

It is assumed for this model that the autonomous technical progress increases the value of $U^*$ by 0.75 % per annum and the value of $P_L^*$ by 1.5 % per annum in all production sectors. It is also assumed that the effects of privatization on the total increase of the value of parameters $U^*$ and $P_L^*$ amounts to 5 % over 5 years in sector $A$, $C$ and $F$, over 8 years for the sector $I$, and over 10 years for the sector $M$. The differences between the length of these time periods reflect the uneven rates of privatization in particular sectors.

Under such assumptions the analysis of the long-term efficiency of production is based on the investigation of the capital-to-labor ratio relative to the trajectories of $U^{**}$, $U^*$ (the limits of the substitution range) and $U_p$.

The actual production rate is defined as the minimum of an optimum production rate and production capacity. If the optimum production rate is less than production capacity then employment is reduced. If the production capacity is greater than the optimum production rate then the employment is being increased. Both labor reductions and increases cannot exceed an assumed amount. The excess employment causes the higher average production costs.

Now let us consider how authors of the model describe product market (see Gadomski and Woroniecka [15]).

The enterprises maintain the stocks of products in order to provide the continuity of sales. The producers observe the changes of the production costs, stock and the desired level of stock and react to them by adjusting the corresponding prices. The desired level of stock is a linear function of demand and is interpreted as such a filling of the distribution system that enables continuity of sales.

Three categories of prices are used in the model: the freely determined price, minimum price and sales price. Each sector sets the sales price while observing values of the freely determined price and the minimum price. The freely determined price $p_t^s$ is the price which would arise on the market if producers were insensitive to production costs. The freely determined price changes under the influence of stock changes and the divergence of stock from the desired level. The price increase $\Delta p_t^s$, is caused by the decrease of stock and/or by the decrease of the actual stock below the desired level:

$$\frac{\Delta p_t^s}{p_t^s} = \varsigma_i + \frac{\Delta Z_t}{Z_t} + \varsigma_2 \frac{Z_t - Z^*}{Z_t} \qquad (1.35)$$

where: $Z_t$ – the stock level, $Z_t^*$ – the desired stock level, $\varsigma_1$, $\varsigma_2$ – the parameters, $\varsigma_1, \varsigma_2 < 0$.

The minimum price, $p_t^{\min}$, is the level considered by the producer to be the lower limit on the sales price. This level is determined by the supply conditions, i.e., the production costs and the degree of monopolization within a given sector—the higher the monopolization the higher the minimum price:

$$p_t^{\min} = \delta AC_{t-1}, \qquad (1.36)$$

where: $\delta$ – the monopolization coefficient, $\delta > 0$, $AC_{t-1}$ – the average production costs.

In the case of perfect competition, the minimum price is defined by the minimum of average variable costs. In the monopolistic case, the minimum price would provide profit despite the production costs.

The sales price $p_t^d$ is equal to:

$$p_t^d = \max\left(p_t^{\min}, p_t^s\right). \qquad (1.37)$$

There are four sources of investment in the model: the depreciation fund, the net profit, the investment loans, and the foreign investments. The investment decisions are made based on the ratio of expected return to the interest rate.

A constant proportion of profit to the demand for investment loans is assumed. The subsequent decisions concern the structure of demand for investment goods (imported vs. domestic). These purchases are influenced by the ratio of prices of imported and domestic products and by the preferences to the supply source.

The model of the labor market includes the equations describing an employment in the particular sectors and formation of wages. The persistence of long-term unemployment is a prerequisite for the assumption on inherent disequilibrium of the labor market and application of the category of "normal" unemployment. the actual employment is defined as the minimum of supply and demand of labor.

The number of economically active persons determined in the demographic part defines the supply of labor. It is assumed that the number of economically active persons does not depend on the wage levels. The demand for labor is formed in the public and productive sectors. The demand of a given production sector is determined by the optimal employment level and the constraint on the maximum admissible employment change. The total demand for labor is the sum of effective demands from all sectors.

The following categories of wages are distinguished: the freely determined wage, the guaranteed wage, and the minimum wage.

The freely determined wages are the average wage that would result if a labor market was not regulated. The freely determined wage increases with the relative decrease of an unemployment and increases when the unemployment rate falls below the "normal" level:

$$\frac{\Delta w_t^s}{w_t^s} = \chi_i \frac{\Delta N_t}{N_t} + \chi_2 \frac{N_t - N^*}{N_t}, \tag{1.38}$$

where $w_t^s$ – the freely determined wage, $N_t$ – the unemployment, $N_t^*$ – the "normal" unemployment, $\chi_1, \chi_2$ – the parameters, $\chi_1, \chi_2 < 0$.

Equation (1.38) is interpreted in the following way. The unemployment $N$ is the result of the households' supply of the workforce and the demand for it in the sectors. An increase of unemployment lowers the average wage. In a contrary situation, the decrease in unemployment results in the increases of average wage. The changes of average wage are induced also by the divergence of unemployment from the long-run equilibrium (natural unemployment). The low unemployment – below the "natural" level – is a factor contributing to the rise of the average wage.

The guaranteed wage is determined in negotiations between the social partners: the government, employers, and employees. The changes of the guaranteed wage $\Delta w_t^g$ depend upon the rate of GDP growth and upon the rate of consumption price increase:

$$\Delta w_t^g = \nu_i \frac{\Delta p_t^c}{p_t^c} + \nu_2 \frac{\Delta Y_t}{Y_t}, \tag{1.39}$$

where $p_t^c$ is the consumption price index, $Y_t$ is the $GDP$, $\nu_1, \nu_2$ – the parameters, $\nu_1, \nu_2 > 0$.

The dependence of guaranteed wage upon price increase reflects the mechanism of wage indexation, while dependence upon the GDP growth defines the long-run relation of wages and the GDP.

The effective average wage in the economy $w_t^a$ is defined as the maximum of the freely determined and the guaranteed wage levels:

$$w_t^a = \max\left(w_t^g, w_t^s\right). \tag{1.40}$$

Whenever a sector has insufficient means to pay full wages the actual average wage is lowered beneath the normative average sectors' wages. The minimum wage is the lower limit for the case of financial losses in the production sectors and is related to the guaranteed wage. The average wages in particular sectors are determined in definite proportions to the average wage in whole economy.

Another large-scale economic model was evaluated for Poland simultaneously with the above-considered model. This is a simulation econometric model SEMP (see Babarowski et al. [2]). The model consists of 800 equations with more than detailed 1,000 variables. We cannot consider it in details due to a lack of space, therefore let analyze the pricing mechanisms taking into account in this model.

In accordance with the classical theory, determining the equilibrium in the product market at time $t$ can be reduced to computation of the price $p_t^*$, being a solution to the following equation

$$y_{dt}(p_t) - y_{st}(p_t) = 0, \tag{1.41}$$

where $t$ – the discrete time, $y_{dt}$ – the demand for product, and $y_{st}$ – the supply of product expressed in the same units.

If the supply and demand are explicit functions of the price, then the calculation of the equilibrium price does not cause any difficulties. The demand function (expressed in real units), when consumers' money is determined, is of hyperbolic type with respect to the price. More difficult situation is in the case of suppliers because, in general, the supply function is not given explicitly. After a disturbance, the equilibrium is established in a dynamic way. It is not known whether a producer adjusts the production to price changes according to the demand function determined in the previous equilibrium state, or modifies the supply function. It should be noted that it is impossible to observe the supply function values in more than one point. The current production and stocks determine this point.

Various pricing models (Keynesian, monetary, monopolistic, etc.) depending on the chosen hypotheses are built. These models allow to answer the following question: what are the price and supply responses to a rapid change of demand? In general, there are two basic hypotheses of the price formation. The first one assumes that the prices are determined by the demand, the second one states that the prices are determined by the production costs. In the framework of the first hypothesis two subclasses of producers are distinguished: one—with the full elasticity of supply and second—without the elasticity of supply with respect to the price level. Unfortunately, the basic elements of these hypotheses: the demand, supply, and costs depend upon many (often the same) factors. Due to this fact they cannot be considered independently. All of these facts are taken into account when the model is built.

In the considered model it is assumed that in the consumption goods market the nominal demand $Y_{dt}$ is expressed by an amount of money designed for the current consumption (characteristics of household sector), and the supply $y_{st}$ is expressed by a quantity of product in the constant prices distributed between the state and private producers which are determined independently. Both values are calculated with the use of dynamic models. Earlier the values of demand and supply play important role in these models. The changes of supply and demand values depend upon a large number of variables. A prediction of the price change plays an important, but not primary role in these changes. The supply and demand are expressed in different units, hence their comparison is possible when the price is known. If the previous price $p_{t-1}$ is known, then the measure of dynamical disequilibrium is the difference $\Delta Y_t^N = Y_{dt} - p_{t-1} y_{st} = 0$. It is the main reason of the price change $p_{t-1} > p_t$ excluding the case when the price is stimulated by the costs. Three main variants of the price change mechanism are taken into account:

(i) the dynamical disequilibrium brings the price to its equilibrium value; that is, for any previous value $p_{t-1}$ the new value $p_t$ fulfils the equation:

$$\Delta Y_t = Y_{dt} - p_t y_{st} = 0, \tag{1.42}$$

(ii) the price level $p_t$ is computed in the submodel of producers and depends upon the production costs and the preferred profit level:

$$p_t = g_{dt}, \tag{1.43}$$

where $g_{dt} = f(N_t, Z_{dt}, y_{st})$; $N_t$ – the production costs, $Z_{dt}$ – the preferred profit level.

(iii) the price $pt$ changes in the direction determined by the dynamical disequilibrium but does not attain the equilibrium value

$$p_t = p_{t-1}\left[1 + f\left(\Delta Y_t^N\right)\right] \tag{1.44}$$

In addition, more general "mixed" model of the price mechanism has been investigated. In this model, the market disbalance $\Delta Y_t^N$ as well as the production costs can simultaneously influence the price. The following formula has been applied:

$$p_t = \alpha_i p_{t-1} + \alpha_2 f(\Delta Y_t^N, y_{st}) + \alpha_3 f(g_{dt} - p_{t-1}), \tag{1.45}$$

where:

$\alpha_1$ is the "structural" inflation parameter (slightly greater than 1);
$\alpha_2$ is the weight of the relative disequilibrium in the market;
$\alpha_3$ is the weight of the difference between the current price and the price, resulting from the desired profit index (cost-determined-price).

The simulation model using the Polish economy information in the 1990s has demonstrated that Eqs. (1.42) and (1.45) describe the processes of inflation with approximately equal accuracy and they are more presize than Eqs. (1.43) and (1.44). The question, what kind of inflation, costs, or demand was dominated at that time, is still open. Probably, several costs and demand factors, which are correlated each with other has been implied the price growth. It should be noted that such factors as the retance, the following macromodel was proposed for Bulgaria in [26]. This model considers the difference in behavior of newly organized and former state-owned private enterprises as one of the main factors of high costs inflation in Bulgaria in the beginning of 1990s.

Bulgaria started the economic reforms in February 1991 with a *big bang* type of the price and trade liberalization (a detailed analysis of the reform process in Bulgaria is presented in [1]). The markets for goods and services, as well as the financial and the labor markets, were liberalized to a great extent, making space for competition and the appearance of a private sector. A free foreign exchange market with a floating rate was established and the tax system was changed considerably and became closely to the market standards.

The target of these significant institutional changes introduced at the onset of the reform process was to demonopolise the domestic and foreign trade, and to promote the competition in goods and financial markets allowing for a new private sector to appear. But even after these reforms, the state-owned enterprises still accounted for most of the production of goods and services in Bulgaria (as it was also in Romania and in the republics of the former Soviet Union – FSU). The weight of the private sector in GDP as of mid-1994, as estimated and published in Transition Report, EBRD, October 1994, was as follows: Bulgaria 40 %, Romania 35 %, most of the FSU republics – between 15 and 20 %. The weight was even higher in Hungary, 55 %, Poland, 55 %, Latvia, 55 %, and the Czech Republic, 65 %.

Reforms affected negatively the performance of state-owned enterprises. The first step undertaken in 1991 was to split the existing big state-owned companies and conglomerates into their component plants and factories. So, the traditional trade and technological relations were disrupted to give economic and financial independence to the separate small and medium size enterprises. This decentralization of the decision making process to a firm's level was a precondition for privatization.

Deprived of the upper level administration shelter the small and medium size firms had to establish their own trade relations resorting to the new tiny private firms appearing almost instantaneously and in large numbers. A private sector interlinked with the state-owned sector was born. The state firms found themselves trapped in-between private firms that performed both as suppliers of raw materials and as distributors of the final goods, ready to capture an increasing part of the rents by imposing price differentials. The state firms had to accept the new rules keeping producer prices quite low and lagging behind the general price level dynamics. On the other hand, the trade unions tried to keep real wages constant through indexation.

These two forces were sufficient to wipe out firms' operating profits and transform net profits into losses (see Commander and Coricelli [6]).

The disappearance of operating profits in the state-owned firms had a twofold impact on the overall performance of the economy: on one hand tax collections from the state-owned enterprises fell sharply and a fiscal deficit appeared. On the other hand, indebtedness of state-owned firms with banks increased abruptly through interest arrears which were honored with new loans.

The Central Bank issued money to finance both the growing fiscal deficit and the rediscounting of the new loans granted to enterprises.

To capture this twofold impact, the model presented in the paper includes the losses of the state-owned enterprises into the budget constraint of the Treasury and assumes that they are covered by money creation. Price and wage setting policies of the state-owned firms are explicitly considered in the model to allow for the analysis of different patterns of behavior.

The budget deficit in period $t(D_t)$ is defined as government expenditures $(G_t)$ less taxes $(T_t)$ plus losses of state-owned enterprises $(E_t)$:

$$D_t = G_t - T_t + E_t. \tag{1.46}$$

In the last two cases, a situation of shortage $Y_t > 0$ (a part of demand greater than supply is put out for the future or is covered by import) as well as of overproduction $Y_t < 0$ (increase of stocks) is possible.

Assuming that the nominal government expenditures and tax revenues are proportional to the nominal GDP $(Y_t)$ through constants $\overline{g}$ and $\overline{h}$, Eq. (1.46) was transformed into

$$D_t = \left(\overline{g} - \overline{h}\right) Y_t + E_t. \tag{1.47}$$

The Olivera–Tanzi effect (is an economic situation involving a period of high inflation in a country which results in a decline in the volume of tax collection and a slow deterioration of real tax proceeds being collected by the government of that country) is ignored to simplify the model. A model dealing with similar problems and including the Olivera–Tanzi effect is presented in Heyman and Canavese [12].

The supply and demand behavior of goods and services produced by state-owned enterprises is introduced in Eqs. (1.46) and (1.47). Econometric estimations show that the price elasticity of demand for goods produced by state-owned enterprises is quite low (see Nenova [25], where it was estimated that the price elasticity of demand for goods produced by a sample of 3,720 state-owned firms is equal to –0.1) and so it is assumed that the demand depends only on the real income $(y_t)$ through a constant $\overline{s}\overline{S}$

$$\text{quantity demanded} = \overline{s}\, y_t. \tag{1.48}$$

The supply side is presented by a fixed coefficient production function with only one input, labor $(L)$

$$\text{quantity demanded} = \frac{1}{l} L_t, \tag{1.49}$$

where $l$ is the labor-output ratio.

It is also assumed that the quantity produced by the state-owned enterprises is always equal to the demanded quantity. From (1.48) and (1.49), the losses of state-owned enterprises can be presented in a detailed form as a function of the output, employment, labor-output ratio, prices and wages:

$$E_t = w_t L_t - p_{s,t} \overline{s}\, y_t = w_t l\, \overline{s}\, y_t - p_{s,t} \overline{s}\, y_t, \tag{1.50}$$

where $w_t$ and $p_{s,t}$ are wages paid and prices of goods and services produced by state-owned enterprises.

By introducing Eq. (1.50) into (1.47), $D_t$ can be obtained as

$$D_t = \left(\overline{g} - \overline{h}\right) Y_t + w_t l\, \overline{s}\, y_t - p_{s,t} \overline{s}\, y_t. \tag{1.51}$$

The general price level $\pi$ is defined as a geometric mean of prices set by the private firms ($p_t$) and the prices set by the state-owned enterprises, hence, the rate of inflation is

$$\hat{\pi}_t = \alpha \hat{p}_{i,t} + \beta \hat{p}_{s,t}, \tag{1.52}$$

where $\alpha$ and $\beta$ are constant weights such that $\alpha + \beta = 1$ and the hat over a variable denotes its rate of change over time.

Relative prices and real wages are introduced to analyze the impact of real changes in the prices and wages on losses

$$\overline{p}_{s,t} = p_{s,t} / \pi_t \tag{1.53}$$

and

$$\overline{w}_t = w_t / \pi_t. \tag{1.54}$$

Since the real output can be represented by the nominal output $y$ and the general price level $\pi$ ($y = Y / \pi$), the relative prices (1.53) and real wages (1.54) can be explicitly inserted into the total deficit equation (1.51)

$$D_t = (\overline{g} - \overline{h}) Y_t + \overline{w}_t l \overline{s} Y_t - \overline{p}_{s,t} \overline{s} Y_t. \tag{1.55}$$

The losses of state-owned enterprises are covered by the loans or by the accumulation of arrears to the commercial banks. These debts are rediscounted at the central bank and this transaction is equivalent to the direct financing of losses by issuing money.

If the total fiscal deficit is financed by printing money, the rate of growth of money supply is:

$$\frac{D_t}{M_t} = \frac{\Delta M_t}{M_t} = \hat{M}_t, \tag{1.56}$$

where $M_t$ is the money supply and $M_{t-1}/M_t$ is linearly approximated by $1 - \hat{M}_t$.

To emphasize the link between the rate of inflation and the pricing behavior of state-owned enterprises, a constant income velocity of money is assumed ($v = M/Y$). Equations (1.55) and (1.56) can be combined to get the equality

$$\hat{M}_t = (\overline{g} - \overline{h})v + \overline{w}_t l \overline{s} v - \overline{p}_{s,t} \overline{s} v \tag{1.57}$$

or

$$\hat{M}_t = d + s(\overline{w}_t l - \overline{p}_{s,t}), \tag{1.58}$$

where

$$(\overline{g} - \overline{h})v = d, \tag{1.59}$$

$$\overline{s} v = s. \tag{1.60}$$

Equation (1.58) is the classical expression presenting inflation as the product of the fiscal deficit-to-GDP ratio and the income velocity of money. It also shows explicitly that the losses of public enterprises weighted by their share in total output are a part of the deficit financed by issuing money.

The close relationship between the growth in money supply and the losses of state-owned enterprises has been introduced in Eq. (1.58). The pricing behavior of the private sector affects the inflation rate in a different way. It is assumed that prices in the private sector are set following a mark-up rule on costs. Hence, their dynamics depend on changes in wages in the private sector and in prices set by the state-owned enterprises which are inputs for private sector production. It is also assumed that they are sensitive to the changes in real income as a proxy for the behavior of demand. Equation (1.61) shows the rate of change of goods prices and services produced by the private sector using as inputs labor and the goods supplied by the state-owned enterprises:

$$\hat{p}_{i,t} = \mu_i \hat{w}'_t + \mu_2 \hat{p}_{s,t} + \mu_3 \hat{y}_t, \tag{1.61}$$

where $w'$ is the level of wages in the private sector.

The coefficients $\mu_1$ and $\mu_2$ should add up to unity as they represent the weights in the private sector costs of labor and the goods produced by state-owned enterprises. The constant $\mu_3$ is a measure of the mark-up sensitivity and the flexibility of private sector prices to changes in aggregated demand.

It is also assumed that wages in the private sector are fully indexed to the past inflation. The fact that the majority of private sector firms are small-scaled, organized within the members of the family, upholds this assumption.

The next equation presents the private sector rate of wage change:

$$\hat{w}'_t = \hat{\pi}_{t-1}. \tag{1.62}$$

Equations (1.61) and (1.62) can be inserted in (1.52) to get the inflation rate

$$\hat{\pi}_t = \alpha\mu_i\hat{\pi}_{t-1} + (\alpha\mu_2 + \beta)\hat{p}_{s,t} + \alpha\mu_3 Y_t - \hat{\alpha}\mu_3\hat{\pi}_t \\ - (\alpha\mu_2 + \beta)\hat{\pi}_t + (\alpha\mu_2 + \beta)\hat{\pi}_t, \qquad (1.63)$$

where the approximation of the real growth rate $\hat{y}_t = \hat{Y}_t - \hat{\pi}_t$ has been used.

The constant income velocity of money assumption implies $\hat{Y}_t = \hat{M}_t$ and so

$$\hat{\pi}_t = A\hat{\pi}_{t-1} + B(\hat{p}_{s,t} - \hat{\pi}_t) + Cd + Cs(\overline{w}_t l - \overline{p}_{s,t}, \qquad (1.64)$$

where

$$A = \mu_1/(\mu_1 + \mu_3), \qquad (1.65)$$

$$B = (\alpha\mu_2 + \beta)/\alpha(\mu_1 + \mu_3), \qquad (1.66)$$

$$C = \mu_3/(\mu_1 + \mu_3). \qquad (1.67)$$

Equation (1.64) is the reduced form of the model. It highlights the impact on inflation of the inertia due to the past inflation indexation of private sector wages, the wedge between the general price level dynamics and the changes in prices set by the state-owned enterprises and the size of the fiscal deficit and the losses of state-owned enterprises.

Important conclusions about the price dynamics and interdependence between the inflation and losses of former state-owned enterprises were made using the above-considered model (1.64)–(1.67).

Despite of peculiarities for different models of the cost inflation in a transition economy all of them consider monopolism and high-costed technologies as the key factors of price growth. Therefore, this inflation was not only the result of pre-reform disproportion in the sphere of finances and it cannot be overcomed only by tools of monetary policy. Deep structural reforms in the real sector are necessary to normalize the processes of money turnover.

## 1.4 Modeling of Inflationary Expectations

The inflation and disbalance processes belong to the most typical and the most serious problems of the transition period. Their investigation is rather complex and needs the simultaneous consideration of different economic levels (microeconomics; intermediate level: branches, regions; macroeconomics). The cost and demand inflation, "artificial" inflation of monopolies and specially formed inflationary expectations produce their joint impact on general economic destabilization characterized by means of the growth of "black market" turnover and lack of a market infrastructure. Therefore, an integrated system study of all aspects of pricing is vitally necessary for the transitional period. Mathematical modeling can be a basis for such study. A model can help us to make general interpretations of phenomena. It is important because some policy variables such as taxes can be rapidly changed and this produces difficulties for the detailed studies. Of course, any study requires numerically sound approaches and real data illustrations. The dynamic aspects of models are essential as far as dynamic and unstable processes are concerned. Several models to this end are considered in this section.

It should be noted that modeling of inflation is far from a complete framework even for the market economics. Some results in this area, for instance, presented in Straffa [32], have mainly asymptotical values.

The situation in the area of postinflation analysis is quite the same. The fiscal deficits have increased inflation in Bulgaria and Romania but not in the case of Russia (see Komulainen and Pirttila [18]). Many faces and interdependencies between the inflation and stocks and bonds, nominal and real interest rates, GDP deflator, etc. by means of statistical analysis are studied in Kiani [16].

Despite of this a great deal of experience for decision-making has been collected in several papers, including Johansson [13], Naslund [24], Wyzan [38]. The application of this experience for the economic transition is also the subject of this Section. The situation in Ukrainian economy was taken as the example of transition economic reforms in Eastern Europe referring to their difficulties, contradictions, and problems which cannot be easily solved.

The section has the following structure. First a dynamical macromodel is described. Using this model, we consider the macroeconomic processes connected with the market disbalances and disproportions between supply and demand for goods, services on the consumer market. But such "classical" inflationary processes are not the main source of price growth in a transition economy. In this Section we show that the root of inflation is connected with structural disproportions in industry and monopoly impact on prices. The strong monetary policy been typical as antiinflationary measure transforms these disproportions into payment crisis, the same as "freezing" price policy transforms the demand inflation into permanent deficit of goods.

Then we analyze the results of model runs based on real data concerning Ukrainian economy. Background of recommendations about measures which can prevent high inflation is elaborated there.

The simultaneous analysis also has been done for the postinflation period. Industrial decline, payment crisis and other typical phenomena of this period were illustrated from the viewpoint of the mentioned model. The mentioned phenomena created the environment for wide implementation of the "soft budget constraints." Analysis of their peculiarities concludes this Section.

The proposed model, which is a generalization of macromodel (1.2) describes some inflationary processes at the early stages of economic transition. Let us consider a closed economic system which can be described in terms of aggregated indicators. Insofar as it is a dynamic system, all its indicators are continues functions of time $t$. Three types of inflation typical for a transition period are taken into account in the model. They are:

(a) the demand inflation on the consumer goods market;
(b) the demand inflation in the productive sector of economy (industry, agriculture, etc.);
(c) the costs inflation in the productive sector.

It should be noted that the consumer market and productive sphere are served by different money flows in a transition economy (cash, as a rule, for consumers and cashless for enterprises). The principles of this flows creation are also different. The finances of state-owned (or formerly state-owned) enterprises are not fully separated from the state finances in a transition period. The state subsidies, credits, and other types of the direct financial state support cover the essential part of expenditures of enterprises. As a result the demand of enterprises cannot be considered as the function of their incomes, it is the function of the indicators of the current financial and monetary policy of the state. The changes in money supply are considered in aggregated macromodel as the main such indicator. The consumer's demand is mainly the function of real incomes of consumers.

The scenario of cost inflation in a transition economy was considered in Sect. 1.3. An exponential growth of cost is typical for this scenario. The total cost also depends on prices for imported goods, which are the functions of the world market prices in hard currency and exchange rates.

The proposed macromodel is based on the above-mentioned assumptions and uses the following notation. Let $x(t)$ be the gross social product (GSP) of the system in constant prices at time $t$, $a$ the share of total costs in GSP, therefore $1 - a$ be the share of national income ($NI$); $R(t)$ the part of $NI$ (in constant prices) used for consumption; $\tilde{w}$ is the consumption rate, $w$ is the accumulation rate; $w_0$ is the minimal value of the accumulation rate which is needed to keep a constant level of production; $S(t)$ is the cash-paying demand of consumers; $D(t)$ is the money balances held by consumers at time $t$; $p(t)$ is the consumer price index relative to time $t = 0$; $p_1(t)$ is the component of this index be determined by supply–demand pricing; it corresponds to the demand inflation on the consumer goods market; $p_2(t)$ is the component determined by changes of cost; it corresponds to the costs inflation; $p_3(t)$ is the component to be determined by the changes of money supply and inflationary exceptions; it corresponds to the demand inflation in the productive sector of economic; $\overline{p}(t)$ is the world market price index relative to time $t = 0$; $\hat{p}(t)$ is the relative exchange rate of national currency calculated to the time instance $t = 0$; $b$ is the reciprocal of the rate of return on investments; $q$ is the share of consumer incomes in national income.

Then our model takes the following form:

$$\frac{dx}{dt} = b^{-1}(1-a)(w-w_0)x; \quad R = (1-a)\tilde{w}x \tag{1.68}$$

(the dynamic balance of GSP and consumption fund in constant prices);

$$\frac{ds}{dt} = \frac{d}{dt}\left(\frac{D}{p}\right) \tag{1.69}$$

(the equation of demand dynamics);

$$\frac{dD}{dt} = pq(1-a)x - p\min(S, R) \tag{1.70}$$

(the equation of consumer money balances);

$$\frac{dp_i}{dt} = m(S - R) \tag{1.71}$$

(the Walras–Samuelson's equation for supply–demand pricing, $m$ is the coefficient of price elasticity);

$$p_2 = F\left(\overline{p}, \hat{p}\right) e^{\mu t} \tag{1.72}$$

(the equation of dynamic costs, empirical dependence $F(\overline{p}, \hat{p})$ and reflects the impact of the external aspects of costs growth, $\mu$ is the rate of costs growth as a result of structural inflation crisis, its value can be estimated using approach of the previous Section and data of multibranch balances);

$$\ln\frac{M}{p_3} = \alpha E + \gamma \tag{1.73}$$

(the Cagan's equation for "monetary" pricing, see [4, 5]); $M$ is the value of money supply, $\alpha$ and $\gamma$ are the given coefficients, the inflationary expectations $E$ are obtained as:

$$E = \int\limits_{t-\tau}^{t} \beta(t)\frac{dp}{dt}p^{-1}dt, \tag{1.74}$$

where $\tau$ is the time interval of accumulation of expectations, $\beta(t)$ is the waged function for this interval);

$$p = \max\left(p_i, p_2, p_3\right) \tag{1.75}$$

(the dependence of total prices on their separate components).

The initial conditions for the variables of Eqs. (1.68)–(1.75) are given in Cauchy form: $x(0) = x^0$, $p_i(0) = p_i^0$, $i = \overline{1,3}$; $S(0) = S^0$, $D(0) = D^0$.

The value of $E$ is assumed to be known for the time interval $[0; \tau]$.

The another dependence between $p$ and $p_i$ $(i = \overline{1,3})$ in (1.75) can be used as model variants. For instance, it can be

$$p = y_1 p_1 + y_2 p_2 + (1 - y_1 - y_2)\, p_3 \tag{1.76}$$

where $y_1$ and $y_2$ are given coefficients.

The total income of producers $H = p \min(S, R)(1 - a)^{-1} - pax$ is also calculated in the model. It should be noted that the value of $H$ can be negative. It means the growth of total producers' debt and, as a result, a deep payment crisis. The detailed analogue of the model (1.68)–(1.75) will be also elaborated. It takes into account several branches and groups of consumers, relative prices can also be different for different sectors of economy. But the equations in a such model have no principal differences from (1.68) to (1.75) equations.

The real-data runs were made on the basis of the model (1.68)–(1.75) to estimate the impact of different inflationary factors during Ukrainian high inflation in 1992–1994.

The initial data for model runs were taken from Multibranch Balance [23] and Ukrainian Economics in Figures [34]. Judgements of experts from Ukrainian National Institute of Economics Programs were used to estimate $m$ and $w_0$ parameters. The structure of the model gives us the possibility for separation analysis of impact of different inflationary factors. For instance, assuming $F(\overline{p}, \hat{p}) \equiv \text{const}$ we eliminate the impact of the external price shock. If we assume that $M = C$, $E = 0$, we eliminate the impact of "monetary" pricing. We obtain several such variants; the impact of every factor can be estimated as a result of compared variants.

Parameters $\alpha, \gamma, \tau$ of Eqs. (1.73)–(1.74) were estimated on the basis of statistical data (see Table 1.5). Our analysis of the data for Ukraine in 1992–1993 has shown that the value of $\tau$ was practically zero. This was so because the prices were mainly increased by administrative means at that time and it was impossible to predict the time and the magnitude of price increment by analyzing the statistical data for a large time interval. The economics' agents behaved accordingly.

**Table 1.5** Money supply increment and expectations during high inflation

| Month, 1993 | Money supply (M1 aggregate, bln.crb) | Consumer price index (% to previous month) | Expected annual inflation (%) |
|---|---|---|---|
| January | 2,575 | 208.89 | 13,066 |
| June | 6,099 | 171.7 | 8,604 |
| July | 9,527 | 137.6 | 4,512 |
| August | 16,101 | 121.7 | 2,610 |
| September | 22,513 | 180.3 | 9,631 |
| October | 28,757 | 166.13 | 7,936 |
| November | 34,828 | 145.27 | 5,428 |

In view of above, Eq. (1.73) takes the form

$$\ln \frac{M}{p_3} = \alpha \frac{dp}{dt} p^{-1} + \gamma. \quad (1.77)$$

Our comparison of Ukrainian values of $\alpha$ and $\gamma$ with the corresponding values for hyperinflations in countries with a developed market economy in the 1920–1950s (see [32]) has shown that for the 1993 hyperinflation in Ukraine the unit impact of inflationary expectation $\alpha$ was an order of magnitude lower, while the direct effect $\gamma$ of changes in money supply was 20–40 % higher. This supports the conjecture that the cost inflation in productive sphere (in industry, agriculture, etc.) predominated within the inflationary processes in transition economies of FSU countries.

The model (1.68)–(1.75) runs show the following impact of main factors of inflation. The price growth in 1991–1992 was mainly the result of demand inflation. The share of consumer incomes in national income dramatically increased in 1989–1990 due to particular liberalization of foreign trade in the former Soviet Union and absence of mechanisms which can create the additional consumer incomes into real investments. The industrial decline started at the end of the 1980s. Therefore at the moment of price liberalization in January 1992 the overdemand at the consumer goods market was twice greater than total supply (see Table 1.6). The chaotic price growth created a deep price disequilibrium in different branches of industry. Thus the cost inflation is started. The external oil/gas price shock and internal mechanism of price growth as a result of low technological level and high monopolism were the main reasons of the cost inflation in 1993. This fact was illustrated by the results of model runs, presented in Table 1.7.

**Table 1.6** The modeling results and consequences of high inflation in 1992–1994

| Date | GDP deflator (1.1.91 = 1) | Consumers money stock (bln.crb) | Overdemand (+) Oversupply (−) (% to total supply) |
|---|---|---|---|
| January 1992 | 2.21 | 250.67 | 109.42 |
| February 1992 | 6.02 | 434.7 | $-6 \times 10^{-3}$ |
| April 1992 | 5.12 | 464.21 | 13.8 |
| June 1992 | 8.48 | 551.02 | 6.52 |
| August 1992 | 11.75 | 581.7 | −2.65 |
| October 1992 | 23.76 | 704.5 | −13.13 |
| January 1993 | 71.31 | 942.19 | −24.5 |
| July 1993 | 397.03 | 3,144.0 | −25.3 |
| September 1993 | 1,276.0 | 5,290.0 | −26.9 |
| March 1994 | 1,504.0 | 6,388.0 | −26.6 |
| July 1994 | 2,428.0 | 10,184.0 | −22.9 |
| October 1994 | 2,633.0 | 11,134.0 | −21.5 |

**Table 1.7** The real and forecasted values of GDP deflator in 1993

| Month, 1993 | Real values (1.1.91=1) | Forecusted values | | |
|---|---|---|---|---|
| | | Variant 1 | Variant 2 | Variant 3 |
| January | 73.11 | 20.06 | 48.03 | 71.31 |
| February | 94.153 | 21.5 | 68.31 | 77.51 |
| March | 114.98 | 22.9 | 97.12 | 97.13 |
| April | 142.16 | 24.3 | 138.1 | 138.11 |
| May | 181.38 | 25.7 | 196.37 | 196.37 |
| June | 311.43 | 27.09 | 279.22 | 279.23 |
| July | 428.53 | 28.5 | 397.03 | 397.03 |
| August | 521.7 | 29.2 | 480.25 | 507.1 |
| September | 940.41 | 29.9 | 564.5 | 1,276.0 |
| October | 1,259.9 | 31.37 | 801.73 | 1,454.0 |
| November | 1,930.11 | 31.81 | 1,141.0 | 1,504.0 |

The values of Weighed Price Index (GDP deflator) were forecasted by runs of the above-mentioned model under several assumptions.

For Variant 1 only supply demanded pricing was taken into account. The values of Weighed Price Index (WPI) were calculated as a result of cost pricing for Variant 2. The joint impact of costs and monetary pricing was the basis for forecasting by means of Variant 3. Our comparison of forecasted values of WPI with observed values of GDP deflator have been demonstrated the accuracy of modeling and the real impact of each of the inflationary factors. More than 80 % of the price growth was the result of costs inflation. The monetary factors such as large monetary emissions produce essential effect only in January and September of 1993. The part of additional price growth at the end of 1993 was the result of monopolistic pricing together with an administrative overestimation of the possible costs increase. The real rates of costs inflation at the beginning of 1994 (about 10–25 % monthly) were smaller than the forecasted ones (about 25–35 %) as a result of such overestimation. The impact of external and internal aspects of costs growth was approximately equal.

The high inflation in Ukraine was ended in 1994. The rate of inflation decreased from more than 3,000 % in 1993 to 120 % in 1995 and 46 % in 1996, but the situation in industry and finances didn't become better. The industrial decline is continued and the payment crisis becomes deeper and more destructive. Let us analyze this phenomena in detail.

The inflation was stopped mainly by the restrictive monetary and credit policy of the National Bank of Ukraine. The enterprises could not sell their production in the situation when the total demand is sufficiently smaller than the total supply (see Table 1.6) under current prices which are calculated on the basis of costs pricing. As a result the pricing mechanism (1.75) was substituted by (1.76) mechanism with small values of $y_2$ coefficient. In this situation the difference between the

**Table 1.8** The computational results for postinflation period

| Month, 1996 | Price index (1.1.96 = 1) | Cost price index (1.1.96 = 1) | Total debts of enterprises (% to GSP) |
|---|---|---|---|
| January | 1.0 | 1.0 | 0 |
| February | 1.05 | 1.12 | 0.24 |
| March | 1.09 | 1.17 | 0.53 |
| April | 1.11 | 1.25 | 0.8 |
| May | 1.13 | 1.39 | 5.35 |
| June | 1.15 | 1.53 | 9.6 |
| July | 1.19 | 1.68 | 13.6 |
| August | 1.26 | 1.83 | 17.2 |
| September | 1.29 | 1.91 | 18.1 |
| October | 1.33 | 1.99 | 20.4 |
| November | 1.40 | 2.13 | 23.4 |
| December | 1.46 | 2.28 | 26.1 |

current prices and production costs must be covered from some financial source. The subsidies from the state budget were first used as the mentioned source, but such practice created a large deficit of the budget and was stopped mainly in 1995.

The bankruptcy legislation is not very strong in Ukraine and even the existing bankruptcy procedures do not work always. The enterprise managers used the debts in such situation as a source of selfsubsidies to cover the high costs of their production. They did not pay for previously consumed energy, raw materials, and other products of the intermediate consumption. This artificially (and not legally) has decreased the costs of the finite products and gives them possibility to reduce prices without losses in profits. Such practice has created the deep payment and budgetary crisis in 1995–1997 years which dominated during a postinflation period.

We have used the model (1.68)–(1.76) to estimate the value of such selfsubsidies and their share in the total debts in Ukrainian economy. The part of the prices which are covered by such selfsubsidies is equal to the difference between costs price (1.75) and combined price (1.76). Therefore the total debts of enterprises $B$ at the given time interval $[t_0, t_1]$ will be equal to

$$B = \int_{t_0}^{t_1} (p_2 - p)\, x dt. \tag{1.78}$$

The computational results are presented in Table 1.8. The values of costs price index (1.75) and combined price index (1.76) are calculated together with $B$ values for $y_1 = 0.45$ and $y_2 = 0.55$. It should be noted that the values of price index (1.76) are close to the real values in 1996 and the calculated value of $B$ reflects only a part of total debts. If we take into account the debts which were made before 1996 and debts resulted from payment crisis then the share of total debts in GSP in 1996 must be more than 33 % which corresponds to reality. But selfsubsidies created the main part of these debts (about 80 %). The high rates of total debts growth are typical for this stage of postinflationary processes. Deep technological changes which can decrease the cost of production is the main way to overcome such problems.

The next model, suggested by Piontkivsky [29], is based on Cagan's modified equation and can be applied for a dynamics description of inflationary seniorage and the rate (speed) of supplanting the national currency as a mean of accumulation (such process being called "dolarization").

Unlike the described above, in the modified equation there are not only internal factors of inflation (inflationary expectations) taken into account, but also the external ones (the expected tempos of national currency devaluation). As a consequence the Cagan's equation gets the form:

$$\ln \frac{M}{p} = \alpha_1 E_1 + \alpha_2 E_2 + \gamma, \tag{1.79}$$

where $E_1$ is the expected inflationary level; $E_2$ – the expected devaluation tempos; $\alpha_1, \alpha_2$ – some constants.

Supposing the stability of real GNP and of real percentage rate during the period of simulation, the value $M/p$ can be considered to be the real demand on money $m$. In a similar way, the real demand on foreign currency will be determined with a correlation:

$$\ln f = \alpha + \gamma_1 \pi_e^* + \gamma_2 E_2, \tag{1.80}$$

where $\pi_e^*$ is the expected inflationary level in the country of currency origin, $\gamma_1$ and $\gamma_2$—some constants.

Under such conditions the real profit from the inflationary seniorage $S$ can be estimated as the multiplication of nominal rates of monetary volume increase $\overline{\mu}$ by the volume of real demand for money $m$:

$$S = \overline{\mu} m = \overline{\mu}\, e^{\alpha_1 E_1 + \alpha_2 E_2 + \gamma}.$$

The expected devaluation rates can be considered to be equal to the difference between the rates of inflation in the country with a transition economics and in the country of foreign currency origin: $E_2 = E_i - \pi_e^*$. For that it is enough to suppose that the parity of currency relative to the consumer ability is being kept. It follows:

$$S = \overline{\mu}\, e^{(\alpha_1 + \alpha_2) E_1 - \alpha_2 \pi_e^* + \gamma}. \tag{1.81}$$

The described above Eqs. (1.80)–(1.81) are used for the estimation of inflationary expectations.

Taking into account the made assumptions, the value of unpayments $W$ can be considered as a linear function from seniorage $S$, real monetary volume $m$ and demand on foreign currency $f$:

Here the value of $S$ is determined by Eq. (1.81); $m$ – Eq. (1.79); $f$ – Eq. (1.80).

The empirical estimation of parameters $\alpha_1, \alpha_2$ and $\gamma_1, \gamma_2$ using the monthly data starting from December 1992 till December 2000 was carried out in Piontkivsky [29], the value of aggregate $M2$ (broad monetary volume) minus deposits in the foreign currency being considered as $M$. The consumer price index was chosen as the price index $p$. The inflationary expectations $E_1$ were estimated as rates of monthly increase of consumer price index, and devaluation expectations $E_2$ – as rates of changing of coupon-karbovanets exchange rate (till August 1996) and UAH exchange rate (after August 1996) against US dollar. Qualitative variables $D93$ and $D95$, taking value 1 after December 1993 and April 1995, were added into Eqs. (1.79) and (1.80). These variables reflect changes that took place in Ukrainian economy after getting out from strong inflation 1993 and starting carrying out the stabilization monetary policy in 1995. In these equations the autocorrelation processes of the first row to regressants are also taken into account.

After estimating the parameters with the help of method 1-MLS the transformed Eqs. (1.79) and (1.80) have the following form:

$$\begin{aligned}
\ln m_t = \; & \underset{(0.011)}{0.125} + \underset{(0.87)}{0.00006}\, E_2 - \underset{(0.000)}{0.013}\, E_i + \underset{(0.000)}{0.19}\, D93 \\
& + \underset{(0.023)}{0.007}\, E_i D95 + \underset{(0.000)}{0.93} \ln m_{t-1}; \\
\ln f_t = \; & -\underset{(0.27)}{0.018} + \underset{(0.000)}{0.004} E_2 + \underset{(0.000)}{0.98} \ln f_{t-1} - \underset{(0.002)}{0.18}\, D93,
\end{aligned}$$

where

$m_t$ – the value of real demand on money at the moment of time $t$;
$m_{t-1}$ – the value of this indicator at the previous moment of time $(t-1)$;
$f_t$ – the value of demand on foreign currency at the moment of time $t$;
$f_{t-1}$ – the value of this indicator at the previous moment of time $(t-1)$.

The values of $p$–statistics are indicated in brackets under coefficients.

It's necessary to note that according to received correlations, the weakly value of expected devaluation $E_2$ and not significant influences on the demand on national currency, and the value of inflationary expectations $E_1$ do not have any influence on it. The influence of inflationary expectations on the value $m$ essentially decreased after April 1995. A positive coefficient with $D93$ in the first equation means that the real demand on money in Ukraine has been constantly decreasing after getting out from strong inflation, and quick increase of non-payments at that time can't be explained with insufficient monetary base. The decrement of nominal money supply changed from 100 to 120 % in some months of 1993 till 7–8 % in 1995 has also essentially reduced inflationary seniorage and has decreased the motivation of economical subjects to actions, directed onto its redistribution. Nevertheless, non-payments at that time has essentially increased. The reasons for such a process being dealt with the costs inflation mechanism.

In a transition economy some firms and enterprises posses the large bargaining and lobbing power due to paternalistic relations between them and the state. These firms operate under conditions of soft budget constraints: the part of their expenditures (sometimes, the main part) is covered by the state finances. They have some special conditions for payments to their partners in business. Such situation creates the core of a "shadow" (illegal) economy, but it also impacts on a payment crisis. Several models of soft budget constraints influence appear in the countries with a transition economy. Further, we will consider one of the. It was proposed by Vyshnya [35]. The econometric model analyzes the interrelation between debt dynamics and the presence of soft constraints using the official statistical data and nonofficial estimates of some indicators of "shadow" economy. The model was developed for Ukraine, but it also can be applied to other post-communist countries with widespread corruption.

The aim of this model is to investigate whether the soft budget constraints are granted subject to bargaining and lobbying. Specifically, our attention is concentrated upon the explanation of tax arrears to the state and local budgets, which are assumed to be one of the forms of SBC in transitional economies. In other words, the model examines the relationship between the amount of tax indebtedness of Ukrainian enterprises on one hand and their bargaining power on the other hand.

This concept can be expressed in the following form:

***Tax Arrears to the Consolidated Budget = f (Bargaining Power)***

What are proxies for independent and dependent variables? Capture theories extension of rent-seeking prompts that firms may use their profit to lobby government. Therefore, it is possible to assume that more profitable enterprises can use larger amount of money for lobbying. In other words, their bargaining potential seems to be higher. At the same time, it is necessary to pay attention mainly to after tax profit, since this money can be used by managers for any non-economic activity. Summing up, it is offered to use *after tax profit* of enterprises as a proxy for their bargaining power.

At the same time, the choice of dependent variable appears to be obvious—it is necessary to collect data about the overdue tax indebtedness of domestic firms to the state and local budgets.[1] However, it is impossible to acquire these data because Ukrainian Ministry of Statistics offers only the data about tax arrears of *industrial enterprises* to the consolidated budget and does not calculate the same statistics for other economic participants (i.e., agriculture, transportation, etc.). Alternative sources, for instance TACIS, offer annual aggregated data about tax indebtedness of all domestic economic participants. However, they do not show overdue tax arrears to the budget for each sector of the economy. As a result, tax indebtedness of domestic firms to the state and local budgets cannot be included directly in the model.

[1] We should concentrate upon the overdue tax arrears because they are equivalent to "flow of tax liabilities." It is necessary to remember that the difference between the flow and the stock of tax arrears is crucial, and that only the flow of tax liabilities is an implementation of current soft budget constraints.

A good proxy for the dependent variable is the overdue arrears of domestic firms, including tax and intraenterprise indebtedness. It seems to exhibit high positive correlation with tax arrears and it is covered completely in official and alternative statistical sources. Moreover, since enterprises' indebtedness to the budged is concerned, it is necessary to look at *overdue credit arrears* of domestic economic participants.

It is necessary to add that, in order to minimize the influence of an industry size upon its profit and arrears, it is offered to divide both sides of the equation by quantity of employees. It is assumed that larger industries have more employees. This statement seems to be true even for the post-communist countries, where the hidden unemployment is widespread. Therefore, the relationship between *per employee overdue credit arrears* of industries on one hand and their *per employee after tax profit* on the other hand is analyzed.

Moreover, in order to eliminate the inflation it is offered to express the data related to the dependent and independent variables in constant prices. The traditional Ukrainian base year 1990 is used. It is also necessary to stress out that a more detailed explanation of all proxies to be offered.

As a result, the previous model is transformed into the following form:

***Per Employee Overdue Credit Indebtedness=f (Per Employee after Tax Profit)***

or

$$Y_t = a_0 + \sum a_i^* X_{t-i} + u_t, \tag{1.82}$$

where

$Y_t$ is the per employee overdue credit indebtedness of firms of particular sector of the economy in year $t$ expressed in 1990 year prices;
$X_{t-i}$ is the per employee after tax profit of a particular industry in year $t - i$ expressed in 1990 year prices.

The model has a linear form. It is chosen because, on one hand, there exist no theoretical investigations examining the actual form of relationship between tax arrears and after tax profit of firms, and, on the other hand, other functional forms seem to be less appropriate for this particular case.

First, the reciprocal functions presume inverse relationship between dependent and independent variables. Second, the logistic form is appropriate when dependent variable takes values between 0 and 1. Third, the quadratic form seems to have no clear economic interpretation, because it presumes that arrears would decrease for small profits up to a certain point. Nevertheless, it is presumed that normally arrears should decrease if profit is generated, or it may increase if profit is used for rent-seeking activity. However, in the latter case an increase in arrears is generated more likely by higher profit.

The same argument is valid for the interaction form, as far as it is extremely difficult to give a clear economic interpretation for the interaction variable. Finally, the logarithmic functions are used when changes of both dependent and independent variables are very small. However, our statistical data demonstrate that an increase in arrears is significant during the latest couple of years. Therefore, the linear model seems to exhibit the clearest relationship between the total credit indebtedness and after tax profit of firms.

Another question to cover is why a dynamic model is used. The obvious answer is that the effects of economic and other variables are rarely instantaneous, and it takes some time for economic agents to respond. By the same logic, producers seem to be reluctant first to evaluate their bargaining potential generated in the previous periods and than to use it for lobbying a government to acquire soft privileges in the next periods.

It is also important to discuss the choice of dependent variables. Traditionally, $X_t$ is also included in the model. Nevertheless, in this situation it seems reasonable to assume that firms cannot use the profit generated in period $t$ for rent-seeking activity at the same period of time, because this profit first has to be estimated and reported and only after that, usually at the end of the year, a firm can use it for its non-economic activity. Therefore, it is reasonable to presume that the rent-seeking activity is affected by the profits generated in the previous periods.

Now we are ready to answer the following three questions. First, why the credit indebtedness rather than the net arrears of firms are chosen to be a dependent variable? To answer this question, let us recall that the subject of investigation is the amount of enterprises' arrears to the State and local budgets or, in other words, their credit indebtedness. At the same time, the State is also indebted to companies. As a result, the credit arrears of domestic firms might be caused be growing State debt. Therefore, it seems reasonable to pay attention to the *net* indebtedness of companies. However, in Ukraine it is impossible to calculate net arrears directly because the credit indebtedness is expressed in terms of the sale prices while the debit indebtedness is based on costs. In other words, the debit indebtedness excludes the profit of firms that has to be covered by their debtors. Therefore, the governmental statistics artificially undervalues the amount of the State debt to domestic companies and overvalues firms' arrears to the Budget. Thus, the difference between the credit and debit indebtedness of domestic enterprises (or their net arrears) would be artificially enlarged. Summing up, it is better to use the total overdue credit arrears as a proxy for the Soft Budget Constraint (SBC) because in such a case both the dependent variable (per employee overdue credit arrears) and the independent variables (per employee after tax profit) would be expressed in the market-based prices.

The second question is to discuss the sample range for both dependent and independent variables. It is necessary to stress that tax arrears ($Y_t$) are analyzed for the period 1996–2002, and after tax profit of companies ($X_{t-i}$) covers the period of 1995–2001 respectively. These ranges are explained by two reasons. First, the phenomenon of tax arrears enlarged drastically only since 1995. Second, The Ministry of Statistics started declaring the debit and credit indebtedness for each sector of the economy only since 1996. Previously, the total debit and credit indebtedness of Ukrainian firms were divided into "indebtedness to domestic firms, NIS companies, and other foreign enterprises." (The New Independent States (NIS) which comprise the 15 former Soviet Union republics except the three Baltic States—namely Armenia, Azerbaijan, Belarus, Georgia, Kazakhstan, Kyrgyzstan, Moldova, Russia, Tajikistan, Turkmenistan, the Ukraine, and Uzbekistan—have vast commercial, economic, and scientific potential.) Therefore, the model covers the relationship between per employee credit arrears and per employee after tax profits in all available periods.

The last problem is covered, why per employee statistics is used to mitigate the influence of an industry size upon its indebtedness and profit. As it has been already mentioned, it is assumed that larger enterprises hire more labor. This statement seems to be true, since it is generally agreed that Ukrainian enterprises use extremely depreciated equipment and that old-guard managers are not reluctant to spend money upon technological innovations. Therefore, our assumption seems to sound reasonable.

Moreover, it is possible to argue that there exists a huge hidden unemployment in Ukraine. Employees work only two or three hours per day in this situation. As a result, the quantity of employees is not very important. Nevertheless, this tactics is traditional for all domestic enterprises. Therefore, in any case, it is possible to state that larger firms still have more employees. Moreover, the data about really worked hours per employee is available only since 1998. Thus, it cannot be applied to our analysis.

It also seems reasonable to choose an output of each particular industry as a weight. However, in a country where hidden sector is higher than the official (legal) one, this variable appears to be even more questionable. As a result, the quantity of employees seems to be the best available weight.

To sum up, the model is aimed to demonstrate whether the soft budget constraints acquisition is caused by any bargaining and lobbying. The proxy for the dependent variable (SBC) is the amount of per employee overdue credit arrears of firms of all sectors of domestic economy expressed in 1990 year prices. The proxy for the independent variable (bargaining potential) is per employee after tax profit of companies of all sectors of Ukrainian economy measured in constant prices (1990). The model is expressed in dynamic form because the effects of independent variables upon the dependent one are not instantaneous. Finally, the investigation covers the period 1995–1998 when tax arrears expended drastically.

After our detailed description of the model, it is useful to examine our investigation results.

It is necessary to remember that our model is expressed in the following form:

$$Y_t = a_0 + \sum a_i^* X_{t-i} + u_t,$$

where $Y_t$ is the per employee overdue credit indebtedness of firms of particular sector of the economy in year $t$ expressed in 1990 year prices;

$X_{t-i}$ is the per employee after tax profit of firms of particular sector of the economy in year $t - i$ expressed in constant prices.

**Table 1.9** Relationship between per employee credit arrears and per employee after tax profits of firms in 1997 year

LS // Dependent variable is **LK1997**
Date: 06/04/99 Time: 15:29
Weighting series: W97
Sample (adjusted): 1 16
Included observations: 16 after adjusting endpoints

| Variable | Coefficient | Std. Error | $t$-Statistic | Prob. |
|---|---|---|---|---|
| LPROF1996 | 0.001183 | 0.000358 | 3.307766 | 0.0052 |
| LPPROF1995 | 1.30E−05 | 1.45E−06 | 9.007214 | 0.0000 |
| Weighted statistics | | | | |
| $R$-squared | 0.985620 | Mean dependent var. | | 5.40E−08 |
| Adjusted $R$-squared | 0.984593 | S.D. dependent var. | | 1.34E−07 |
| S.E. of regression | 1.66E−08 | Akaike info criterion | | −35.71154 |
| Sum squared resid. | 3.86E−15 | Schwarz criterion | | −35.61497 |
| Log likelihood | 264.9893 | $F$-statistic | | 959.5538 |
| Durbin–Watson stat. | 2.033405 | Prob($F$-statistic) | | 0.000000 |
| Unweighted statistics | | | | |
| $R$-squared | 0.123971 | Mean dependent var. | | 5.93E−08 |
| Adjusted $R$-squared | 0.061397 | S.D. dependent var. | | 6.50E−08 |
| S.E. of regression | 6.29E−08 | Sum squared resid. | | 5.55E−14 |
| Durbin–Watson stat. | 2.115456 | | | |

The estimation procedure covers the relationship between per employee overdue credit arrears in 1997 and 1998 on one hand and per employee after tax profit in 1995–1997 respectively on the other hand for all sectors of Ukrainian economy. In other words, the model takes the form:

$$\begin{aligned} Y_{1997} &= a_0 + \sum a_i^* X_{1997-i} + u_{1997}, \\ Y_{1998} &= a_0 + \sum a_i^* X_{1998-i} + u_{1998} \end{aligned} \tag{1.83}$$

Table 1.9 presents the estimated regression coefficients and associated statistics for per employee credit arrears in 1997.

According to Table 1.9, all coefficients are significant for 5 % test. Actually, this model has 14 d.f., and $t_{14}^*(0.05/2) = 2.145$. Thus, for each regression coefficient to be significantly different from zero, the $t$-statistics given in Table 1.9 must be greater than 2.145. It should be pointed that the regression coefficients for LProf 1996 and LPprof 1995 are significant because 3.307766 > 2.145 (for LProf 1996), and 9.007214 > 2.145 (for LPprof 1995). The same conclusion can be made based on the $p$-value approach. According to this criterion, each regression coefficient is significantly different from zero if its $p$-value is less than the level of significance. In our example the level of significance is equal to 5 % or 0.05, and $p$-value for LProf 1996 and LPprof 1995 are less than 0.05: $0.0052 < 0.05$ and $0.0000 < 0.05$.

Moreover, according to $F$-statistics, it is possible to conclude that all regression coefficients have a significant joint effect on $Y$. Actually, for this model, $k - 1 = 2$ and $T - k + 1 = 16 - 3 + 1 = 14$. The degrees of freedom $F$-statistics are therefore 2 for the numerator and 14 for the denominator. From the $F$-table, the critical value for a 5 % test is $F_2, 14^*(0.05) = 3.74$. Since the $F$-value in Table 1.6 is 959.5538 ($959.5538 > 3.74$), we reject the null hypothesis that all the regression coefficients are zero.

The value of adjusted $R2$ also exhibits a good general relationship between dependent and independent variables. However, it can be slightly overvalued because lagged independent variables may cause the multicollinearity problem. However, even in the case of multicollinearity, the Ordinary Least Squares (OLS) estimators are still unbiased, efficient and consistent, and the $t$-test is still valid. The only negative outcome of multicollinearity is a possibility to undervalue $t$-statistics. Nevertheless, the previous investigations demonstrate that all the coefficients are significant; therefore, it is not necessary to care about this problem.

Recall that OLS makes the assumption that the variance of the error term is constant. If the error terms do not have constant variance, they are said to be heteroskedastic, i.e., with differing variance. The OLS estimates are not BLUE (Best Linear Unbiased Estimator) if among all the unbiased estimators, OLS does not provide the estimate with the smallest variance.

It is necessary to mention that these results are acquired after the adjustment for heteroskedasticity. The estimation was provided with the help of WLS procedure. Therefore, these estimates are more efficient than OLS estimates. Moreover, WLS (Weighted Least Squares) estimates are well-behaved or BLUE.
Moreover, special analysis demonstrates the absence of serial correlation (Fig. 1.10).

Figure 1.10 exhibits that the tendency for successive residuals to cluster on one side of the zero line or the other does not exist. It is the graphical representation of the absence of serial correlation.

Another possibility to verify this statement is to examine the Durbin–Watson statistics. For arrears in 1997 the Durbin–Watson (DW) statistics is $d = 2.033405$. The DW statistic is a test statistic used to detect the presence of autocorrelation (a relationship between values separated from each other by a given time lag) in the residuals (prediction errors) from a regression analysis.

The number of observations is 16 and the number of explanatory variables, $k'$, is 2. For a one-tailed test the critical values are $d_l = 0.982$ and $d_u = 1.539$. Since $d > d_u$, it is possible to conclude that the first-order serial correlation in the residuals is absent at the 5 % level.

The underlying analysis demonstrates that our model seems to provide correct explanation between the dependent and independent variables. Statistics demonstrate that there exits no autocorrelation in the residuals, and estimated coefficients

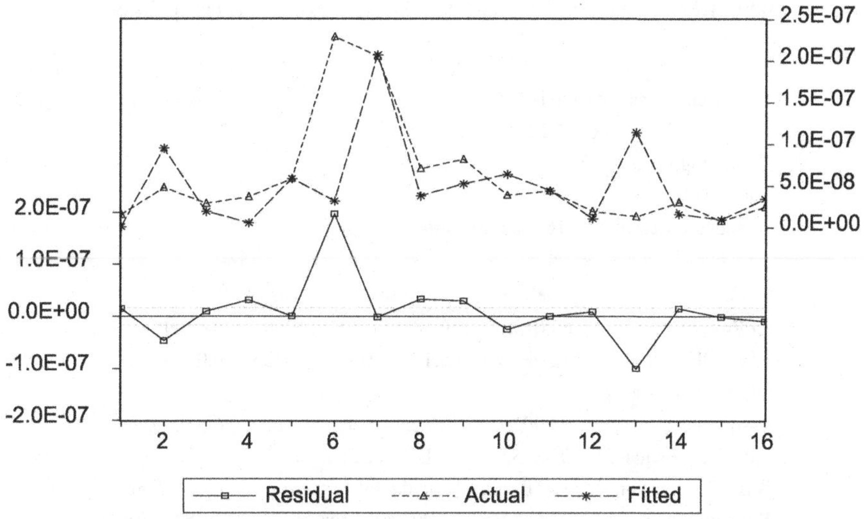


**Fig. 1.10** Autocorrelations in residuals for credit indebtedness of firms in 1997

are significant and BLUE. The same analysis is conducted for the relationship between per employee overdue credit arrears of Ukrainian firms in 1998 on one hand and their per employee after tax profits in previous years on the other hand.

The estimated regression coefficients and associated statistics for per employee credit arrears in 1998 are given in Table 1.10.

It is obvious that there exist no constant term in Table 1.10. In the intermediate stage its $p$-value exceeded even 0.1. It means that this coefficient was insignificant even at 10 % level. However, the coefficients for dependent variables are significant both in the intermediate and final stages. According to Table 1.10, they are significant for 5 % test. Actually, this model has 14 d.f., and $t_14^*(0.05/2.) = 2.145$. Thus, for each regression coefficient to be significantly different from zero, the $t$-statistics given in Table 1.10 must be greater than 2.145. We note that the regression coefficients for LProf1996 and LPprof1995 are significant because 7.160684 > 2.145 (for LProf 1996), and 23.34526 > 2.145 (for LPprof 1995). The same conclusion can be made based on the $p$-value approach. In our example the level of significance is equal to 5 % or 0.05, and its $p$-value for Prof 1997 and Prof 1996 are less than 0.05: 0.0000 < 0.05 and 0.0000 < 0.05 respectively.

According to $F$-statistics, it is possible to conclude that all regression coefficients have a significant joint effect on $Y$. Actually, for this model, $k - 1 = 2$ and $T - k + 1 = 16 - 3 + 1 = 14$. The degrees of freedom $F$-statistics are therefore 2 for the numerator and 14 for the denominator. From the $F$-table, the critical value for a 5 % test is $F_2, 14^*(0.05) = 3.74$. Since the $F$-value in Table 1.10 is 1,423.892 (1,423.892 > 3.89), the hypothesis that all the regression coefficients are equal to zero was rejected.

**Table 1.10** Relationship between total credit arrears and after tax profits of firms in 1998

LS // Dependent variable is **LK1998**
Date: 06/04/99 Time: 15:22
Weighting series: W98
Sample (adjusted): 1 16
Included observations: 16 after adjusting endpoints

| Variable | Coefficient | Std. Error | $t$-Statistic | Prob. |
|---|---|---|---|---|
| LPROF1996 | 0.001934 | 0.000270 | 7.160684 | 0.0000 |
| LPPROF1995 | 1.20E−05 | 5.13E−07 | 23.34526 | 0.0000 |
| Weighted statistics | | | | |
| $R$-squared | 0.990264 | Mean dependent var. | | 6.35E−08 |
| Adjusted $R$-squared | 0.989568 | S.D. dependent var. | | 1.42E−07 |
| S.E. of regression | 1.45E−08 | Akaike info criterion | | −35.97786 |
| Sum squared resid. | 2.96E−15 | Schwarz criterion | | −35.88128 |
| Log likelihood | 267.1198 | $F$-statistic | | 1,423.892 |
| Durbin–Watson stat | 2.209110 | Prob($F$-statistic) | | 0.000000 |
| Unweighted statistics | | | | |
| $R$-squared | 0.207316 | Mean dependent var. | | 7.65E−08 |
| Adjusted $R$-squared | 0.150695 | S.D. dependent var. | | 9.80E−08 |
| S.E. of regression | 9.03E−08 | Sum squared resid. | | 1.14E−13 |
| Durbin–Watson stat | 2.150764 | | | |

The value of adjusted $R2$ also exhibits a good general relationship between dependent and independent variables. Moreover, since the analysis was provided with the help of the Weighted Least Squares (WLS) procedure (to eliminate heteroskedasticity), the estimators are BLUE.

It is also possible to conclude that there is no serial correlation in residuals. Figure 1.11 exhibits that the tendency for successive residuals to cluster on one side of the zero line or the other do not exist. It is the graphical representation of the absence of serial correlation. This conclusion is strengthened by the DW (Durbin–Watson) statistic. For arrears in 1998 the DW statistics is $d = 2.209110$. The number of observations is 16 and the number of explanatory variables $k'$ is 2. For a one-tailed test the critical values are $d_l = 0.982$ and $d_u = 1.539$. Since $d > d_u$, it is possible to conclude that there is no first-order serial correlation in the residuals at the 5 % level.

To sum up, the underlying models provide a good explanation of the relationship between the total overdue credit arrears of Ukrainian firms in 1997 and 1998 years on one hand and their after tax profits in 1995–1997 on the other hand. The estimated coefficients are significant and BLUE. Therefore, the model can be used to analyze the actual relationship between the dependent and independent variables.

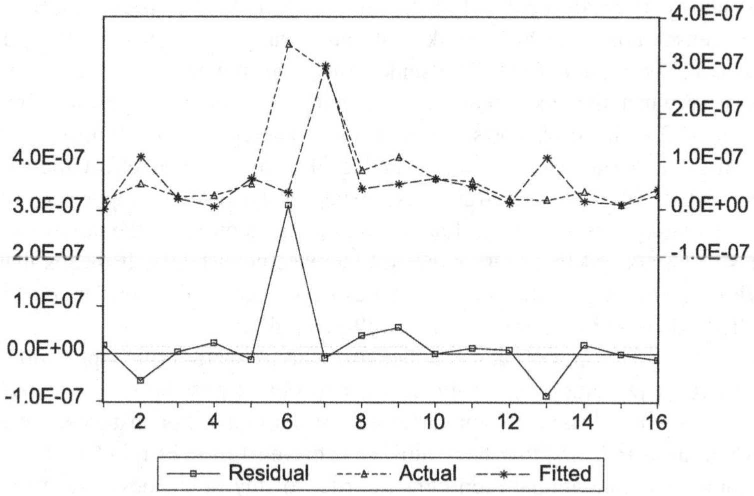


**Fig. 1.11** Serial correlation in residuals for credit indebtedness of firms in 1998

Data presented in Tables 1.9 and 1.10 can be briefly summarized by the following formulas:

$$\text{Per Employee Overdue Credit Arrears}_{1998}\text{ (estimated)}$$
$$= 0.0019336393^{*}(\text{ Per Employee After Tax Profit})_{1996}$$
$$+1.979588^{*}10 - 5^{*}(\text{ Per Employee After Tax Profit})_{1995};$$

$$\text{Per Employee Overdue Credit Arrears}_{1997}\text{ (estimated)}$$
$$= 0.0011830715^{*}(\text{Per Employee After Tax Profit})_{1996}$$
$$+1.3039236^{*}10 - 5^{*}(\text{ Per Employee After Tax Profit})_{1995}.$$

Both equations demonstrate that there exist a strong positive relationship between the profitability of enterprises and enlargement of their tax arrears. It may mean that the original statement that firms spend their profits generated in previous periods to acquire the possibility not to pay taxes in the future seems to be true. For instance, according to the first equation, for an average Ukrainian enterprise, a per employee increase in tax indebtedness in 1998 by one 1990 ruble is financed by approximately 845 (= 1/0.0019336393) 1990 rubles of per employee profit obtained in 1996 and about 76,692 (= 1/1.979588*10 – 5) base year rubles of per employee profit generated in 1995. At the same time, in 1997 a one 1990 ruble increase in per employee tax arrears costs an average domestic enterprise about 517 (= 1/0.0011830715) 1990 rubles of per employee profit obtained in 1996 and about 50,516 (= 1/1.3039236*10 – 5) base year rubles of per employee profit generated in 1995.

The underlying results demonstrate that rent-seeking activity is extremely costly. These expenses comprise huge sunk costs and yearly premiums for the official (governmental) representatives. The sunk cost is the money paid by industries in 1995 a year when the tax arrears start growing. According to results obtained by Vyshnya [35] the sunk costs are enormous. Therefore, only limited quantity of industries, for instance, agriculture and coal mining, could effort themselves to become tax debtors. Moreover, these expenses have to be supported by the additional annual premiums to prolong the arrears agreement for the next year.

It has to be noticed that these expenses increase persistently. It means that the industries willing to join the tax arrears business later have to pay much higher prices. This idea can be supported by the following figures.

According to these graphs, new-comers, like energy sector or transportation, pay higher share of per employee profit in order to obtain possibility not to pay the taxes in the future. However, their expenses are not equal. For instance, the firms of transportation industry that have already achieved full debt relief in 1997 pay rather some moderate annual premiums in order to stay in business. Nevertheless, these premiums are much higher per ruble (each time per "national currency" might be submitted by any of them, for example, "Ukrainian Grivna" introduced later on) of indebtedness than for other sectors that have been traditionally supported by the government (like coal mining).

At the same time, the energy sector that has started acquiring some privileges from the Government only a year ago and that has never obtained possibility not to pay the taxes, faces enormous rent-seeking costs.

Figures 1.12 and 1.13 demonstrate that there exist sectors of the domestic economy, called other industries, which have huge tax arrears and pay minimum

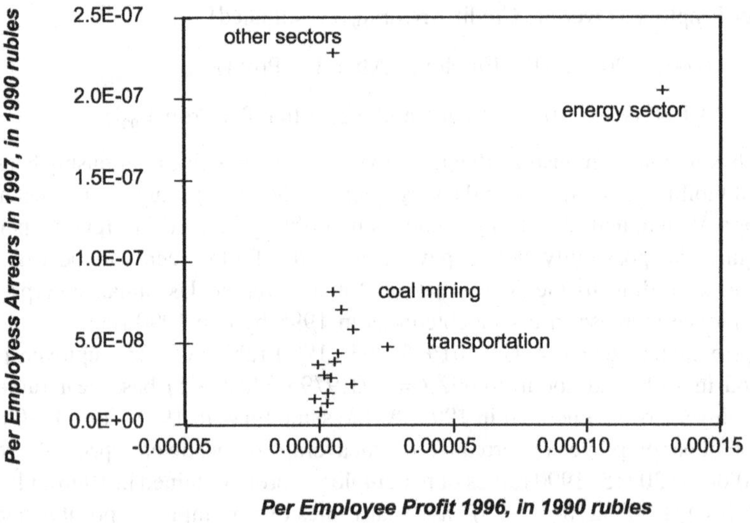


**Fig. 1.12** Cross-industry relationship between arrears in 1997 and profit 1996

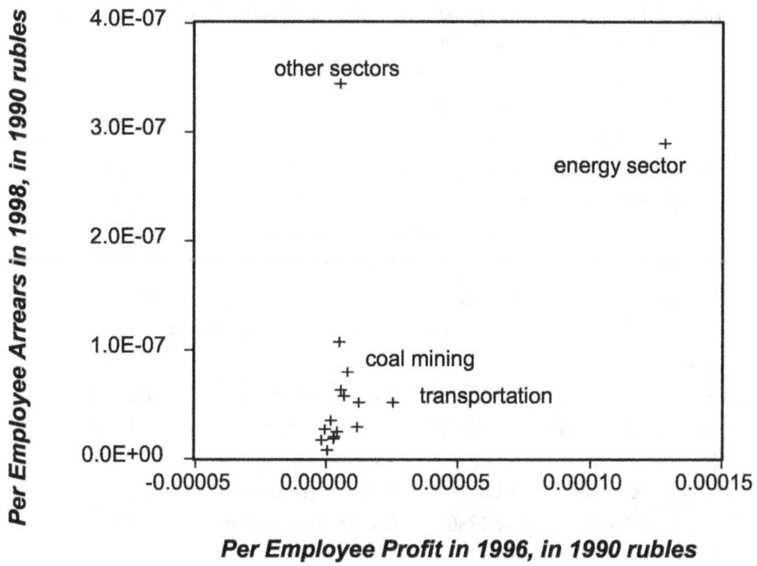


**Fig. 1.13** Cross-industry relationship between arrears in 1998 and profit 1996

rent-seeking costs. They include mainly the state non-profitable industries, like forestry or fishery. As far as the share of the state capital for them is prevailing, they do not have to pay for soft budget constraints.

Anyway, it is necessary to take into account that the rent-seeking activity is financed by the profits generated in several periods. It means that firms expect that they would acquire possibility not to pay the taxes in the future, and they plan their expenditure in advance. Therefore, it is possible to conclude that expectations play an important role in the enlargement of tax arrears. Moreover, these expectations are strengthened by a real possibility to obtain SBC privileges. It means that, in announcing a struggle with tax debtors, the Ukrainian government fails in advance, since previously it has created a strong tradition for the tax privileges acquisition through bargaining and bribes.

To sum up, it is possible to assume that the enlargement of nonpayments to the State and local budget are caused by bargaining and bribes. Moreover, the rent-seeking costs are extremely high. Therefore, only a few industries can effort themselves not to pay the taxes regularly. At the same time, the expectations and the failed Government commitment to handle this problem seem to strengthen the tradition of tax arrears.

Nevertheless, a complete understanding of the relationship between the rent-seeking and soft budget constraints is impossible without clarifying the problem of causality. Therefore, it seems reasonable to investigate whether this positive relationship is mutual.

**Table 1.11** The relationship between per employee profit of industries in 1997 and their per employee arrears in 1996

LS // Dependent variable is **LPROF1997**
Date: 06/04/99 Time: 15:44
Weighting series: WPR97
Sample (adjusted): 1 6
Included observations: 16 after adjusting endpoints

| Variable | Coefficient | Std. Error | $t$-Statistic | Prob. |
|---|---|---|---|---|
| C | 4.88E−06 | 2.52E−07 | 19.35462 | 0.0000 |
| Weighted statistics | | | | |
| $R$-squared | 0.982305 | Mean dependent var. | | 4.37E−06 |
| Adjusted $R$-squared | 0.982305 | S.D. dependent var. | | 8.17E−06 |
| S.E. of regression | 1.09E−06 | Akaike info criterion | | −27.31388 |
| Sum squared resid. | 5.90E−12 | Schwarz criterion | | −27.34859 |
| Log likelihood | 74.42802 | Durbin–Watson stat. | | 1.716805 |
| Unweighted statistics | | | | |
| $R$-squared | 0.013142 | Mean dependent var. | | 4.77E−06 |
| Adjusted $R$-squared | 0.011142 | S.D. dependent var. | | 1.03E−05 |
| S.E. of regression | 1.03E−05 | Sum squared resid. | | 5.30E−10 |
| Durbin–Watson stat | 2.522524 | | | |

The model discussed above demonstrates that an increase in the industry's rent-seeking potential eases an acquisition of the soft budget constraints. However, this conclusion will be meaningless if we find out that an expansion of the tax arrears also has a positive influence upon industry profitability. Therefore, this subsection is aimed to investigate the inverse relationship.

The following model traces the connection between the per employee profit of domestic industries in 1997 and their per employee tax indebtedness in 1996. The model covers only these years because of the lack of data. However, if the model demonstrates that the relationship between these variables is negative or zero, it will be possible to conclude that an expansion of arrears does not cause an increase in industries' profitability in general.

The results of the model are summarized in Table 1.11.

Table 1.11 exhibits that, after an adjustment for heteroskedasticity typical for cross-section data, the changes in per employee arrears of industries have no impact on their per employee profitability. Alternatively, the per employee profitability is determined independently of industries' arrears. This conclusion is strengthened by the value adjusted $R2$ for the weighted results. At the same time, Table 1.11 demonstrates that there is no serial correlation in residuals.

Summing up, the underlying model shows that per employee profit of industries generated in 1997 is formed independently from their tax arrears. This conclusion gives us the possibility to mention that an expansion of tax indebtedness of the domestic firms does not lead to an increase in their profit in general. Therefore, the previous model, which stated that an acquisition of the soft budget constraints is affected by the rent-seeking power, provides valid results.

The investigations allow us to conclude that strong bargaining power based on firms' profit eases an acquisition of the soft budget constraints in Ukraine. This result seems to support the main conclusion of capture theories which stress that firms may use their profit for the rent-seeking activity. Moreover, the investigations provide a support for one of the postulates of soft budget constraints concept proclaiming that SBC are obtained subject to lobbying and bargaining. Nevertheless, since investigations were provided only for Ukraine, the extension of these conclusions to more general cases requires experiments to be provided in different countries.

Moreover, it is highly desirable that later experiments cover larger period of time. For instance, monthly data might be used.

At the same time, it seems reasonable to choose weights other than the total quantity of employees per industry. As it has been already mentioned, hidden unemployment in some transitional economies makes this weight not very strong. Therefore, it would be much better to use the value added per industry as the weight for both the total tax arrears and total profit of a particular industry.

Summing up, it is possible to say that the underlying investigations demonstrates a positive relationship between an increase in industries' profit on one hand and an enlargement of their tax arrears in Ukraine on the other hand. Moreover, according to our analysis the rent-seeking activity is costly. Therefore, only a few sectors of the economy could effort it to themselves. These results seem to support the capture theories conclusion and SBC postulates.

## Concluding Remarks

The investigations of peculiarities of inflation and post-inflation phenomena (such as payment crisis) in the transition economy allow us to state the following conclusions:

1. The specificity of a transitional economy produces the special forms of pricing and inflationary processes. A high level of expenditures typical for the system of technologies of the former of Soviet Union countries together with sufficient influence of monopolies lead to the costs growth. Sooner or later costs pricing will be dominated under such conditions and costs inflation became the main reason for the price growth.
2. In order to develop an anti-inflation policy for a transition economy it is important to distinguish inflation that directly results from or does not directly result from monetary circulation. The anti-inflationary monetary measures are restricted by the specificity of costs inflation in a transition economy. The price growth will be substituted by payment crisis and industrial decline, if cardinal changes in

technologies don't take place. Therefore, a monetary policy must be combined with policies of far-reaching structural changes in the economy, for instance, with technological projects of the government.

3. The optimum money supply in a transition economy is the function of costs pricing and has been calculated for every time period. If money supply is less than the optimum, the deficit of money is increased and the economic decline is worsened.
4. If the nominal money supply is larger than its optimal value, the deficit of money can be reduced only for restricted time interval. Then the real money supply, recalculated according to changed prices, will not be greater than the optimum.
5. Subsidies for some industries are necessary during the high external price shock, but their structure and values must be analyzed together with the general monetary policy and budget dynamics.

## References

1. Avramov, R., Antonov, V.: Economic Transition in Bulgaria. Agency for Economic Coordination and Development, Sofia (1994)
2. Babarowski, J., Gutenbaum, J., Inkielman, M.: Simulation of the inflation process in a macroeconomic model. The case study of Poland. In: Kulikowski, R., Nahorski, Z., Owsinski, J. (eds) Modelling of Economic Transition Phenomena, pp. 43–57. University of Information Technology and Management, Warsaw (2001)
3. Bruno, M., Fisher, S.: Seigniorage, operating rules and the high inflation trap. Q. J. Econ. **105**(2), 353–374 (1990)
4. Cagan, P.: The monetary dynamics of hyperinflation. In: Friedman, M. (ed) Studies in the Quality Theory of Money, pp. 25–117. University Press, Chicago (1956)
5. Christiano, L.: Cagan's model of hyperinflation under rational expectations. Int. Econ. Rev. **28**, 33–49 (1987)
6. Commander, S., Coricelli, P.: Unemployment, Restructuring, and the Labor Market in Eastern Europe and Russia. The World Bank, Washington, DC (1995)
7. Czerwinski, Z., Gedymin, W., Kiedrowski, R., Panek, E.: Makroekonomiczny sredniookresowy model gospodarki Polski KEMPO 94. Ogolna charakterystyka i rownania modelu (Macroeconomic medium-term model of Polish economy KEMPO 94. General characteristics and equations of the model; in Polish). In: Budowa i implementacja modeli makroekonomicwych. Institute of Development and Strategic Studies, Warsaw (1996)
8. Danh Thi Hieu La: Vietnam: Transition to Market. Economics, Moscow (1999, in Russian)
9. De Clementi, M., Morciano, M., Orlandi, A., Perella, R.: Cumulative inflation and dynamic input-output modelling. In: Ciaschini, M. (ed) Input-Output analysis: Current Developments. International Studies in Economic Modelling Series, pp. 149–165. Routledge, London/New York (1988)
10. Frenkel, I.A.: The forward exchange rate, expectations and the demand for money: The German hyperinflation. Am. Econ. Rev. **67**, 653–670 (1977)
11. Hahn, F.: Money, Growth, and Stability. McGraw-Hill, New York (1985)
12. Heyman, D., Cavanese, A.: Tarifas Publicas y Deficit Fiscal: Compromisos entre Inflacion de Corto y Largo Plazo. Montevideo: Revista de Economia del Banco Central del Uruguay **3**(3) (1989)
13. Johansson, K.H.: On Market Dependencies of Agents' Learning for a Hyperinflation Model. WP-93-47, IIASA, Laxenburg (1973)
14. Gadomski, J.: Production in a non-market economy. Control Cybern. **21**(2), 38–52 (1992)

15. Gadomski, J., Woroniecka, I.: Poland under transition: A systems dynamics model. Owsinski, J. (ed) Final Report, vols. 1 (part IV), 2 (parts V, VI), and 3. Working Group on Dynamic Macro-Economic Modelling of Polish Economy in Transition "DYMOPET". Polish Operations and Systems Research Society, Warsaw (1995)
16. Kiani, K.M.: Inflation in transition economies: An empirical analysis. Trans. Stud. Rev. **16**(1), 34–46 (2009)
17. Gandolfo, G.: Economic Dynamic Methods and Models. North-Holland, Amsterdam (1980)
18. Keynes, J.M.: The General Theory of Employment, Interest and Money. Palgrave Macmillan, London (1936)
19. Kornai, J.: From Socialism to Capitalism: What Is Meant by the Change of System. The Social Market Foundation, London (1998)
20. Marchuk, G.I.: Methods of Calculative Mathematics. Nauka, Moscow (1977, in Russian)
21. Mikhalevich, V.S., Mikhalevich, M.V.: Dynamic pricing macromodels for a transition economy. Cybern. Syst. Anal. **31**(3), 409–420 (1995)
22. Mikhalevich, M.., Podolev, I.: Modeling of Selected Aspects of the State's Impact on Pricing in a Transitional Economy. WP-95-12. IIASA, Laxenburg (1995)
23. Arno Kappler, Stefan Reichart: Multibranchial Balance of Ukrainian Economy for 1993. Minstat, Kiev (1995, in Ukrainian)
24. Naslund, B.: Decision-making under inflation. In: Haley, K. (ed) Operational Research, pp. 849–859 (1978)
25. Nenova, M.: The Bulgarian State Firms: Results from a Financial Dataset. Agency for Economic Coordination and Development (1995)
26. Nenova, M., Cavanese, A.: Inflation stabilization and the state-owned enterprise behaviour in transition economies. The case of Bulgaria. In: Kulikowski, R., Nahorski, Z., Owsinski, J. (eds) Modelling of Economic Transition Phenomena, pp. 58–68. University of Information Technology and Management, Warsaw (2001)
27. Nikaido, H.: Convex Structures and Economic Theory. Academic, New York (1967)
28. Petrov, A.A., Buzin, A.Yu., Krutov, A.P., Pospelov, I.G.: Estimates of the Consequences of Economic Reform and Large-Scale Engineering Projects for the Economy of the USSR. Computer Center, Acad. Sci. USSR, Moscow (1990, in Russian)
29. Piontkivsky, R.V.: Impact of the currency substitution on the money demand and the income from seignorage in Ukraine. Naukovi zapiski, Kyiv-Mohyla Academy **19**, 27–30 (2001, in Ukrainian)
30. Roland, G.: Transition and Economics: Politics, Markets, and Firms. MIT, Cambridge (2000)
31. Sargent, T.I., Wallace, N.: Rational expectations and the dynamics of hyperinflation. Int. Econ. Rev. **14**, 328–350 (1973)
32. Straffa, P.: Monetary inflation in Italy during and after the war. Camb. J. Econ. **17**, 7–26 (1993)
33. Tarasevich, L.S., Grebennikov, P.I., Leusskyi, A.I.: Microeconomics. Uright, Moscow (2006, in Russian)
34. Koshlaǐ, L.B.; Mikhalevich, M.V.; Sergienko, I.V. Simulation of employment and growth processes in a transition economy. (English. Russian original) Cybern. Syst. Anal. 35, No.3, 392–405 (1999); translation from Kibern. Sist. Anal. 1999, No.3, 58–75 (1999)
35. Kuznetsov, Y.A.: Elements of Applied Bifurcation Theory. Springer-Verlag New York (1998).
36. Waugh, F.V.: Inversion of the Leontief matrix by power series. Econometrica **18**, 142–154 (1950)
37. Welfe, W., Welfe, A., Florczak, W.: Makroekonomiczny minimodel gospodarki polskiej (Macroeconomic minimodel of Polish economy; in Polish). In: Budowa i implementacja modeli makroekonomicznych, pp. 25–37. Institute of Development and Strategic Studies, Warsaw (1996)
38. Wyzan, M.L.: Monetary independence and macroeconomic stabilisation in Macedonia: An initial assessment. Communist Econ. Econ. Transform. **5**(3), 351–368 (1993)
39. Zhang, W.B.: Economic Dynamics - Growth and Development. Springer, Berlin (1990)

# Chapter 2
# Modeling of Structural, Institutional and Technological Changes

Using the notion main direction when talking about transition might in some sense be misleading. It may provide an impression that well-defined plans were designed at the beginning of transition and that those have been followed like a road map. First of all, because of the many distinctions among states (countries) with the non-market economies in terms of the so-called pre-conditions of transition there is nothing like a road map in transition. It means that the actual paths of reforms can also turn out to be very different from the initial plans. The transition paths can be as much adjusted (tuned) responses to unexpected events which have been overlooked in the initial schedule.

## 2.1 Main Directions of the Structural Reforms in Economics

Upon the achievement of the financial stabilization, the countries, which have chosen the path to market reforms, faced a complex of significant problems, induced by the alternation of the existing systems of productive and other social and economic relations targeted at the best way leading to the market economy standards. These problems are solved through the implementation of structural, institutional, and technological changes (reforms) (see Sergienko et al. [38]). Their planning is associated with considerable difficulties, since unlike for the initial financial stabilization there are no universal methods. The plan of the reforms should take into account the country's peculiarities and available resources. A large amount of semistructured and unformalized tasks arises in the course of the structural transformations. However, the application of the mathematical models to economy can be an efficient and supportive tool of the decision making in certain aspects of the mentioned reforms.



I.V. Sergienko et al., *Optimization Models in a Transition Economy*,
Springer Optimization and Its Applications 101, DOI 10.1007/978-1-4899-7544-7_2

It should be noted that despite of the specific character of their implementation, the structural reforms in the countries with a transitional economy should be concentrated on several similar directions. An accelerated demonopolization of the economy, i.e., the creation of competitive markets is the first of them. As it was mentioned in Chap. 1, the presence of the imperfect competition, especially in the economic instable environment, can significantly falsify the effectiveness of the market self-regulation mechanisms. By estimating the degree of monopolism in a transitional economy, one should take into account not only the formal, but also informal relations between firms and enterprises. The antimonopoly legal norms, transferred from the legislation of the countries with market economy, would not completely solve the emerged problems under such circumstances, but sometimes they still have a positive effect. Some extra criteria of monopolism are required, first of all, for the price setting analysis. Our mathematical models allow to forecast the price change under different assumptions about the forms of imperfect competition on the given market segment. Comparing the forecasted results with the real world practice, it is possible to determine under which assumptions the most accurate results were obtained, and evaluate the existing situation more adequately on this basis.

The decentralization and disaggregation of the large-scale enterprises are traditionally viewed as an important way toward the development of the competitive market. However, one form of imperfect competition – monopoly – is often replaced by another – oligopoly in such a way. To estimate, to which extend this replacement would comply with the public interest, and to forecast its consequences, one may use the models of the oligopoly pricing. The consequences of the imperfect competition also may be restrained by tools of taxation and customs-tariff policies; our application of this tool is also connected with mathematical modeling.

The second important direction of the structural reforms deals with the creation of the market infrastructure – the development of the stock and money exchange system, insurance market, banking and consulting structures, etc. Our mathematical models can be used to verify the managerial decision and forecast their consequences for the state's economic policy in this direction, as well as to support the management in the newly created economic structures. The development of the market infrastructure demands thoroughgoing alternations of the taxation, budgetary, customs-tariff policies and a fiscal policy as a whole. Such alternations should be systematically based on the quantitative methods of the analysis, including modeling, are widely used in the development of the reform plans in the indicated sphere.

The transformation of a fiscal system is traditionally viewed as one of the most important directions of the structural reforms. The taxation system of post-communist states needs a permanent analysis and perfection, especially in such matters as tax rates, preference grants and their abolition, scheme of taxes payment, etc. The changes in the taxation policy must be connected with budgetary policy. A forecast of the changes in budget incomes, according to which the expenses should be calculated, is required. Some well-known mathematical models are widely used to meet this challenge in the countries with the stable market economy.

The models of KESSU series (see Hetemaki [16]), deserve a special attention – since early 1980s they have been used to support the decision making in the budgetary policy in Finland under the conditions which in some aspects corresponding to the peculiarities of transition economy. Note that Finland has the so-called mixed economy, i.e., an economic system in which both the private sector and state direct the economy, reflecting characteristics of both market economies and planned economies (see, e.g., Bradley [2]).

Most mixed economies can be described as market economies with strong regulatory oversight, and many mixed economies feature a variety of government-run enterprises and governmental provision of public goods.

It should be pointed that, during the development of the fiscal policy, the necessity not only for a checking-up of existed recommendations, but also for the elaboration of new propositions may arise. Optimization models, which were applied in the countries with transitional economy, are being used for this purpose. Their further development needs taking into account the specific features of every country.

The stimulation of changes in existed production technologies is the next important direction of the structural reforms. As it was mentioned in Chap. 1, a large power and material intensity of the production is one of the main reasons for the cost inflation and non-payment crises. These effects can be fully gotten over by significantly reduced production costs per unit, and rejecting the old, outdated technologies. The planning of such changes needs considering of input–output models with variable direct cost matrix. Models of this type started to develop in the 1970s in connection with elaboration of the DISPLAN planning system (see Glushkov [13]) in the former USSR. Their application for the transitional economy requires both a significant revision of the structure of the models, and perfection of the principles for modeling results implementations.

Often, while planning the structural reforms, managers' decisions have to be made by weighting alternative options by means of various indicators (criteria). The mathematical models and methods can also be used here to support the decision making, in particular, the Analytic Hierarchy Process (AHP), proposed by Saati [32], and the methods of fuzzy set theory, developed by Bellman and Zadeh [1]. A choice of alternative for the administrative reform, which is also one of the directions of structural changes in transition economy, is the field of possible application of the above-mentioned approaches.

The current chapter reflects all directions described above. Section 2.2 presents our mathematical models of monopoly and oligopoly pricing, both universal and special-purposed, which allow us to take into account the peculiarities of certain market segments. Section 2.3 presents a mathematical model of the state budget implementation process and a "Budget" modeling complex, created on its basis. The latter has been applied, in particular, to analyze the Project of Ukrainian State Budget. Our results of the conducted analysis are also listed in Section 2.3. A special emphasis is made on the optimization of the fiscal policy standards and solution of the emergent problems of the multi-objective optimization and optimal control with discrete time.

Section 2.4 presents the optimization models for planning and analysis of the changes in production technologies, and reports our results of the calculations made with real world data. These calculations are connected with the solution of complex nondifferentiable and discrete optimization problems which required the development of specialized numerical methods (see Sergienko [35]).

Section 2.5 deals with a huge of alternatives for planning reforms in the transition economy by means of modern information decision support technologies. These technologies are represented by new mathematical models and computational algorithms including their theoretical justification and software development.

The fuzzy set theory and ordinal regression methods are introduced and applied in Sects. 2.6 and 2.7, respectively. These methods taking into account incomplete, weakly structured information applied to define the limits and status of suburban area around the capital of Ukraine – Kiev.

## 2.2 Modeling of Antimonopoly Measures

We will consider several models of antimonopoly measures based both on tools of financial and industrial policy in this section. Financial tools are first introduced.

A high level of monopoly is one of the reasons for inflation in a transitional economy. In the conditions of unsaturated markets, a monopolist can obtain additional profits by reducing production, which leads to the price increases.

It was shown in Mikhalevich [23] that one possible way to stop such an activity is to increase the interest rate and, connected with it, loans. The loss of profits resulting of production decline cannot be compensated by additional future profits if the interest rate is higher than the inflation rate. Otherwise, the higher interest rate generates a large amount of money in circulation, and hence, makes the market unsaturation deeper and gives the monopolist an additional motivation to increase prices. Such a situation also stimulates savings, but does not direct them to the production purposes. Therefore, the optimum value of the interest rate exists, and this corresponds to the maximum of productive activity of a monopolist. Let us consider the model to obtain this value.

Let us assume that there is a producer who makes only one kind of goods and this producer is a monopolist in his sector. His goal is to maximize his real (with taking into account of discounting) profits in the time period $[0; T]$. Being a monopolist, he can control prices by varying the value of this production, i.e., influencing the supply–demand disequilibrium. We also assume that prices change according to the Walras equation (see Nikaido [30]).

Let us denote the price for the good at the time $t$ as $p(T)$, the share of the producer's profit in the price as $\hat{q}$, the value of the good to be produced at time $t$ as $u(t)$, the producer capacity as $\overline{u}$, the demand for the good as $S(t)$, the initial price at the time instant $t = 0$ as $p_0$, the coefficient of the pricing equation as $m$, the discounting coefficient being further interpreted as the interest rate as $\nu$. For the given assumptions, the producer's activity can be described by the following problem of optimal control:

$$\int_0^T e^{-\nu t} p(t)\,\hat{q} \min\Big(S(t), u(t)\Big) dt \to \max,$$

$$\frac{dp(t)}{dt} = m\Big(S(t) - u(t)\Big),$$

$$0 \le u(t) \le \overline{u}, \quad t \in [0; T], \tag{2.1}$$

$$p(0) = p_0.$$

Let us denote an optimal solution to this problem obtained for the given $\nu$ and $S(t)$ as $(p^*, u^*)$. Now let us consider the model of the stimulation of a production activity of a monopolist by the state which establishes the value of $\nu$ parameter.

The minimization of the supply–demand disequilibrium is the state goal. The demand for goods is assumed to be the proportional to the ratio of consumers' money stock $D$ and the price $p$. Two sources of consumers' income are considered in our model:

1. the income which consumers receive independently from producer profits (payments from the budget, for instance);
2. the fixed portion $q$ of profits of the producer–monopolist. Let us also assume that the consumers' money stock is discounted with the same coefficient (interest rate) $\nu$.

If the intensity of the consumers' income is denoted as $Q(t)$, the following problem is obtained:

$$\int_0^T \max\Big(S(t) - u^*(t), 0\Big) dt \to \min,$$

$$\frac{dD(t)}{dt} = \nu D(t) + (1 + \nu) \times \bigg(Q(t) + \hat{q} q p^*(t) \min\Big(S(t), u^*(t)\Big) - p^*(t) \min\Big(S(t), u^*(t)\Big)\bigg),$$

$$\frac{dS(t)}{dt} = \frac{d}{dt}\Big(D(t) \,/\, p(t)\Big), \tag{2.2}$$

$$D(0) = D_0, \quad S(0) = S_0,$$

where $D_0$, $S_0$ are the given values of $D$ and $S$ variables for the time instant $t = 0$.

Equations (2.1)–(2.2) are a rather complicated two-level optimal control problem. However, this problem can be simplified.

Let's show that $u^*(t) \leq S(t)$ for every $t \in [0;T]$. Indeed, the producer has no reason to produce goods, which cannot be sold, never bring him any profits and decrease prices, which he tries to increase. Therefore, the equations of the problem (2.2) can be substituted by linear ones:

$$\frac{dD(t)}{dt} = \nu D(t) + (1+\nu)\Big(Q(t) + (\hat{q}q - 1)\, p^*(t)u^*(t)\Big),$$

$$\frac{dS(t)}{dt} = \frac{d}{dt}\left(\frac{D(t)}{p^*(t)}\right).$$

By substituting their solution $S(\nu, p^*, u^*)$ which is obtained for the given $\nu$ in problem (2.1), we transform the latter to the former which allows the application of algorithm proposed in Mikhalevich [23] for its solution. If we further substitute the solution of the latter to the objective function of the problem (2.2), we obtain a rather complex, but quite solvable, one-dimensional optimization problem. Its solution $\nu^*$ will be the optimum value of the interest rate $\nu$.

Antimonopoly policy can also use tools directly influencing the real sector. Development and application of such measures need the modeling of their consequences. The following model was directed to the solution of this problem.

The goal of the model is to derive the policy prescriptions to be used by a government faced with the problem of regulating the natural monopoly controlled either by the local business groups or by the foreign direct investors. A simplified version of this model suggested by Eaton and Lipsey [9] with the entry deterrence is used for these purposes. Assuming that the market can sustain only one firm in the long run and complete depreciation of capital after two periods, Shy [44] came to the conclusion that when the entry is allowed incumbent overinvests into productive capital to deter entry of any other firm. However, deriving the results of the model Shy [44] implicitly assumed that both incumbent and potential entrant belong to local businesses. Here two cases can be considered. First, a natural monopoly is owned by local business groups and is exposed to a potential entry of either foreign investors or local company. Second, a natural monopoly is owned by the foreign investors and it also faces potential entry. Under the assumption that the goal of the government is to maximize the welfare of society as the whole, the government will asymmetrically treat investments done by local and foreign businesses. In a sense, when the locals own the enterprise, the standard conclusions follow: it might be better for the state if the government blockades the entry to such an industry to prevent the overinvestment of the local firm into unproductive activities. However, when foreigners control the local monopoly, the standard conclusion may not be the best for the state. The government could consider the overinvestment of multinational corporations into the country as a positive phenomenon since it may actually increase the welfare of the host country. It might be better to permit entry into the industry that would lead to overinvestment by the foreigners.

Adapting the assumptions of Shy's [44] simplified version of Eaton and Lipsey [9] model to their framework, it state that there are two parties in the game. Incumbent natural monopoly owned either by one of local business groups or by foreign investor and potential entrant which could be any firm interested in serving the market. They assume that incumbent and potential entrant produce homogenous product and the cost structure of both players is the same. Without loss of generality, the variable costs of production of both firms can be normalized to zero. The other assumptions are as follows (the denotations of model authors are saving):

1. Each firm can produce in any period $t$ if it has capital in this period $t$. To build the capital firm has to invest amount $B$ that includes buying new equipment, advertising, promotion, etc.
2. The investment lasts for two periods without depreciation and then completely depreciates, thus to continue producing each firm has to invest again after two periods.
3. If only one firm produces and serves the market, it collects the monopolistic revenue $H$.
4. If two firms operate in the market then each collects revenue $L$.
5. The inequality $2L < B < H$ held, which means that the market cannot sustain two firms in the long run, which is a way to impose assumption of natural monopoly.
6. Let $0 < p < 1$ be the same discount factor for both firms.

In that framework the problem for the incumbent monopoly arises from the fact that potential entrant could step into the market at any time. Consequently, the existence of a threat of entry for the incumbent may affect the frequency of its investment.

Action taken by the firm $i$ in time $t$ is denoted as $a_t^i \in \{B, NB\}$ where $B$—invest and $NB$ not invest in period $t$ and $i$ {incumbent, potential entrant}. Incumbent is already in the industry (by definition) and thus it is assumed to invest in period $t = -1$ and game starts at period $t = 0$, both players can invest in any period $t = 0, 1, 2, \ldots$.

Each firm maximizes the sum of its discounted profits:

$$\ddot{I}^{i} = \sum_{t=0}^{\infty} \rho^{t} \cdot \left(R_t^i - C_t^i\right),$$

where $R_t^i = H$ if one firm has invested and solely operates in the market and $R_t^i = L$ if both firms have invested and together serve the market in period $t$. $C_t^i = B$ if firm invests at $t$ and $C_t^i = 0$ otherwise.

The introduction of foreign direct investor to the model leads to two distinct cases which have opposite conclusions with respect to the government actions.

Consider the first case when local business group controls the natural monopoly and faces potential entry into the industry. According to our assumptions, if local business group operates in the industry alone it gets the revenue $H$ and if a potential entrant steps into the industry both firms get $L$. Here the properties of equilibrium strategies that each firm follows to maximize its total profits are determined by the following theorem, which consists of two parts.

**Theorem 2.1.** *1. If a potential entrant is not allowed entering the market then incumbent invests only once in two periods to renew depreciated capital (a potential entrant clearly stays out of the industry). Hence, the strategy for the incumbent is as follows:*

$$a_t^{inc} = \begin{cases} B, & for \quad t = 1, 3, 5, \ldots \\ NB, & for \quad t = 0, 2, 4, 6, \ldots \end{cases}$$

*2. If a potential entrant is allowed entering, and if the discount factor is high enough satisfying*

$$\frac{B}{H} < \rho < \sqrt{\frac{B-L}{H-L}} (A).$$

*Then the following strategies constitute a subgame perfect equilibrium for the game:*

$$a_t^i = \begin{cases} B, & if \quad a_{t-1}^j = NB \\ NB, & if \quad a_{t-1}^j = B \end{cases}$$

*$i, j$ = incumbent, potential entrant $i \neq j$, $t = 0, 1, 2, \ldots$.*

*Proof.* The subgame perfect equilibrium, i.e., the situation where no player finds it beneficial to deviate form his strategy outlined in formulation of Theorem 2.1, is considered. The theorem is proved by contradiction.

**First Proposition.** Since capital lasts for two periods, the best incumbent can do is to invest in each odd period and minimize its costs. The fact that $B < H$ by assumption makes that investment scheme profitable for incumbent and eliminates incentives to deviate from it.

**Second Proposition.** Suppose that a potential entrant deviates from the strategy specified in the second part of the theorem then it gets the following profit:

$$\Pi^2 = (1+\rho)(L-B) + \rho^2 \cdot \frac{H-B}{1-\rho}$$

If it will not deviate (not enter) then the profit should be less than zero, Hence, if $\rho$ satisfies the considered below condition, potential entrant will not step in:

$$(1+\rho)(L-B) + \rho^2 \cdot \frac{H-B}{1-\rho} < 0 \Rightarrow \rho < \sqrt{\frac{B-L}{H-L}}. \tag{2.3}$$

Now for incumbent not to deviate from its optimal strategy profit obtained in one period should be less than all future discounted profits:

$$H < \frac{H - B}{1 - \rho} \Rightarrow \rho > \frac{B}{H}. \tag{2.4}$$

Hence, if $\rho$ satisfies condition (2.4) incumbent will not deviate.

Together (2.3) and (2.4) form the condition which ensures that the entry is deterred by incumbent through overinvesting, i.e., investing every period instead of once in two periods. The Theorem is proved.

Let analyze now the most important and remarkable practical implications of the obtained results.

First note that under quite general conditions specified in assumptions and restrictions on, the model generates the following patterns of investment for incumbent natural monopoly the first five periods.

Essentially, what the theorem tells us is that *if* the potential entrant is allowed to step into the industry (e.g., entry is not blockaded by a government) *then* incumbent will have to carry out excess investment to deter the entry. Since capital lasts for two periods investment in each period ensures that the incumbent will operate in the next period and thus makes the threat of incumbent remaining in industry for one more period credible. However, the excess investment leads to wasting of society resources in a sense that some resources could be invested for production purposes rather than entry deterrence. It causes the reduction in the overall welfare leading to the need of government intervention and prohibition of entry.

According to Table 2.1 local business carries out costly activities to deter the entry. However, the additional resources spent on deterrence could be directed to productive purposes if the government prevents the entry in the first place. Hence, if local business controls the incumbent monopoly it is better for society to prohibit entry in an industry. The threat to the incumbent will be eliminated and it will not misallocate resources. We summarize this with the following policy prescription.

**Policy Proposition 1.** If the natural monopoly industry is controlled by *local business* (*that invest only domestically*), then the government may improve welfare of the country by blockading the entry of other firms into this industry.

Consider now the second case when a foreign investor owns the local incumbent monopoly and is exposed to possible entry of either local firms or other foreign investors into an industry. Theorem 2.1 tells us that if entry is blockaded foreign investor will invest only in odd periods following pattern number 1 in Table 2.1. The outcome will be efficient amount of investment from the foreign investor's point of view. However, in that case government misses the opportunity to improve social welfare in a country though making foreigner incur additional investment. If the entry is allowed the foreign investor will spend money on its deterrence

**Table 2.1** Investment patterns under different policy rules

| Period | −1 | 0 | 1 | 2 | 3 | 4 | 5 |
|---|---|---|---|---|---|---|---|
| 1. Entry is blockaded | B | 0 | B | 0 | B | 0 | B |
| 2. Entry is not blockaded | B | B | B | B | B | B | B |

and thus it will invest additional money in a country every even period following pattern number 2 in Table 2.1. Consequently, investments in a country spent on maintaining monopoly status by foreigner will double in our model. Since local government welcome more investment it will perceive overinvestment as positive phenomenon, which leads to the redistribution of monopoly profits in favor of locals and increases overall welfare. Therefore, the government should, in contrast to the first case, behave in completely different way—it should not prohibit the entry as it was in the case with the local business control and make foreign investors carry out additional investment activities. This can be summarized with the following policy prescription.

**Policy Proposition 2.** If the natural monopoly industry is controlled by *foreign* business, then the government may improve welfare of the country by making the *entry free* for other firms into this industry.

Using the simplified version of Eaton and Lipsey [9] model, it is demonstrated that the government should provide asymmetric regulation of natural monopolies under different ownership structure. If local business group controls natural monopoly, government should blockade entry to an industry preventing the misallocation of resources to unproductive activities. On the other hand, government should allow entry to an industry if the owner is foreign investor. In that case, social welfare will be increased through additional foreign investments to buying new equipment, advertising and promotion in a country and thus increasing employment in it. Further, investments in promotion may increase consumer surplus and welfare through decrease in monopoly prices.

Of course, there are possible dangers of using the above proposed policy prescriptions without further considerations of each particular case. The most important one is the possibility of increase in direct bribing of local politicians and governmental officials by foreign investor to maintain its monopoly status. Another one is the possibility of violation of assumption of identical technologies. If the entry is blockaded by the government, the old production technology might be chosen by local businesses. Without competitive pressure and possibility of entry of investors with better technologies a country could be worse off in the long run; this is a subject of another study. Some aspects of technology development in transition economy are considered in detail in Sect. 2.4.

Models of real sector with the account of imperfect competition at some of its segments and externality impact are considered in Cavalletti [5]. They also can be applied for the development of anti-monopoly policy in transition economy.

A two-sector model of economy is considered. A competitive sector ($S_X$) produces an appropriable good ($X$) by means of a single-output return to scale technology, using as primary inputs labor and capital. In the non-competitive sector ($S_G$) two firms – a private and a public owned one – produce a differentiated product, sharing the same constant return to scale technology. Each unit of this market good produced by the private and the public firm generates a unit of public characteristic, which enters the utility function of consumer. Consumers maximize utility over characteristics to their constrains.

Market goods are denoted as $X$, $G_z$ and $G_h$; the primary factors are $L$ and $K$. Market prices for produced good are indexed by $i = x, z, h$ while factor prices are denoted as $w$ and $r$ for labor and capital services, respectively.

There are two productive sectors or industries denoted as $S_X$ and $S_G$, producing the good $X$ and goods $G_z$ and $G_h$, respectively. The market goods $G_z$ and $G_h$ can be interpreted as the production of a differentiated product by two firms active in the same industry.

Let $X$ is produced by means of a single-output, Cobb–Douglas [6] technology whose unit cost function can be written as:

$$C_x(p, w, r, 1) = \phi_x^{-1}\left(\frac{r}{\beta}\right)^{\beta}\left(\frac{w}{1-\beta}\right)^{1-\beta} \tag{2.5}$$

where $p$ is a vector of good prices.

There are two firms in sector $S_G$, each producing a differentiated product denoted as $G_z$ and $G_h$ for the public and the private firm, respectively. Each unit of goods , $G_z$ and $G_h$, generates one unit of public characteristic $Y$ which represents the positive external effect on the utility of the consumers.

Firms share the same technology, which for a given level of inputs produces multiple outputs in the fixed proportion. A simple representation of the technical constraint is obtained for a technology of the following form:

$$F\left(G_i, Y_i, K_i, L_i\right) = g\left(G_i, Y_i\right) - f\left(K_i, L_i\right); \quad i = z, h. \tag{2.6}$$

In this case the functions $g(\cdot)$ and $f(\cdot)$ are linear homogeneous in outputs and inputs respectively, hence, it is possible to separate the producer's problem into two separate optimizations of conditional outputs and inputs. In the considered model, $g(\cdot)$ represents a fixed-coefficient combination of outputs, $G_i$ and $Y_i$; $f(\cdot)$ is a Cobb–Douglas aggregate of factors. For any given output vector $z_i = [G_i, Y_i]$, the unit cost function is defined as follows:

$$C_i\left(w, r, z, (p, w, r, 1)\right) = \phi_i^{-1}\left(\frac{r}{\alpha_i}\right)^{1-\alpha_i}\left(\frac{w}{1-\alpha_i}\right)^{\alpha_i}; \quad i = z, h. \tag{2.7}$$

The second important characteristics of the commodity produced in sector G is that firms do not act as the price takers. Each firm's behavior is then defined in terms of some pricing rule, which maps the relation between prices and quantity decisions. The quantity produced by each firm – and therefore the total supply of the sector – depends on a number of factors: the strategy adopted by the firms, the degree of product differentiation, the number of active firms in each sector and the objective pursued by the firms. In our model we assume that private firms act as price competitors a la Bertrand and firms charge an ad valorem mark-up rate on marginal cost $\mu_{gi}$. The mark-up rate is a function of the elasticity of individual demand $\xi_{gi}$:

$$\mu_{gi} = \frac{1}{|\xi_{gi}| - 1}; \quad i = z, h, \tag{2.8}$$

which, in turn, depends on the price-elasticity of aggregate demand at the equilibrium and other factors such as the number of firms active in the sector considered, the strategy adopted by the firms, their degree of product differentiation, the structure of preferences.

Expenditure decisions are taken by a representative consumer $j$ and in our model we use the Lancaster's [21] characteristic approach. Two are the goods available for consumption, namely $X$ – a private appropriable good – and $G$ an impure public good. Each unit of $X$ – traded at price $P_x$ – generates one characteristic $x$ and has no effect on the utility of other individuals. Each unit of $G$ – traded at price $P_g$ – is generating $\alpha$ units of a private characteristic $g$ and $\beta$ units of a public characteristic $Y$. The utility function of this consumer is defined over the space of characteristics $x$, $g$, $Y$; it is strictly increasing, quasi-concave, continuous and everywhere twice differentiable, and can be written as follows:

$$U_j = u_j\left(x_j, g_j, Y\right) \tag{2.9}$$

where the trade off between the different characteristics is determined by the choice of $u$ which summarizes individual preferences. Even without his personal contribution, each consumer can consume units of the public characteristic generated by the consumption of the others. Consumers' demands derive from the solution of a standard maximization problem. At the top level utility is a Cobb–Douglas function of the private good ($X$), the private characteristic of the impure public good ($G$) and the public component ($Y$):

$$U_j = X_j^{\alpha_{xj}} G_j^{\alpha_{gi}} Y_j^{\alpha_{yi}}, \tag{2.10}$$

where $Y_j$ is treated as a public good, i.e., the public characteristic depends, for each consumer, on the total consumption of good $G$. To represents this characteristic, the consumption of good $Y$ is aggregated by a Leontief technology (a production function in which no substitution between inputs is possible) which transforms one unit of $Y$ into one unit of $Y_j$, for $j = 1, \ldots, n$, so that:

$$Y = Y_1 = \ldots = Y_n \quad \text{and} \quad P_y = \sum_{j=1}^{n} P_i. \tag{2.11}$$

At the bottom level the utility function is defined as a constant elasticity of substitution aggregate of private and public production of the impure public good:

$$G_j = \left(\delta G_{zj}^{\sigma-1/\sigma} + (1-\delta)\, G_{hj}^{\sigma-1/\sigma}\right)^{\sigma/\sigma-1}, \tag{2.12}$$

where $\delta$ is the share parameter and $\sigma$ is the inter-industry elasticity of substitution.

The budget constraint for consumers is given by the sum of his earnings: consumers supply a fixed amount of labor $\overline{L}_j$ at the price $w$, earn capital rents $r\overline{K}_j$ and collect the profits distributed by the private firm. His income $I_j$, gross of the externality effect, is then defined as:

$$I_j = r\overline{K}_j + w\overline{L}_j + \theta_j \Big(\sum_{i=z,h} \pi_i\Big)_i, \tag{2.13}$$

where $\pi_i$ is defined as $\pi_i = P_{gi} G_i - rk_i - wL_i$ and $\theta_j$ is the share of total profits redistributed to the consumer $j$.

The consumption pattern of each individual does not depend only on his decisions since he can also consume the public characteristic of the quantity of good $G$ bought by other consumers. This is the case of positive externality since the total positive external effect $Y = Y_z + Y_h$ enters the utility function of the representative consumer, but the consumer does not compensate the producers for the value of the external effect produced. This creates an interesting problem from an Applied General Equilibrium (AGE) modeling point of view (see e.g., Cardenete et al. [4]). The codes for an AGE model in fact are based on balanced Social Account Matrices where consumers have to pay for each commodity they receive and firms ought to be paid a price according to their marginal cost of production. A positive external effect ($Y$) is produced by sector $S_G$ and enters the utility functions of the consumers, but the consumers do not compensate sector $S_G$ for the value of the external effect produced.

To represent this situation a fictitious tax is levied on the output of sector $S_G$. The rate of this tax is chosen to make its revenue equal to the value of the external effect. The revenue is used to subsidize the consumer for the 'purchase' of $Y$. In this way the externality effect causes a net transfer of income from the producers (which are charged with the fictitious taxation) to the consumers (which are subsidized for the same amount).

The modeling of the subsidy to the consumer takes explicitly into account the fact that the externality affects directly his utility by modifying his unit expenditure function (utility price).

Let us define the unit expenditure function of the $j$-th consumer as:

$$e_j\,(w, r, p) \quad = \quad \phi^{-1}\left(\frac{p_x}{\alpha_x}\right)^{\alpha_x}\left(\frac{p_Y}{\alpha_y}\right)^{\alpha_y}\left[\frac{\left(\delta^{\sigma} p_{g_z}^{1-\sigma} + (1-\delta)^{\sigma} p_{g_h}^{1-\sigma}\right)^{1/1-\sigma}}{\alpha_g}\right]^{\alpha_g} \tag{2.14}$$

The unit expenditure function can be used to define income in terms of indirect utility as:

$$e_j V_j = r\overline{k}_j + w\overline{L}_j + \theta_i \sum_{i=z,h} \pi_i + p_j Y_j \tag{2.15}$$

As shown in Eq. (2.15) the consumer is refunded for his purchase of good $Y$: the positive external effect causes the unit expenditure function to be lower than it would have been without externality; if $s_j$ is defined as $p_j Y_j / V_j e_j$, the consumer's income, net of the eternality effect, is equal to

$$M_j = e_j = \left(1 - s_j\right) V_j = r\overline{k}_j + w\overline{L}_j + \theta_i \sum_{i=z,h} \pi_i, \tag{2.16}$$

where $e_j(1 - s_j)$ is the unit expenditure function net of the external effect (the net price of utility).

The explicit considerations of combined market failures such as oligopoly and externality in applied analysis can greatly improve our understanding of specific policies concerning the provision of public goods and the correction of externalities in non-competitive markets.

The literature suggests subsidizing the price of the impure public good for an amount which corresponds to society's evaluation of the public good produced. In the specific case presented above, $G_z$ and $G_h$ produces the public characteristic at the same rate, hence they should be subsidized at a uniform rate.

On the other hand, the literature on market regulation to reduce the inefficiencies caused by imperfect competition suggests that the public firm to whom the regulator can impose marginal pricing rules should increase its production in order to reduce the monopolistic power of the private competitor. In both cases and when only one market failure applies to a specific institutional context the use of these rules are expected to improve welfare.

The model (2.5)–(2.16) allows to verify if these rules are still valid when several market failures, namely the presence of oligopoly (hence a monopolistic rent) and a merit good affect the same market. For our numerical exercise we have aggregated consumers into one single representative consumer; the value of the exogenous parameters and elasticities are summarized in Table 2.2. The observed unit price for a price competitor a la Bertrand depends on the calibrated value of the substitution elasticity between differentiated goods in the oligopolistic sector (i.e., the calibrated value of $\sigma$). It seems that these values turned out to be 8.0 for 0.2 of observed unit profit of the private firm and 6.0 of the individual elasticity of demand.

**Table 2.2** Calibrated elasticities for the model (2.5)–(2.16)

| Calibrated elasticities | Value |
|---|---|
| Elasticity of input substitution of firm $X$ | 1.0 |
| Elasticity of input substitution of firm $z$ in sector $G$ | 1.0 |
| Elasticity of input substitution of firm $h$ in sector $G$ | 1.0 |
| Elasticity of substitution between $G_z$ and $G_h$ | 8.0 |
| Elasticity of individual demand of firm $h$ in sector $G$ | 6.0 |
| Unit average profit of firm $h$ in sector $G$ ($\mu$) | 0.2 |

In this first exercise the cost of the public characteristic of the impure public good is financed by general taxation and the optimal production is induced through a subsidy to sector $G$.

In particular, the model is calibrated for an initial equilibrium in which no charge is made on the consumer for the use of the public good (the externality effect is a net income transfer from producers to the consumer). Starting from this equilibrium, the optimal subsidy schemes $S_z$ and $S_h$ which internalize the external effect associated with the total consumption of the impure public good are obtained. The cost of the subsidy is financed by a proportional tax on total income, which due to the assumption of fixed labor supply can be considered as a first-best instrument. The results of this first simulation are presented in Table 2.3.

Two countervailing equilibria are presented for alternative scenarios regarding the behavior of the private firm. In the first case, it is assumed that the private firm producing $G_h$ can act as a Bertrand competitor where the value of the mark-up is sustained by the elasticity of substitution between $G_z$ and $G_h$ calibrated in the initial equilibrium; in the second case, the firm is regulated since it has to charge a price equal to its marginal cost, i.e., the firm has not a profit. This second case differs from a perfectly competitive market since the two qualities of goods produced $G_z$ and $G_h$ are not a perfect substitute. The introduction of the subsidy (case 1) is slightly welfare improving, as shown by the equivalent variations (EV) in Table 2.3.

**Table 2.3** Correcting the externality with a subsidy[a]

| | $\sigma = 8.0$ | | |
|---|---|---|---|
| | Bench | 1 | 2 |
| $X$ | 1 | 0.888526 | 0.873164 |
| $G_h$ | 1 | 0.829519 | 2.64792 |
| $G_z$ | 1 | 1.20237 | 0.615821 |
| $w$ | 1 | 1 | 1 |
| $r$ | 1 | 0.90032 | 0.88755 |
| $P_x$ | 1 | 0.929133 | 0.919888 |
| $P_{gz}$ | 1 | 0.968989 | 0.964846 |
| $P_{gh}$ | 1.2 | 1.16063 | 0.964846 |
| $P_y$ | 0.3 | 0.223293 | 0.21441 |
| $S_z$ | | 0.230439 | 0.222222 |
| $S_h$ | | 0.19239 | 0.222222 |
| $s$ | 0.127659 | — | — |
| $e$ | 1 | 0.822985 | 0.7850748 |
| $\nu$ | 2.35 | 2.35735 | 2.40429 |
| $\mu$ | 0.2 | 0.1977 | 0 |
| Tax revenue | | 0.300933 | 0.306931 |
| EV | | 0.007347 | 0.054291 |

[a] 1 = an optimal subsidy with the private firm being a Bertrand competitor;
2 = optimal subsidy with the private firm regulated with the price = marginal cost

The positive EV is the result of a reduction in the unit expenditure function ($e$) and an increase in the utility index ($\nu$). The demand for $X$ is decreasing as we might expect. In the benchmark, in fact, the price paid by the consumer to buy from sector $G$ was not optimal since it included the cost to produce the public good. The introduction of a subsidy allows the consumer to increase its purchase of $G$, in particular the variety produced by the public sector which is cheaper since there is no mark-up on it. The subsidy rate is not uniform and it is higher for the private producer. If the hypothesis of mark-up pricing (case 2) is removed, i.e., the private producer charges a price equal to its marginal cost, there is a more substantial improvement in welfare, an increase in the variety produced by the private sector and the subsidy rate is uniform. The results presented in Table 2.3 allow us to make the following conclusions:

- the presence of oligopoly is a sufficient condition to make uniform rates a sub-optimal policy; and
- the same conclusion applies to welfare gains and losses, which are represented by the equivalent variations (EV).

The second type of issue can be discussed in relation with the market regulation. The second-best optimal policies to apply in a mixed oligopoly, i.e., a market in which a private firm and a publicly controlled one are coexisting have been intensively studied. The monopolistic power of the private firm derives from the imperfect substitution nature of the goods produced by the two firms and cannot be removed through regulation. The problem for the public firm is to determine an optimal price/quantity policy for its product given the structure of the market in which it operates, i.e., given the reactions of the private producer. In general, the private firm is going to reduce the quantity offered in response to an increase in the quantity provided by the public firm in order to maximize its profit. If, as it is assumed in neoclassical models, all available resources are used, the final composition of the supply will be biased toward public production.

This is a typical second-best problem for which the traditional literature on pricing policies points out that the simple rule price equals the marginal cost (MPR), although being welfare improving, it might not be the optimal second-best policy, and concludes that a price higher than marginal cost could in fact be a superior policy to the marginal cost pricing rule. A priori, one could expect that a marginal pricing rule, in the presence of externalities could be a superior instrument. In this case, in fact, the public production would increase, but since the commodity traded is a merit good, it could still be underprovided because of the externality element.

In the first simulation (case 1), the public producer corrects its price policy taking into account the external effects. A subsidy on the marginal cost of production is levied at a rate $S_z$ such that the subsidy matches the value of the external effect. In this way, the market price of good $G_z$ reflects the unit cost of production of the appropriable characteristic only. The subsidy is financed with a general tax on income while, in order to gain some understanding from the comparison with the previous exercises, the tax revenue is kept at the original level of the previous simulations and results are shown in Table 2.4.

**Table 2.4** Simulations using regulation[a]

| | $\sigma = 8.0$ | |
|---|---|---|
| | Bench | 1 |
| $X$ | 1 | 0.876397 |
| $G_h$ | 1 | 0.239923 |
| $G_z$ | 1 | 1.38588 |
| $w$ | 1 | 1 |
| $r$ | 1 | 0.87633 |
| $P_x$ | 1 | 0.91173 |
| $P_{gz}$ | 1 | 0.743125 |
| $P_{gh}$ | 1.2 | 1.11032 |
| $P_y$ | 0.3 | 0.218045 |
| $S_z$ | 0 | 0.226855 |
| $s$ | 0.127659 | – |
| $e$ | 1 | 0.82159 |
| $\nu$ | 2.35 | 2.28550 |
| $\mu$ | 0.2 | 0.15517 |
| Tax revenue | 0 | 0.300933 |
| EV | 0 | −0.0645 |

[a] 1 = marginal cost pricing rule for the public producer

It should be noted that the simple marginal pricing rule in this context produces a welfare loss, i.e., it creates a deadweight loss that is greater than the inefficiency it tries to correct. This is an important result since such a rule, although not being perhaps optimal, it is said to produce welfare gains in a monopolistic market context. This result is due to the presence of the externality element. The negative sign is caused by an excess of public production compared to both the private production of the same good and the private production of the numeraire good, which, in turns, results from the way private and public characteristics interacts in the model. The public characteristic is produced at the same rate by $G_z$ and $G_h$; the optimal rule in this market should then be to buy this good only from the producer that sells at the lowest cost. However, on the private market the appropriable characteristic is not a perfect substitute hence both goods are produced. The final equilibrium will depend on a balance between these two points, both of which are far from optimal. This in turn causes the welfare loss. These effects are obviously inversely related to the degree of substitution between the private and the public produced merit good.

The model (2.5)–(2.16) allows us to derive important conclusions on optimal second best policy in the presence of market power and externalities. Now we want to demonstrate that the use of subsidy and regulation instruments depends on the environment in which the policy has to be taken.

For this reason, the results of the simulations previously presented in Table 2.3 are displayed for alternative levels of $\sigma$ in Table 2.5; the special attention is paid to the analysis of changes of the degree of substitution between $G_z$ and $G_h$ which in turn influences $\mu$.

**Table 2.5** Sensitivity analysis

| | $\sigma = 10$ | $\sigma = 8^{a}$ | $\sigma = 5.7$ | $\sigma = 4.5$ | $\sigma = 1.4$ |
|---|---|---|---|---|---|
| $X$ | 0.88864 | 0.80885 | 0.88830 | 0.88113 | 0.88605 |
| $G_h$ | 0.28265 | 0.82951 | 0.83189 | 0.83390 | 0.89392 |
| $G_z$ | 1.20264 | 1.20237 | 1.20186 | 1.20144 | 1.19682 |
| $w$ | 1 | 1 | 1 | 1 | 1 |
| $r$ | 0.90041 | 0.90032 | 0.90013 | 0.89972 | 0.89825 |
| $P_x$ | 0.92920 | 0.92913 | 0.92899 | 0.92888 | 0.92763 |
| $Pgz$ | 0.96902 | 0.96898 | 0.96892 | 0.96887 | 0.96832 |
| $P_{gh}$ | 1.11654 | 1.16063 | 1.25640 | 1.35219 | 2.01436 |
| $P_y$ | 0.22336 | 0.22329 | 0.22316 | 0.22305 | 0.22183 |
| $S_z$ | 0.23050 | 0.23043 | 0.23031 | 0.23021 | 0.22909 |
| $S_h$ | 0.20004 | 0.19239 | 0.17761 | 0.16491 | 0.01101 |
| $S_h$ | 0.82242 | 0.82298 | 0.82420 | 0.82542 | 0.91238 |
| $\mu$ | 0.154 | 0.2 | 0.3 | 0.4 | 20 |
| Tax revenue | 0.30120 | 0.30093 | 0.30036 | 0.29983 | 0.27025 |
| EV | 0.00940 | 0.00734 | 0.00294 | −0.00136 | −0.7236 |

[a] Simulation of Table 2.3

The welfare gains deriving from the subsidy on goods $G_z$ and $G_h$ are larger the smaller the mark-up is. It is worth remembering that in the considered model the mark-up is determined by the product differentiation, hence a small mark-up implies that the two varieties of $G$ are relatively high substitutes so that the consumer has a smaller reduction in his utility from the oligopoly. A higher mark up implies that the distortion caused by oligopoly is relatively more important than the presence of externality. The subsidy can only remove the second source of market failure hence its effectiveness will be positively related to the relative importance of the failure it is correcting. For the same reasons, the welfare gain is greater the greater is $\sigma$. In the presented exercise, the sign of the equivalent variation changes for a value of $\sigma$ is equal to 4.5. This means that if the degree of substitution between the two varieties is less than or equal to 4.5 giving a subsidy is not a first best policy. The following conclusions implies from obtained results:

- the optimal subsidy rates are inversely related to the degree of monopoly power which, in turn, depends on the degree of product differentiation, captured in the model by $\mu$, given the number of active firms on the market; and
- the same trend occurs in the magnitude of welfare gains and losses, which are represented by the equivalent variations (EV).

Table 2.6 reports the results of the simulations presented in Table 2.4 for alternative levels of $\sigma$. It should be pointed out that, if $\sigma$ is relatively high ($\sigma = 10$), the two goods are not highly differentiated, so that the increase in the production of the public variety of good $G$ creates a distortion that is smaller than the market failure created by the positive externality it is correcting. The increase in public production brought about by this type of regulation is then welfare improving.

**Table 2.6** Sensitivity analysis

| | $\sigma = 10$ | $\sigma = 8^a$ | $\sigma = 5.7$ | $\sigma = 1.4$ |
|---|---|---|---|---|
| $X$ | 0.85830 | 0.87639 | 0.87471 | 0.87429 |
| $G_h$ | 0.15673 | 0.23992 | 0.38857 | 0.86809 |
| $G_z$ | 1.85543 | 1.38588 | 1.32548 | 1.14218 |
| $w$ | 1 | 1 | 1 | 1 |
| $r$ | 1.065 | 0.87633 | 0.86866 | 0.85670 |
| $P_x$ | 1.04552 | 0.91173 | 0.90614 | 0.89738 |
| $P_{gz}$ | 0.83110 | 0.74312 | 0.74074 | 0.73543 |
| $P_{gh}$ | 1.13832 | 1.11032 | 1.19081 | 1.87881 |
| $P_y$ | 0.18816 | 0.21804 | 0.21789 | 0.21922 |
| $S_z$ | 0.18460 | 0.22685 | 0.22730 | 0.22963 |
| $e$ | 0.80019 | 0.82159 | 0.82082 | 0.86932 |
| $\mu$ | 0.11680 | 0.15517 | 0.24218 | 18.6804 |
| Tax revenue | 0.30120 | 0.30093 | 0.30036 | 0.27025 |
| EV | 0.28402 | -0.0645 | −0.08164 | −0.71161 |

[a] Simulation of Table 2.4

When the two goods are highly differentiated, the first effect (substitution of private variety with public variety) has worse consequences than correcting for the externality effect, i.e., this policy is Pareto-inferior.

These results are confirmed by the sensitivity analysis on $\varphi$, the scale parameter for the production of the impure public good. The value of $\sigma$, the elasticity of substitution between private and public production constant, is kept out and the value of $\varphi_h$, the share of the private production of the impure public good, is alternatively increased and decreased.

If we compare the results in Tables 2.5 and 2.6 with those presented in Tables 2.3 and 2.4, we can note that there is an inverse correlation between the scale parameter and the signs of the equivalent variations. An increase in the market share, other things being equal, has the same impact on welfare as a decrease in the benchmark elasticity of substitution: the producer has a greater monopolistic power on the variety, which results, other things being equal, in a higher benchmark value of $\mu$ compatible with a pre-determined value of $\sigma$. Subsidy and regulation are not welfare improving policies in this case, but the regulation has an even worse impact, i.e., it is a Pareto-inferior instrument.

## 2.3 Models for Fiscal System Optimization

The economic situation in many post-communist countries (first of all, in CIS countries) does not fit the assumptions on which market and centrally planned economies are based.

The characteristic of economy in transition, such as a dominance of formerly state-owned property, strong monopoly influence, the absence of the centralized control of industry and prohibitive technology costs result in the complexity of the problem of supporting decisions in the budgetary, monetary and financial policy. Both experience of experts (this mainly refers to the pre-reform period) and analogies with market economies are not suitable in this case. A detailed investigations of transitional economy as a special economic system using quantitative and qualitative methods is necessary. These research require the mathematical modeling of transitional processes in financial and budget spheres.

The models of financial sphere and its essential components (such as pricing) were elaborated by Cagan [3], Sargent and Wallance [33], Johansson [19]. Input–output models were applied in the analysis of influence of the existing technological structure on prices and profitability of the different sectors of economy in such papers as Moroney and Toevs [29], Gandolfo [12] and Hahn [15]. It should be noted that all mentioned authors deal with a well-developed market economy and their results cannot be applied to the transition period without modification.

The models for the state budget consist of a special class of financial models. The procedures of development and realization of budget are determined by well-developed national legislation; therefore they are very special in every country. General ideas and details of models are the subject of such publications, which have been collected in Wallis and Whitley [49], Hetemaki and Kaski [17].

The problem of fiscal policy parameters optimization is especially important in such conditions. The model which can connect the budget, tax system, and other main macroeconomic parameters (processes) is necessary for these investigations. However, several problems appear when an optimization approach is applied to budgetary modeling. The construction of goal (objective) function is one of them. Therefore, the complex consideration of models, optimization methods, and necessary computer tools seems to be urgent.

The development of rational budget is one of the vital issues of economic stabilization. This possesses some special features in transition economy. The structure of expenditures is one of them. The financial stability of formerly state-owned enterprises (i.e., soft private sector) cannot be ensured without large state subsidies if the current prices are rapidly changing and consumer's demand is decreasing. Therefore, such subsidies become the main budget expenditures (about 40–59 % of total expenditures in the Ukraine in 1993 and 20–25 % in Russia at the moment of writing this text). Transformation of industrial structure and implementation of new technologies also need the state investments with long lag of revenues. Values of such subsidies and investments are strongly determined by the current prices for manufactured and consumer goods production. Control of prices and stimulation of activity in the main industrial sectors are other important goals of direct and indirect financial support from the state budget. Beside the direct impact the budget also produces a fundamental and indirect influence on prices through money supply increases to cover the deficit, through taxes which can change the cost of production and through social payments from the budget which change consumer demand. Therefore, the accurate forecast of the development of budget needs cannot be done without information about the structure and values of main expenditures and revenues of the budget. These two problems must be solved simultaneously.

The budget deficit has been traditionally considered among the most important reasons of inflation. It should be noted that the reduction of the budget deficit is important not only for the period of planning as a whole, but also for every subperiod. For example, the budget deficit in Ukraine in 1993 covered about 15 % of total expenditures. But for several months of this year it was more than 50 % of total income (Mikhalevich, Podolev [25]). Of course, it was the strong reason for hyperinflation. The analysis of dynamics of the budget is necessary for accounting of such effects.

The well-balanced budget is necessary, but not sufficient for financial stability in the transition economy. The dominance of enterprises which are managed independently, but bear no responsibility for their results, and the absence of a working bankruptcy mechanism together with the pressure of managers on state bodies, create possibilities for large out-of-budget outlays of money and credits, such as took place in Ukraine in 1993 and in the first half of 1994 (for all details see Mikhalevich, Podolev [25]).

The estimation of the impact of the budget on production is also important. High taxes intensify the industrial decline and decrease the real budget incomes. This produced a higher budget deficit than had been planned. The joint analysis of all these aspects needs modern methods, including mathematical modeling and system researches.

It should be noted that the modeling of state budget is far from being complete even for market economies (see Wallis and Whitley [49]). The construction of such models for transition economy needs special approaches. It is difficult to construct econometric models for such purposes due to the lack of appropriate statistical data. Input–output models seem to be more suitable for such purposes.

One of such models is considered in this section. It can be used to analyze the consequences of the tax policy and decisions about budget outlays. To provide an explanation of the impact of the budget deficit or surplus on the dynamics of prices and, consequently, on the budget revenues, is the main goal of the model. It takes into account all the above-mentioned peculiarities of the problem of budget development in transition.

Note that the effect of the government budget on prices in a transition economy is very complex and multifaceted. It includes the imbalance of supply and demand caused by budgetary reallocation of producer and consumer incomes, the effect of taxes on production costs, and the consequences of emissions intended to cover the budget deficit. It is difficult to allow for all these factors in a relatively simple analytical model. We accordingly propose a simplified simulation model for analyzing the consolidated state budget, which is based on following assumptions.

1. We consider a system of $n$ material production industries plus a nonproductive sector, $m$ consumer groups, and the government budget.
2. The revenues of the government budget derive from tax receipts from the industries, the income tax paid by consumers, fixed payments from industries and consumers, and other sources.
3. The model considers three types of taxes on industries:

(a) value added tax charged at a rate $q_j^{(1)}$, which is possibly differentiated by industry;
(b) tax on profit charged at a basic rate $q_j^{(2)}$, which also may be differentiated by industry;
(c) excise taxes on some industry products, which constitute a fraction $q_j^{(3)}$ of the final price.

4. The actual budget revenue from these taxes is proportional to the volume of sales of the industries in current prices.
5. The revenue from income tax on consumers is proportional to the nominal income of consumers with proportionality coefficient $q_j^4$, which is differentiated by consumers groups.
6. The budget revenue from fixed payments and other sources is assumed given.
7. The main budget expenditures include the following:

   (a) subsidies to industries;
   (b) indexed and unindexed payments to consumers;
   (c) other expenditures.

   The expenditures of the first group are proportional to the production volume of the industries eligible for subsidies. Payments to consumers are indexed in proportion to the weighted price index calculated from data on the volume of sales of industry products in current and constant prices. Other expenditures are assumed given.
8. The model is based on the following pricing mechanism. Price is determined by the cost and monetary factors according to the econometric model presented in Mikhalevich, Podolev [25]. Three types of product prices are considered:

   (a) fixed price, when products are sold at a given (constant) price and the difference between industry costs and sales revenues is made up from the budget;
   (b) controlled prices, when the budget compensates a part of industry costs through subsidies;
   (c) free prices, when no subsidies are paid.

   We denote the set of industries with fixed prices by $\Omega_1$, the set of industries with controlled prices by $\Omega_2$, and the set of industries with free prices by $\Omega_3$.
9. The time in the model is discrete with an increment $\Delta t$ in the interval $[0, T]$. Let us give the detailed description of the model starting from model inputs and variables which are recalculated during the model runs.

$a_{ij}$ is the direct input in constant prices of industry $i$ product consumed to produce a unit of industry $j$ product ($i, j = 1, \dots, n$);

$\overline{q}_j$ is the share of labor costs in the price of industry $j$ product ($j = 1, \dots, n$);

$\widehat{q}_j$ is the standard profit margin in the price of industry $j$ product $(j = 1, \ldots, n)$;

$q_j^+$ is the share of other value added components in the price of industry $j$ product $(j = 1, \ldots, n)$;

$q_j^{(1)}$ is the rate of value added tax for industry $j$ $(j = 1, \ldots, n)$;

$q_j^{(2)}$ is the rate of tax on profit for industry $j$ $(j = 1, \ldots, n)$;

$q_j^{(3)}$ is the share of excise taxes in the price of industry $j$ product $(j = 1, \ldots, n)$;

$x_j^{(t)}$ is the predicted output of industry $j$ product at time $t \in [0, T]$ in constant prices; this quantity may be viewed as the upper bound on the sales volume produced by the available assets and capacities $(j = 1, \ldots, n)$;

$B_j^{(t)}$ is the production cost per unit industry $j$ product at time $t$, which does not depend directly on internal prices $(j = 1, \ldots, n,\ t \in [0, T]$;

$W_j(t)$ is the excess profit per unit output of industry $j$ at time $t$ $(j \in \Omega_2 \bigcap \Omega_3,\ t \in [0, T]$;

$\overline{W}_j(t)$ are the subsidies per unit output of industry $j$ at time $t$;

$H(t)$ is the budget revenue from fixed payments at time $t$;

$\widehat{H}(\mathrm{t})$ is the budget revenue from other sources at time $t$;

$\overline{p}_j(t)$ is the fixed price of industry $j$ product at time $t$ $(j \in \Omega_1, t \in [0, T])$;

$g_j$ is the proportion of industry $j$ product used for consumption $(j = 1, \ldots, n)$;

$\beta_{jk}$ is the proportion of industry $j$ product in the bundle of goods and services used by consumers of group $k$ $(j = 1, \ldots, n,\ k = 1, \ldots, m)$;

$Q_k^{(1)}(t)$ is the amount of indexed payments from the budget to consumers of group $k$ at time $t$ in constant prices $(k = 1, \ldots, m, t \in [0, T]$;

$Q_k^{(2)}(t)$ is the amount of unindexed payments from the budget to consumers of group $k$ at time $t$ in constant prices $(k = 1, \ldots, m, t \in [0, T])$;

$G(t)$ are other budget expenditures at time $t$;

$c_{jk}$ is the share of group $k$ consumers in the labor costs of industry $j$;

$q_k^4$ is the income tax rate for group $k$ consumers;

$D_k^{(0)}$ are the money balances for group $k$ consumers $(k = 1, \ldots, m)$;

$\overline{M}(t)$ is the predicted quantity of money at time $t$ assuming zero budget deficit in the time interval $[0, T]$;

$p_j(t)$ is the relative price ( measured in relation to the base price) of industry $j$ product at time $t$ $(\ j \in \Omega_2 \bigcup \Omega_3, t \in [0, T])$;

$\tilde{p}_j(t)$ is the price component determined by production costs;

$y_j(t))$ is the industry $j$ product in constant prices used for consumption $(j = 1, \ldots, n, t \in [0, T])$;

| | |
|---|---|
| $z_j(t)$ | is the sales volume of industry $j$ product at time $t$ in constant prices; |
| $\Pi_j(t)$ | is the income of industry $j$ at time $t$; |
| $\bar{\Pi}_j(t)$ | is the income earned by consumers of all groups from work in industry $j$ at time $t$; |
| $\widehat{p}$ | is the weighted average price index of goods and services; |
| $S_j^k(t)$ | is the cash-paying demand of group $k$ consumers for industry $j$ product ($j = 1, \ldots, n, k = 1, \ldots, m$); |
| $M(t)$ | is the quantity of money at time $t$ (allowing for emission to cover the budget deficit); |
| $D_k^{(t)}$ | are the money balances of group $k$ consumers at time $t$ ($k = 1, \ldots, m, t \in [0, T]$); |
| $A_0^{(t)}$ and $B_0^{(t)}$ | are the budget revenue and expenditure at time $t$, respectively. |

Before starting the model simulation, we calculate the value added percentage $q_j$ and the price influence coefficient $\overline{a}_{ij}$ from the formulas (see Mikhalevich V.S. et al. (1993)):

$$q_j = \frac{\overline{q}_j + q_j^+}{1 - q_j^{(1)}} + \frac{\widehat{q}_j}{\left(1 - q_j^{(1)}\right)\left(1 - q_j^{(2)}\right)} + q_j^{(3)}, \quad j = \overline{1, n},$$

$$\overline{a}_{ij} = \frac{a_{ij}}{1 - q_j}, \quad i, j = \overline{1, n}.$$

We also calculate the elements $\overline{a}_{ij}$ of the inverse of the matrix $(E - A)$, $A = \{a_{ij}\}$, $E$ is the identity matrix. Then we do the following for each time instant $t = 0, \Delta t, 2\Delta t, \ldots$.

**1. Calculate the relative prices of the industry products:**

$$\tilde{p}_j(t + \Delta t) = \sum_{i \in \Omega_1} \overline{a}_{ij} \overline{p}_i(t) + \sum_{i \in \Omega_2 \bigcup \Omega_3} \overline{a}_{ij} p_i(t) + B_j(t + \Delta t)$$

$$+ \frac{1 - q_j^{(2)}(1 - q_j^{(1)})}{1 - q_j^{(1)}} W_j(t + \Delta t) - \nu_j(t + \Delta t),$$

$$j \in \Omega_2 \bigcup \Omega_3,$$

where

$$\nu_j(t + \Delta t) = \begin{cases} \overline{W}_j(t + \Delta t), & \text{if } j \in \Omega_2, \\ 0, & \text{if } j \in \Omega_3, \end{cases}$$

$$p_j(t + \Delta t) = \max\Big(\tilde{p}_j(t + \Delta t), \ \exp\big((\ln p_j(t) - \ln M(t) + \gamma)\times$$

$$\times e^{-\alpha^{-1}\Delta t} + \ln M(t) - \gamma\Big)\Big), \qquad j \in \Omega_2 \bigcup \Omega_3.$$

The last equality implies from above-mentioned pricing mechanisms.

2. **Calculate the subsidies needed to sustain fixed prices:**

$$\overline{W}_j\,(t+\Delta t) = \sum_{i\in\Omega_1} \overline{a}_{ij}\,\overline{p}_i(t) + \sum_{i\in\Omega_2\bigcup\Omega_3} \overline{a}_{ij}\,p_i(t) + B_j\,(t+\Delta t)$$

$$+\frac{1-q_j^{(2)}(1-q_j^{(1)})}{1-q_j^{(1)}}W_j\,(t+\Delta t) - \overline{p}_j\,(t+\Delta t)\,, \quad j \in \Omega_1.$$

3. **Calculate the volume of each industry product used for consumption:**

$$y_i\,(t+\Delta t) = (x_i\,(t+\Delta t) - \sum_{j=1}^{n} a_{ij}x_j\,(t+\Delta t)\,g_i; \quad i = \overline{1,n},$$

and also the cash-paying demand of consumers:

$$S_j^k\,(t+\Delta t) = \frac{D_k(t)\beta_{jk}}{p_j\,(t+\Delta t)}; \quad j = \overline{1,n}, \quad k = \overline{1,m}.$$

4. **Determine the sales volume of industry products:**

$$z_i\,(t+\Delta t) = \sum_{j=1}^{n} \widehat{a}_{ij}\left(\min\Big(\sum_{k=1}^{m} S_j^k\,(t+\Delta t)\,, y_j\,(t+\Delta t)\Big) + \tilde{z}_j\,(t+\Delta t)\right),$$

$$i = \overline{1,n},$$

where $\tilde{z}_j\,(t+\Delta t)$ are the uses of product for purposes other than consumption and production.

5. **Calculate the weighted price index:**

$$\widehat{p}\,(t+\Delta t) = \frac{\sum_{j\in\Omega_1} \overline{p}_j\,(t+\Delta t)\,z_j\,(t+\Delta t) + \sum_{j\in\Omega_2\bigcup\Omega_3} p_j\,(t+\Delta t)\,z_j\,(t+\Delta t)}{\sum_{j=1}^{n} z_j\,(t+\Delta t)}.$$

6. **Determine the industry income and profits received by consumers:**

$$\Pi_j\,(t+\Delta t) = p_j\,(t+\Delta t)\,z_j\,(t+\Delta t) - \Big(\sum_{i\in\Omega_1} \overline{a}_{ij}\,\overline{p}_i(t) + \sum_{i\in\Omega_2\bigcup\Omega_3} \overline{a}_{ij}\,p_i(t)$$

$$+B_j (t + \Delta t) + \frac{1 - q_j^{(2)} (1 - q_j^{(1)})}{1 - q_j^{(1)}}$$

$$\times W_j (t+\Delta t) - v_j (t+\Delta t) \Big) x_j (t+\Delta t), \quad j \in \Omega_2 \bigcup \Omega_3,$$

$$\Pi_j (t + \Delta t) = \widehat{q}_j z_j (t + \Delta t) \overline{p}_j (t + \Delta t), \quad j \in \Omega_1,$$

$$\overline{\Pi}_j (t + \Delta t) = \begin{cases} \overline{p}_j (t + \Delta t) \overline{q}_j z_j (t + \Delta t), & j \in \Omega_1, \\ p_j (t + \Delta t) \overline{q}_j z_j (t + \Delta t), & j \in \Omega_2 \bigcup \Omega_3. \end{cases}$$

**7. Calculate the consumer money balances:**

$$D_k (t + \Delta t)$$

$$= D_k(t) + \sum_{j=1}^{n} c_{jk} \overline{\Pi}_j (t + \Delta t) + \widehat{p} (t + \Delta t) Q_k^{(1)} (t + \Delta t)$$

$$+Q_k^{(2)} (t+\Delta t) - \sum_{j \in \Omega_1} \overline{p}_j \min \left( S_j^k (t+\Delta t), y_j (t+\Delta t) \frac{S_j^k (t+\Delta t)}{\sum_{i=1}^{m} S_j^i (t+\Delta t)} \right)$$

$$- \sum_{j \in \Omega_2 \bigcup \Omega_3} p_j (t+\Delta t) \min \left( S_j^k (t+\Delta t), y_j (t + \Delta t) \frac{S_j^k (t+\Delta t)}{\sum_{i=1}^{m} S_j^i (t+\Delta t)} \right).$$

**8. Determine the budget revenues:**

(a) from value added tax:

$$h^{(1)} (t + \Delta t) = \sum_{j \in \Omega_1} q_j^{(1)} q_j \overline{p}_j (t + \Delta t) z_j (t + \Delta t)$$

$$+ \sum_{j \in \Omega_2 \bigcup \Omega_3} q_j^{(1)} q_j p_j (t + \Delta t) z_j (t + \Delta t);$$

(b) from tax on industry profits:

$$h^{(2)} (t + \Delta t) = \sum_{j=1}^{n} q_j^{(2)} \Pi_j (t + \Delta t);$$

(c) from excise taxes:

$$h^{(3)}(t+\Delta t) = \sum_{j\in\Omega_1} q_j^{(3)}\overline{p}_j(t+\Delta t)\,z_j(t+\Delta t) + \sum_{j\in\Omega_2\bigcup\Omega_3} q_j^{(3)}p_j(t+\Delta t)\,z_j(t+\Delta t);$$

(d) from income tax on consumers:

$$h^4(t+\Delta t) = \sum_{k=1}^{m} q_k^4 \times \left(\sum_{j=1}^{n} c_{jk}\overline{\Pi}_j(t+\Delta t) + \widehat{p}(t+\Delta t)\,Q_k^{(1)}(t+\Delta t) + Q_k^{(2)}(t+\Delta t)\right).$$

**9. Calculate the total budget revenues:**

$$B^0(t+\Delta t) = \sum_{l=1}^{4} h^{(l)}(t+\Delta t) + H(t+\Delta t) + \widehat{H}(t+\Delta t).$$

**10. Determine the expenditure on subsidies to industries:**

$$\overline{W}(t+\Delta t) = \sum_{j\in\Omega_1\bigcup\Omega_2} \overline{W}_j(t+\Delta t)\,x_j(t+\Delta t).$$

**11. Calculate the total budget expenditure:**

$$A^0(t+\Delta t) = \overline{W}(t+\Delta t) + \sum_{k=1}^{m}(\widehat{p}(t+\Delta t)\,Q_k^{(1)}(t+\Delta t) + Q_k^{(2)}(t+\Delta t)) + G(t+\Delta t).$$

**12. Determine the change of money quantity:**

$$M(t+\Delta t) = M(t) + (\overline{M}(t+\Delta t) - \overline{M}(t)) + (A^0(t+\Delta t) - B^0(t+\Delta t)).$$

The values of the model variables at time $t = 0$ are determined as follows. The values of $D_k(0)$, $k = \overline{1,m}$, $p_j(0)$, $j \in \Omega_2 \bigcup \Omega_3$, $\overline{W}_j(0)$, $j \in \Omega_1$, are assumed given. The values of $y_i(0)$, $S_j^k(0)$, $\Pi_j(0)$ are calculated from the formulas

$$y_j(0) = \Big(x_i(0) - \sum_{j=1}^{n} a_{ij} x_j(0)\Big) g_i, \quad i = \overline{1,n},$$

$$S_j^k(0) = \frac{D_k(0)\beta_{jk}}{p_j(0)}, \quad k = \overline{1,m}, \quad j \in \Omega_2 \bigcup \Omega_3,$$

$$S_j^k(0) = \frac{D_k(0)\beta_{jk}}{\overline{p}_j(0)}, \quad j \in \Omega_1;$$

$$\Pi_j(0) = p_j(0) z_j(0) - \Big(\sum_{i \in \Omega_1} \overline{a}_{ij} \overline{p}_i(0) + \sum_{i \in \Omega_2 \bigcup \Omega_3} \overline{a}_{ij} p_i(0) + B_j(0)$$

$$+ \frac{1 - q_j^{(2)}(1 - q_j^{(1)})}{1 - q_j^{(1)}} W_j(0) - \nu_j(0)\Big) x_j(0), \quad j \in \Omega_2 \bigcup \Omega_3.$$

The other model variables are calculated from the previously given formulas setting $t + \Delta t = 0$.

The considered model has been implemented in the decision support system (DSS) UKRAINIAN BUDGET (further—BUDGET), which models the effect of the state budget proposal on the main macroeconomic and interindustry characteristics and develops recommendations concerning efficient macroeconomic policies. The system is realized by PC using "Windows" environment. It does not demand the special knowledge in computer programming and coding. User can operate in multiscreen mode, open and close "Windows" with necessary information. The BUDGET system is equipped with multifunctional table and graphical editor and provides the SQL inquires. The results of model simulations (including various alternatives of macroeconomic policy) can be presented in tabular or graphic from. BUDGET produces a forecast of price dynamics, domestic sales volumes, budget revenues and expenditures. The modeling results can also explain the deviations of the predicted values from initial assumptions of the draft budget. Using hierarchical menus and computer graphics, the user of system (an expert in macroeconomic analysis) can select for each branch the pricing mechanisms and other model assumptions that are most essential in his view. Eliminating some factors, the user can estimate their influence at observed processes. An important function of the system is checking the effectiveness of various strategies of direct and indirect subsidies to producers and developing recommendations for reasonable taxation of production. A set of different optimization models can be applied for this purpose. Some problems concerning with their development is the subject of further consideration.

The development of simulated budget processes depends on several parameters which we can change in the model. First of all, the tax rates $q_j^{(1)}, q_j^{(2)}, q_j^{(3)}$ and $q_j^4$ can be considered as controlled variables. The subsidies $\overline{W}_j(t)$ and fixed prices $\overline{p}_j(t)$ are also the same. In some cases the budget problems can be solved changing the value and structure of budget expenditures $Q_k^{(1)}(t)$, $Q_k^{(2)}(t)$ and $G(t)$. Therefore the problem of optimal search of these parameters from the set of their admissible values appears. If the objective function is known this problem can be formulated as the optimal control problem for systems with discrete time; see Ermoliev et al. [10]. However, the construction of objective function in mentioned problem is rather difficult because we have to take into account the different nature goal indicators. Among them:

- the total value of budget deficit which is the commonly used indicator of well-developed budgetary policy:

$$F_1 = \sum_{t=1}^{T} A^0(t) - \sum_{t=1}^{T} B^0(t) \to \min;$$

- the values of budget deficit at some time moments $t \in \overline{\Im}$ can essentially impact on general financial situation as it was mentioned above, so they must be taken into account separately:

$$F_2^t = A^0(t) - B^0(t) \to \min, \quad t \in \overline{\Im};$$

- the stimulation of economic growth is one of the main goals of macroeconomic policy; therefore the value of total production $F_3$ and rates of economic growth $F_4$ at the end of observed time interval $[0; T]$ we consider as important economic indicators:

$$F_3 = \sum_{t=1}^{T} \sum_{i=1}^{n} Z_i(t), \quad F_4 = \frac{\sum_{i=1}^{n} Z_i(t)}{\sum_{i=1}^{n} Z_i(1)};$$

- the price stability is necessary condition for social development; taking this fact into account we consider the following criterion:

$$F_5 = \frac{\overset{\frown}{p}(T)}{\overline{p}(1)} \to \min .$$

Another important economic figures such as total incomes of consumers and producers, total cash-paying demand are also criteria for consideration.

Thus we obtain the multicriteria optimization problem. The changes in values of some its criteria cannot be fully compensated by other criteria. Such fact creates some difficulties for the objective function construction. Taking this all into account, we apply the methods of ordinary regression to solve this problem; see Gruber [14].

Let us consider the vector $F = (F_1, \ldots, F_K)$, where $K$ is the number of criteria. We write the following analytical form

$$U(F) = \min_{k=\overline{1,K}} \left( \alpha_k F_k - \beta_k \right) + \sum_{k=1}^{K} \gamma_k F_k,$$

where $\alpha_k$, $\beta_k$, $\gamma_k$ are unknown parameters, satisfied by means of the above mentioned conditions.

The values of these parameters can be identified as the solution of the next inequality system:

$$U\left(F^{(i)}\right) \geq U\left(F^{(j)}\right), \tag{2.17}$$

where $F^{(i)}$ is the value of criteria vector for alternative $i$, $F^{(j)}$ is the value of criteria vector for alternative $j$ and it is known that alternative $i$ is no worse than $j$.

The methods of nondifferentiable optimization (see Shor [40]) have been applied to find the system solution (or pseudo-solution). When we substitute the parameters by their known values we receive the objective function. The best solution in the budget model can be obtained as the maximal point of mentioned function under the constraints from previous section. The gradient method for optimal control problems solution [10] can be used to find this point.

The model simulations were made using the above mentioned approach applied to the basis of real data obtained from [48] *Ukrainian Economics in Figures (1994–1996)* and [7] *Consolidated Interindustry Balances (1990–1995)*. The search of rational tax rates values was the main goal of these experiments. The first step was to construct the objective function. Taking into account the monetary reform in the Ukraine in 1996 which changed the name of national currency, the scale of prices and the mechanisms of financial control, we present the indicators of budget deficit and total production in related from. Thus, the considered objective function depends on the following criteria:

– related value (to total budget expenditures) of budget deficit (in %):

$$F_1 = \frac{\sum_{t=1}^{T} A^0(t) - \sum_{t=1}^{T} B^0(t)}{\sum_{t=1}^{T} A^0(t)} 100\,\% \rightarrow \min.$$

– rates of production total growth/decline related to known basic variant (in %):

$$F_2 = \left( \frac{\sum_{t=1}^{T} \sum_{i=1}^{n} Z_i(t)}{\sum_{t=1}^{T} \sum_{i=1}^{n} \tilde{Z}_i(t)} - 1 \right) \cdot 100\,\% \rightarrow \max,$$

where $\tilde{Z}_i(t)$ is the value of production of branch $i$ at time moment $t$ according to the basic variant. The variant in which the total growth, compared with 1995, was considered as the basic;

– the rates of total economic growth/decline at the last month of the year related to the its first month (in %):

$$F_3 = \left( \frac{\sum_{i=1}^{n} Z_i('')}{\sum_{i=1}^{n} Z_i(1)} - 1 \right) \cdot 100\,\% \to \max .$$

– the rate of inflation:

$$F_4 = \frac{\widehat{p}\ (T)}{\widehat{p}\ (1)} \to \min .$$

The analytical form of function

$$U(F, \alpha, \beta, \gamma) = \min_{i=\overline{1,4}} (\alpha_i F_i - \beta_i) + \sum_{i=1}^{4} \gamma_i F_i$$

considered in the previous section was applied for this function construction by methods of ordinary regression. Unknown parameters $\alpha$, $\beta$, $\gamma$ were identified to construct this function on the basis of comparison of eight alternatives. The values of criteria for these alternatives were presented in Table 2.7. Alternatives I–VI are the result of different approaches to the development of fiscal and structural economic policy. Alternatives VII and VIII were introduced to emphasize the growth of objective function by its $F_2$ variable and its decline by $F_4$ variable for sufficiently large inflation rates. This demand is the natural restriction for the economy after hyperinflation and connected with it essential GDP decline.

**Table 2.7** Criteria values for the considered alternatives

| | Criteria | | | |
|---|---|---|---|---|
| Alternatives | $F_1$ | $F_2$ | $F_3$ | $F_4$ |
| I | 7.6 | −9.1 | −7.3 | 1.42 |
| II | 4.7 | 1.5 | 0.5 | 1.36 |
| III | 7.6 | −9.1 | −7.1 | 1.42 |
| IV | 5.2 | −7.4 | −6.1 | 1.23 |
| V | 7.0 | −9.1 | −7.1 | 1.42 |
| VI | 7.7 | 2.3 | 1.0 | 1.54 |
| VII | 7.6 | −10.0 | −7.3 | 1.42 |
| VIII | 7.6 | −10.0 | −7.3 | 1.55 |

The alternatives were compared by experts from analytical divisions of central governmental bodies. According to their consolidated mind, alternative II is the best one, alternative VI with high inflation but positive rates of economic growth is more preferable than the low inflationary alternative IV.

Alternative $v$ with low budgetary deficit is better than alternatives III and I which deficit values are greater. Therefore all alternatives can be ranked from more to less preferable as follows:

II, VI, IV, V, III, I, VII, VIII.

The identification of parameters was made by special computer program SPOF on the basis of above-mentioned data. The parameters values were obtained as the solution of inequality system (2.17). Thus, constructed objective function is the following:

$$U(F_1, F_2, F_3, F_4)$$
$$= \min\Big(F_1 - 16.209,\ F_2 - 2.038,\ F_3 + 13.588,\ F_4 - 1.387\Big)$$
$$-57.986F_1 + 85.197F_2 + 64.921F_3 - 0.769F_4.$$

The values of function $u(\cdot)$ for the considered alternatives are presented in Table 2.8. It should be noted that constructed function decreases with $F_1$ variable growth and increases with $F_2$ and $F_3$ variables growth. For small values of inflation rates it slowly increases with $F_4$ growth, but for large values it is decreased by this variable. Therefore this function is rather adequate to preferences existed in macroeconomic planning for transition economy.

The rational values of tax and subsidies rates were received as the maximum point of constructed objective function under the restrictions considered in Sect. 2.3. The obtained optimization model was solved as the optimal control problem with discrete time. The special version of quasigradient method [10] was applied for this purpose.

The calculations to be done show the necessity of deep differentiation of VAT and profit tax rates for different kinds of economic activity. This is connected with the great impact of the cost inflation scenario called [25] as the "Structural hyperinflation crisis" on the processes of budget realization. The structural disproportions in the economy – large rates of cost growth in some branches – are the source of such crisis. In this case all other pricing mechanisms would be, sooner or later, substituted by cost pricing. The rates of indirect taxes must be close to minimal to eliminate the

**Table 2.8** Values of constructed objective function for the considered alternatives

| Alternatives | VIII | VII | I | III | V | IV | VI | II |
|---|---|---|---|---|---|---|---|---|
| Values of objective function | −1761 | −1760 | −1684 | −1671 | −1636 | −1321 | −171.5 | −99.23 |

cost growth in branches with largest industrial expenditures. Low taxes in last ones can be compensated by tax increase for economic activities with low cost and quick money turnover. Thus, obtained rates of VAT change from 5–7 % for fuel and energy production branches to 22–25 % for trade and public supply.

Real data runs show the low efficiency of the "frozen" price policy for high cost branches which initiate the structural hyperinflation crisis. The inflationary impact of large money outlays which are necessary to provide subsidies, is no less in this case than the consequences of cost increases.

But the policy of "Low taxes, no subsidies" leads to the high rates of total cost growth in the branches under cost inflation. The demand restriction (usual strong monetary anti-inflation measures) leads to an additional industrial decline. It can decrease the rates of price growth and even made them less than rates of cost growth, but under the absence of effective bankruptcy mechanism enterprises try to subsidize themselves at the expense of their suppliers. Deep payments crisis observed in majority of CIS countries is the result of such activity. The best policy, according to the model simulations, is to subsidize the sector of economy weakly connected with branches which provoke the cost growth. Applying to Ukrainian data for 1996, this subsidies can cover up 17–22 % of the total cost in food industry, agriculture, transport, and telecommunications.

The forecasted values of main macroeconomic indicators for economy with obtained tax and subsidies rates are presented in Table 2.9 as "constructed variant." They can be compared with real values of these indicators for existing tax rates which are presented as "real values." Analyzing the information from this table, it is possible to assume that the proposed changes in fiscal policy would essentially change the general economic situation in the last half of 90th.

The following conclusion can be elaborated by the results of our research.

**Table 2.9** Values of main macroeconomic indicators

| Indicators | Constructed variant | Real values in 1996 |
|---|---|---|
| Related value of budget deficit (%) | 7.0 | 12.1 |
| Rates of total production growth (%) | −4.6 | −9.0 |
| Rates of total sales growth during the year (%) | 2.26 | −6.2 |
| Rates of inflation (GDR deflator) | 0.91 | 1.214 |
| Rates of consumer real incomes growth (real values = 100) | 125.8 | 100 |
| Rates of producer real incomes growth (real values = 100) | 119.5 | 100 |
| Real money supply (real values = 100) | 120.2 | 100 |

1. The development of anti-inflationary policy in a transition economy must take into account the effect of inflationary processes both inside and outside the money sphere. Therefore, pure monetary measures in the framework of this policy should be supplemented by deep structural–technological changes and corrections in fiscal policy, in particular as a part of government-directed program of structural reforms. The main priority of this program is to reduce production costs in branches enmeshed in a structural hyperinflation crisis.
2. Streaming of the state budget is one of the main stabilization mechanisms in a transition economy. However, elimination of budget deficit is not the only goal: it is necessary to take into account the effect of taxes on prices and product sales, positive and negative consequences of differential subsidies to some industries, etc. Hence, the multicriteria approach and construction of objective function are necessary for budget optimization.
3. The changes in the utility born by decrease of some arguments of objective function cannot be fully compensated by means of increase of its other arguments in the model of budget optimization. Therefore the objective function in this model is nondifferentiable. Its construction can be done by methods of ordinary regression.
4. According to model runs the tax system in a transition economy should be different from that under stable conditions. The tax system should rely on taxes assessed independently of business results (taxes on land, property, etc.). The tax burden should be 20–30 % of the profit of an average business entity. The sum of other payments to the budget should not exceed 30–40 % of these taxes.

   The following studies will be made for further investigation:

   - the account of internal impact on national economy and joint optimization the customer tariff and tax rates;
   - the analysis of influence of random processes in considered model;
   - the development of computer tools for model runs and objective function construction;
   - the real-data model runs using the current economic information for different countries with a transition economy.

The application of smooth approximations of objective function and comparison of obtained results also seems to be interesting. For example, quadratic objective functions for our model can be constructed on the basis of ordinal data.

Our analysis of the existing fiscal system in Ukraine was done by means of using the BUDGET system.

It is well known that the existing tax system is one of the major factors of the economic crisis that lasts in Ukraine. Destructive influence on manufacture exercises not so much overestimate of existing taxes rates as their macro and a microeconomic nonagreement, unsystematic character, indirectivity on stimulation of economic growth, ambiguity of charging taxes rules, an opportunity of evasion from their payment and other general systematic lacks. So, reforming of tax system in Ukraine can be considered as the major compound of system-structural transformations, it should be preceded by versatile scientific researches with application of various quantitative and qualitative methods, including mathematical modeling.

During scientific researches on this subject, mathematical modeling has been applied for revealing the mutual influence of existing kinds of taxes and specifications of the taxation, on one hand, and macroeconomic processes—on the other one. For a solution of this problem the simulation dynamic model from the BUDGET system has been applied, which displays movement of means on separate compounds of the system of national accounts. Except for the tax system, the real sector of economy, the state budget, end-consumers (households), the consolidated account of the external economic operations, bank system are considered in the models. These models provide a more detailed description of any of the specified subsystems, for example, sectored division is taken into account in the real sector. Consideration of simultaneous or alternative actions of various mechanisms of the pricing, inherent in transition economy, and various strategies of direct or mediate influences of the state on the prices and final consumer demand are the main features of our models. In particular, our models may allow for subsidizing of manufactures and social payments to consumers from the budget, influence on the prices of different ways of covering the deficit of the state budget and changes in monetary policy, etc.

With the purpose of our analysis of influence of the suggested taxation rates changes on volumes of inland revenues, our alternative computer modeling has been carried out on the basis of the imitating model considered above. Taking into account the available information, complete character of the budget for the period of its adoption and some other factors, the modeling was conducted for data of the state budget of Ukraine for 1998. An overall objective was to estimate the influence of taxation rates reductions on filling of the budget and on the course of macroeconomic processes.

Calculations were carried out according to the following variants.

**Variant I.** Upon specifications acting at that time and conditions of the taxation it was supposed that the production of manufactured volumes and sales will respond to macroeconomic forecasts on the basis of which the budget project has been developed. In particular, 3 % reduction of real GDP were supposed. This variant was considered for an estimation of accuracy of modeling which would be characterized by approximation of values of the indicators received during modeling to values in the project specified.

**Variant II.** Upon specifications acting at that time and conditions of the taxation it was supposed that the production of manufactured volumes and sales are determined by effective consumers' demand. This variant responds to a real world situation which has developed together with fulfilment of the state budget upon fiscal policy acting at that time.

**Variant III.** Under assumptions of a previous variant concerning factors which influence volumes of manufacture and sales of production that changes concerning reduction of tax pressure on the manufacturer and rationalization of system of taxes will be carried out. In particular it is supposed that the rate of the VAT will

**Table 2.10** Results of alternative modeling of fulfilment of the basic indicators of the state budget of Ukraine for 1996 (in % to planned values

| Indicators | Variants | | |
|---|---|---|---|
| | I | II | III |
| Budget incomes from: | | | |
| (a) VAT | 103.9 | 67.1 | 58.2 |
| (b) Profits tax | 81.6 | 46.7 | 45.4 |
| (c) Payment fund taxes | 96.7 | 81.2 | 78.1 |
| Aggregate budget incomes | 97.9 | 77.2 | 85.3 |
| Aggregate budget expenses | 101.1 | 98.4 | 100.5 |

be decreased from 20 % down to 16 %, and the cumulative rate of the taxation of a payment fund will be decreased from 52 % down to 32 %. Also an introduction of the fixed tax to activity of small enterprises and the uniform land tax with the abolishment of other direct taxes for agricultural producers was supposed. This complex of preconditions covers the majority of constructive suggestions concerning changes in the taxation expressed in the middle of the 90th. Our results of modeling according to this variant, displayed in Table 2.10, have shown how the specified suggestions can influence the economic situation in the country.

Let's analyze the received results. The deviation of values of the majority of indicators according to variant from planned in the project of the budget does not exceed 3–4 %. So, there is some ground to consider this size as a rough estimate of modeling accuracy. Comparison of variants and II gives the basis to assume that during macroeconomic forecasting reduction of a final demand which, according to variant II, should serve as the reason of recession of GDP at a rate of 9 % have been inadequately appreciated. (Real reduction of GDP in 1996 has barely exceeded 10 % that completely falls into the modeling accuracy specified above.) Under these conditions, reduction of tax pressure can become one of the major factors of economic stabilization. The fact that in variant III in comparison to variant II inpayments from the profits taxes practically have not decreased under essential reduction of the tax base (owing to the abolishment of this tax for small and agricultural enterprises) testifies of probable increase of profitability in the majority of branches of economy. It is necessary to pay a special attention to the fact that reduction of inpayments from payment fund taxes in variant III are much more smaller in comparison with reduction of rates of these taxes. It testifies of overvaluation of such rates in the working tax laws. Though in variant III essential budgetary deficit remains, it nevertheless would be less, than in variant II. So, it is possible to draw a conclusion that the reductions of taxes mentioned above as a whole would exercise positive influence on process of fulfilment of the state budget.

An even greater positive influence they could exercise on the general macroeconomic situation. Comparison of some macroeconomic indicators according to variants II and III is carried out in Table 2.11.

Thus, there are all grounds to state that absence of positive changes in tax system is one of the major factors of long economic decline. The carried out researches allow to determine such mainstreams of necessary improvement of this system.

**Table 2.11** Value of the main macroeconomic indicators for variants II and III

| Indicators | Variants | |
|---|---|---|
| | II | III |
| Increase (decline) of GDP (%) | −9.0 | −6.1 |
| including for the last quarter | −5.5 | 1.7 |
| Cumulative real incomes of consumers (% to values for var. II) | 100 | 128.7 |
| including group with the lowest incomes | 1000 | 113.1 |
| Incomes of the enterprises (% to values for var. II): | | |
| (a) For group with the highest incomes | 100 | 101.5 |
| (b) For group with the lowest incomes | 100 | 102.8 |
| Consumer price index (% to the beginning of the year) | 1,281.4 | 120.7 |

1. The question concerning expediency of attraction to the state budget of finance in volume of 35–40 % of GDP remains debatable. On one hand, this specification is, certainly, too big under conditions of weak economy, in which market relations are still developing and the public sector starts to function. On the other hand, taking into account the structure of final consumption on needs of the state, the state budget is one of the major factors of creation of permanent effective demand for end production, and consequently, one of the factors of deceleration of economic decline. In addition, Ukraine has developed social infrastructure that is financed from the state budget destruction of which will have unforeseen consequences. In our opinion, the main flaw of fiscal system existing in Ukraine is not the fact that tax rates are too big, but the fact that the existing system of taxation does not assist economic growth, in particular, suppresses the final effective demand.
2. Taking this into account, the most negative influence on economy have indirect taxes on fund of payment. These taxes can be reduced to 10–20 %. If extricated current assets are directed on increase of a level of wages, it will exercise appreciable positive influence both on dynamics and on structure of GDP. Thus, the general inpayments from the specified taxes will decrease much less, than reduction of their rates, and financing of target expenses on which they were directed, will not make especial problems.
3. Introduction and the further development of taxes the volumes of which do not depend on results of economic activities is necessary: the land tax, the property tax, the rent payments, the fixed taxes, etc. Inpayments from these taxes should make at least 20–25 % from volume of the general inpayments to the consolidated budget of the country.
4. The existing custom duties do not exercise such destructive influence on economy as taxes do. The majority of them can be considered close to the optimum under conditions which have developed. Increase of the custom duties concerning some groups of consumer goods and services (polygraphic production, products

of light industry, luxury goods) at the same time is possible. But the part of these kinds of production in the general import is rather small. An increase of import tariffs concerning production of intermediate use is inexpedient as it intensifies processes of cost-push inflation to which there are already some preconditions.
5. The further improvement of tax system demands carrying out of complex economic researches with attraction of experts of different directions.

## 2.4 Optimization Model of Structural and Technological Changes

The experience with overcoming high inflation in Ukraine in 1992–1994 has focused attention on both the unavoidable use and the limited effect of traditional measures, such as reduction of money emission, reduction of the budget deficit, elimination of subsidized enterprise credit, and the active exchange-rate policy of the National Bank. The price increases were a consequence of deep structural distortions in the economy inherited from the earlier decades and significantly amplified in the initial stages of the transition. Under these conditions, monetary stabilization methods not accompanied by radical structural adjustment simply transformed these distortions into a different form, e.g., inflation was replaced with an increase of accounts payable, but the origins of the distortions were not eliminated. Once relative financial stability has been achieved, it is essential to proceed with a structural reform of the economy. The main tasks include restructuring of the fiscal system and the social security system, changes in property relations, development of a market infrastructure, encouragement of small- and medium-size entrepreneurship, and transformation of the technological structure of the economy.

Structural transformations are much more difficult task than initial stabilization of the quantity of money. No standard solutions are available in this area, and much depends on the numerical characteristics describing the features of the economy. This is particularly relevant for the technological structure, which is unique for each country. The elucidation of technological–structural distortions that affect crisis phenomena in the economy and analysis of techniques for the elimination of these distortions require the use of quantitative methods, in particular, mathematical modeling. In this section, these methods are applied to plan technological changes for increased production efficiency.

The existing technological system in Ukraine is the result of several decades of cumulative influence of the following factors:

- low prices of natural resources (especially low energy prices), which encouraged wasteful use;
- low prices of basic consumer commodities combined with low wages, which resulted in a low share of value added in the consumer prices of most goods;

- a pervasive system of budgetary subsidies, which distorted the interindustry relative prices, artificially lowering the prices of some goods while raising the prices of others;
- the closed nature of the socialist economy, which did not require competitiveness of domestic producers. Today the economy is functioning under fundamentally different conditions. The prices of energy imports have risen to the world level, and in some cases actually exceed the world prices. Energy prices were one of the main factors driving the cost inflation in 1992–1994 [25,47]. The inflationary process was characterized by increases of production costs, which led to price increases long before consumer incomes had a chance to catch up. This reduced the consumer demand, lowered the share of value added in the prices of goods, and triggered Keynesian macroeconomic processes [20], when the price is insensitive to demand and the production volume is insensitive to price. Under these conditions, one of two alternative scenarios is observed, depending on the strictness of the monetary policy:
  (a) lax payment discipline leading to an increase of accounts payable and accelerated decrease of production, or
  (b) accelerated cost inflation. To overcome this situation, it is necessary to increase the disposable income of the consumers (primarily households) while at the same time reducing production costs.

Decrease of production costs is thus a major goal of technological–structural change. It can be achieved in a twofold way: first, by changing the technology so as to reduce inputs of commodities, materials, and other manufactured products per unit of output, and second, by gradually reducing the use of energy- and resource-inefficient technologies, until they are completely phased out. Both these approaches involve changes at the interindustry level, which explains the choice of interindustry models as an analytical tool. Note that these changes are manifested in the reduction of the input matrix elements of the aggregated interindustry balance. Indeed, the direct input coefficients $a_{ij}$ representing the quantity of industry $i$ product required to produce a unit of industry $j$ product is representable

$$a_{ij} = \sum_{k_1=1}^{l_i} \sum_{k_2=1}^{l_j} \tilde{a}_{ij}^{k_1 k_2} W_j^{k_2}, \tag{2.18}$$

where

$k_1$ are the indices of the technologies and subindustries included in industry $i$,
$k_2$ are the indices of the technologies and subindustries included in industry $j$,
$l_i, l_j$ is the number of such technologies and subindustries,
$\tilde{a}_{ij}^{k_1 k_2}$ is the consumption of products manufactured by technology $k_1$ in industry $i$ to produce a unit of product by technology $k_2$ in industry $j$,
$W_j^{k_2}$ is the share of technology $k_2$ in industry $j$.

Then the first of the two approaches to technological–structural change mentioned above will reduce the values of $\overline{a}_{ij}^{k_1k_2}$, whereas the second approach will reduce the values of $W_j^{k_2}$ for the largest $\overline{a}_{ij}^{k_1k_2}$. We see from (2.18) that both approaches reduce $a_{ij}$. This explains the choice of interindustry aggregated balance models as the tool for analyzing and planning technological–structural change. One such a model is considered in this section.

Assume that the country's economy consists of $n$ aggregated industries and $A = \left\{a_{ij}\right\}$ is the direct input matrix for these industries. We assume that labor cost is a linear function of production volume in each industry. Let $q_i$ be the share of labor costs in the price of industry $i$ product. Denote by $y_i$ and $x_i$ respectively the net and the gross product of industry $i$ in constant prices. These quantities are linked by the relationship

$$x_i = \sum_{j=1}^{n} a_{ij} x_j + y_i, i = \overline{1, n}, \tag{2.19}$$

and the real consumer income $D$ is given by

$$D = \sum_{i=1}^{n} q_i x_i. \tag{2.20}$$

The problem is to determine what changes in the elements of the matrix $A$ and the vector $q = \left\{q_1, \ldots, q_n\right\}$ will maximize the consumer income $D$ without triggering additional inflationary forces.

Conditions excluding outburst of cost inflation under the impact of endogenous factors are considered in Mikhalevich [23]. Using the coefficients of the matrix $A$, we can write these conditions in the form

$$\sum_{i=1, i \neq j}^{n} \frac{a_{ij}}{1 - a_{jj} - \overline{q}_j} < 1, \quad j = \overline{1, n}, \tag{2.21}$$

where $\overline{q}_j$ is the share of value added in the price of industry $j$ product.

We assume that the net product consists of two parts: a part that depends on $D$ and a part independent of $D$. Assuming linear dependence of the first part on consumer income, we obtain

$$y_i = \alpha_i D + h_i, \quad i = \overline{1, n}, \tag{2.22}$$

where the coefficients $\alpha_i$, mainly reflect the structure of individual consumption and internal investment, while $h_i$ is determined by the export–import balance of the industries and by public consumption.

We use these relations to express $D$ in terms of $A$ and $q$. From (2.19), $x = (E-A)^{-1}y$, where $x = \left(x_1, \ldots, x_n\right)$, $y = \left(y_1, \ldots, y_n\right)$. Therefore, $D = (q, x) = \left(q, (E-A)^{-1}y\right)$. From the last equality and (2.22) we obtain

$$D = \frac{q^T(E-A)^{-1}h}{1 - q^T(E-A)^{-1}\alpha},$$

where $h = (h_1, \ldots, h_n)$, $\alpha = (\alpha_1, \ldots, \alpha_n)$.

Also note that $\overline{q}_j$ is representable in the form

$$\overline{q}_j = l_j q_j + d_j,$$

where $l_j$ and $d_j$ are given coefficients.

Allowing for errors in input data and for the maintenance of a reserve to support a noninflationary increase of the value-added components that are independent of $q$, we replace (2.21) with the conditions

$$\sum_{i=1, i\neq j}^{n} \frac{a_{ij}}{1 - a_{jj} - \left(l_j q_j + d_j\right)} \leq \beta, \quad i = \overline{1, n}, \tag{2.23}$$

where $\beta < 1$ is a prespecified threshold value. Based on these remarks, the model takes the following form. Find the changes $\Delta a_{ij}$ and $\Delta q_j$ ($i = \overline{1, n}$, $j = \overline{1, n}$) in the current elements of the matrix $A$ and the vector $q$ so as to maximize the function

$$f(\Delta A, \Delta q) = \frac{(q + \Delta q)^T \left(E - (A + \Delta A)\right)^{-1} h}{1 - (q + \Delta q)^T \left(E - (A + \Delta A)\right)^{-1} \alpha} \rightarrow \max, \tag{2.24}$$

where $\Delta q = \left(\Delta q_1, \ldots, \Delta q_n\right)$, $\Delta A = \left\{\Delta a_{ij}\right\}_{ij}^{nn}$ subject to the constraints

$$\sum_{i=1, i\neq j}^{n} \frac{a_{ij} + \Delta a_{ij}}{1 - (a_{jj} + \Delta a_{ij}) - \left(l_j (q_j + \Delta q_j) + d_j\right)} \leq \beta, j = \overline{1, n}, \tag{2.25}$$

$$0 \leq q_j + \Delta q_j \leq 1, \quad 0 \leq a_{ij} + \Delta a_{ij} \leq 1, \quad i, j = \overline{1, n}, \tag{2.26}$$

$$a_{jj} + \Delta a_{jj} + l_j \left(q_j + \Delta q_j\right) + d_j \leq 1, \quad j = \overline{1, n}. \tag{2.27}$$

These constraints may be augmented with bounds on allowed changes of direct input coefficients imposed by the existing technologies

$$\gamma_{\_ij} \leq \Delta a_{ij} \leq \overline{\gamma}_{ij}, \quad i, j = \overline{1, n}. \tag{2.28}$$

and bounds imposed by the availability of resources for technological–structural changes,

$$\sum_{j=1}^{n} \sum_{i=1}^{n} b_{kij} \max(0, -\Delta a_{ij}) \leq B_k, \quad k = \overline{1, K}, \tag{2.29}$$

where

$K$ is the number of different resources,
$B_k$ is the quantity of resource $k$ allocated for reduction of production costs,
$b_{kij}$ is the consumption of this resource to implement measures that lead to a unit reduction in the consumption of industry $i$ product to produce a unit of industry $j$ product.

Note that aggregated interindustry models are widely used for planning and analyzing technological–structural change in the economy. In particular, the proposed model may be regarded as further development of [13], where Glushkov has considered the changes in the matrix $A$ that ensure attainment of the production target $(x, y)$ subject to the given productive resources. Although this is a standard problem for a centrally planned economy, it remains relevant for a transition economy highly dependent on critical imports of energy, raw materials, and other industrial inputs.

The optimization problem (2.24)–(2.29) is a difficult mathematical programming problem. The functional (2.24) and the system of constraints (2.25)–(2.29) lead to a number of difficulties even when we try to find a local extremum of problem (2.24)–(2.29). These difficulties impose certain requirements on the choice of the solution method. The simplest requirement is associated with the constraint system (2.25)–(2.29), where constraints (2.12) are nonsmooth, and the group of constraints (2.25) are fractional-linear. In constraint $j$ from (2.25) the denominator is independent of $i$, and by (2.27) we conclude that it is positive. We can thus rewrite the group of constraints (2.25) in the form

$$\sum_{i=1, i \neq j}^{n} (a_{ij} + \Delta a_{ij}) + \beta(a_{jj} + \Delta a_{jj}) + \beta\Big(l_j(q_j + \Delta q_j) + d_j\Big) \leq \beta, \quad j = \overline{1, n}. \tag{2.30}$$

Then the constraint system (2.25)–(2.29) can be replaced with a single convex nonsmooth constraint

$$F(\Delta A, \Delta q) = \max\Big\{f_1(\Delta A, \Delta q), \ldots, f_5(\Delta A, \Delta q)\Big\} \leq 0, \tag{2.31}$$

where

$$f_1(\Delta A, \Delta q) = \max_j \left\{ \sum_{i=1, i\neq j}^{n} (a_{ij} + \Delta a_{ij}) + \beta(a_{jj} + \Delta a_{jj}) + \beta\Big(l_j(q_j + \Delta q_j) + d_j\Big) - \beta \right\},$$

$$f_2(\Delta A, \Delta q) = \max_{i,j} \left\{ a_{ij} + \Delta a_{ij} - 1, -a_{ij} - \Delta a_{ij}, q_j + \Delta q_j - 1, -q_j - \Delta q_j \right\},$$

$$f_3(\Delta A, \Delta q) = \max_j \left\{ a_{jj} + \Delta a_{jj} + l_j(q_j + \Delta q_j) + d_j - 1 \right\},$$

$$f_4(\Delta A, \Delta q) = \max_{i,j} \left\{ \Delta a_{ij} - \overline{\gamma}_{ij}, \gamma\_{ij} - \Delta a_{ij} \right\},$$

$$f_5(\Delta A, \Delta q) = \max_k \left\{ \sum_{j=1}^{n} \sum_{i=1}^{n} b_{kij} \max(0, -\Delta a_{ij}) - B_k \right\}.$$

The functional (2.24) leads to greater difficulties, because it is not continuous for all $\Delta a_{ij}$ and $\Delta q_j$, $i, j = \overline{1,n}$. To avoid discontinuities, we need to satisfy the following conditions:

(a) the matrix $E - (A + \Delta A)$ is nonsingular;
(b) $(q + \Delta q)^T \Big(E - (A + \Delta A)\Big)^{-1} \alpha \neq 1$.

Inequalities (2.21) imply that $A$ is a production matrix, and therefore the first condition is satisfied for $\Delta a_{ij}$ from the feasible region of (2.25)–(2.29) and from some sufficiently large neighborhood of the feasible region. The product $q^T(E - A)^{-1}\alpha$ may be regarded as a multiplier for "increase of production-increase of consumer incomes." Increase of production requires additional cost savings even if resource-efficient technologies are used. Moreover, the newly created value added is never used entirely for consumption, and for actual technologies we always have $(q + \Delta q)^T \Big(E - (A + \Delta A)\Big)^{-1} \alpha \neq 1$. Therefore, in the region where $\Delta A$ and $\Delta q$ satisfy constraint (2.24), the function $f(\Delta A, \Delta q)$ is continuously differentiable and its gradient is computed by the following formulas

$$\frac{\partial f(\Delta A, \Delta q)}{\partial \Delta q_j} = \frac{1}{1 - (q + \Delta q, \tilde{\alpha})}(e_j, \tilde{h}) + \frac{(q + \Delta q, \tilde{h})}{\Big(1 - (q + \Delta q, \tilde{\alpha})\Big)^2}(e_j, \tilde{\alpha}), \quad j = \overline{1,n}, \tag{2.32}$$

$$\frac{\partial f(\Delta A, \Delta q)}{\partial \Delta a_{ij}} = \frac{1}{1-(q+\Delta q,\tilde{\alpha})}(e_i,\tilde{q})(e_j,\tilde{h}) + \frac{(q+\Delta q,\tilde{h})}{\left(1-(q+\Delta q,\tilde{\alpha})\right)^2}(e_i,\tilde{q})(e_j,\tilde{\alpha}), \quad i,j=\overline{1,n}. \qquad (2.33)$$

Here we use the following notation:

$$\left(E-(A+\Delta A)\right)^{-1}h=\tilde{h}, \quad \left(E-(A+\Delta A)\right)^{-1}\alpha=\tilde{\alpha},$$

$$\left(E-(A+\Delta A)^T\right)^{-1}(q+\Delta q)=\tilde{q},$$

$e_i$ is the $n$-dimensional vector whose $i$-th component is 1 and all other components are 0.

When evaluating the gradient of $f(\Delta A, \Delta q)$ with respect to the variables $\Delta a_{ij}$, we use the following differentiation rule for the inverse matrix $X^{-1}$, where $X = \left\{x_{ij}\right\}_{ij}^{nn}$:

$$\frac{\partial X^{-1}}{\partial x_{ij}} = -X^{-1}\frac{\partial X}{\partial x_{ij}}X^{-1} = -X^{1}e_i e_j^T X^{-1}.$$

This is a consequence of differentiating the identity $XX^{-1} = E$.

Yet another difficulty is associated with the large dimension of problem (2.24)–(2.29). Thus, the number of variables in this problem is $n^2 + n$ (if we need to find the changes of all the elements of the matrix $A$, which is fairly large even for $n$ of the order of 10–20.

Thus, to find a local extremum in problem (2.24)–(2.29) we need an efficient extremum-seeking method for nonsmooth functions. The $r(\alpha)$-algorithm [39, 43] meets this requirement. This is an efficient practical method from the family of subgradient type methods with space dilation in the direction of the difference of two successive subgradients. First, it can handle the nonsmoothness of the constraint (2.31). Second, given the specific features of the objective function (2.24), we can use its modified version [42] evaluating the gradient of the function $f(\Delta A, \Delta q)$ only in the region of its continuous differentiability. Third, experience shows that the $r(\alpha)$–algorithm is efficient for functions with several hundreds of variables (including functions with deep narrow valleys). The multi-extremum nature of problem (2.24)–(2.29) can be allowed for (to a certain extent) by running the calculations from different starting points with a small initial step $h_0$.

The $r(\alpha)$–algorithm is the basis of the program MULSTR for PC. This is a Fortran program to find a local extremum of problem (2.24)–(2.29). The program allows control of the following parameters:

(a) the initial approximation,
(b) the penalty multiplier (used to reduce problem (2.24)–(2.29) to an unconstrained optimization problem);

(c) the parameters of the $r(\alpha)$–algorithm, which include the initial stepping multiplier $h_0$,
(d) the maximum number of iterations $maxitn$, stopping parameters specified by accuracy in the argument and the gradient norm $\varepsilon_x$ and $\varepsilon_g$.

Other parameters of the $r(\alpha)$-algorithm are chosen following the recommendations of [42] (see also [39]). In particular, the space dilation factor is taken at $\alpha = 2$, and for the adaptive step control parameter we take $q_1 = 0.95$, $q_2 = 1.1$, $n_h = 3$. The objective function gradients are calculated from formulas (2.32), (2.33) using the numerically stable subprograms DECOMP and SOLVE [11] to solve the systems of linear equations. These programs are based on the Gaussian elimination technique with partial choice of the pivot element.

The program MULSTR has been applied to run test calculations for real-life data with the model (2.24)–(2.29). These test calculations are described below.

Our numerical experiments had three objectives: first, to assess the efficiency of the proposed algorithm when applied to real-life data with problems of real-life size; second, to anticipate difficulties with data and computation procedures that may arise in the model; and third, to analyze the recommendations for management decisions produced by the model.

The program MULSTR has been used at the Institute of Cybernetics to run test calculations using the data from [18]. We consider the aggregated 18-industry balance and make the following changes in the model (2.24)–(2.29).

1. As no reliable information is available for some components of the vector $h$ and other components are unstable, we replace (2.24) with the following objective function:

$$F(\Delta A, \Delta q) = (q + \Delta q)^T \Big(E - (A + \Delta A)\Big)^{-1} \alpha.$$

   The function (2.24) obviously increases with the increase of the function $F(\Delta A, \Delta q)$. In this function, $\alpha$ are the industry shares in the net product components dependent on consumer income (individual consumption and investment). The values of $\alpha$ were calculated using the data of [45].
2. In the constraints (2.25) we took $\beta = 0.85$. Given the accuracy of the input data, this ensured at least a 5 % increase in the share of value added in the price of industry products.
3. Only input coefficients greater than 0.001 were allowed to vary (within 50 % of their initial value); the remaining input coefficients were kept unchanged. The number of variables in problem (2.24)–(2.29) was thus 200 (of which 182 were the changes of input matrix elements and 18 the changes in the components of the vector $q$).
4. As the constraint (2.29) we used the bound on investments attracted for technological–structural change. Alternative calculations were carried out for different investment volumes. The coefficients and the right-hand side of the constraint (2.29) were based on expert estimates provided by various advisory and analytical divisions of the Cabinet of Ministers of Ukraine.

Three scenarios were considered. According to the first scenario, the constraint (2.29) permits reducing the ratio between intermediate consumption and GDP by 10–15 % of the 1995 level. According to the second scenario (which requires double the capital investment), the ratio can be reduced by 15–20 %. The third scenario assumed a fixed share of labor costs in the price of industry products, while keeping all the assumptions of the first scenario. This scenario was regarded as reflecting the impact of labor costs on final outcomes.

The three scenarios led to the following results.

There is a group of "critical" elements in the matrix $A$, which are reduced with any investment volume. These critical elements are the electric power inputs for the production of a unit of product in coal and food industries; inputs of all forms of fuel to produce a unit of product in the power generating industry, in transport, and in agriculture; metal inputs in machine building; and transport costs in most aggregated heavy industries. The reduction of these components ranges from 20 to 50 % of their initial value, depending on the allocated investment resources. The remaining elements of the matrix $A$ can be reduced with a sufficiently large investment, and they are grouped as follows.

**Power Efficiency.** In addition to previously mentioned industries, it is advisable to reduce the electric power costs in trade and catering, in transport, in machine building, in ferrous and nonferrous metallurgy, in chemical and petrochemical industry. The unit inputs of liquid fuel and gas should be reduced in light and food industry, in trade and catering; the unit inputs of coal should be reduced in ferrous metallurgy.

**Resource Efficiency.** It is advisable to reduce (by 20–40 %) the rate of input use from the main material manufacturing industries (ferrous and nonferrous metallurgy, chemical and petrochemical industry) by the power-generating industry, machine building, and transport. It is also required to reduce the rate of input use from machine building and wood and lumber industry in most other material manufacturing industries. At the same time, the second scenario allows these input rates to be increased in construction (this provides evidence that the prices of construction services are maintained at an artificially high level).

**Reduction of Accompanying Costs.** Almost all industries should reduce the rate of use of services from trade and catering and from transport. This signifies, in particular, reduction of trade and transport markups.

Tables 2.12 and 2.13 present the number of coefficients in each block that have to be adjusted under each scenario. Note that scenario 3 requires much greater technological–structural changes than scenarios 1 and 2, and the optimal objective function value for this scenario is approximately half that for the other two scenarios. This indicates that achievement of real economic growth requires not only radical changes of technology, but also deep restructuring of the existing system of labor remuneration.

The observed reduction of material costs makes it possible to increase the share of labor cost in the product price of most industries from the present level of

**Table 2.12** Results of calculations using various scenarios for the aggregated interindustry model

| Scenarios and blocks | Number of varied coefficients | Decrease: Less than 10 % | Decrease: 10–25 % | Decrease: 25–47 % | Decrease: Over 47 % | Increase |
|---|---|---|---|---|---|---|
| **Scenario 1** | 182 | 6 | | | 13 | 40 |
| *Blocks* | | | | | | |
| Energy efficiency | 45 | 2 | 1 | 4 | | 8 |
| Resource efficiency | 54 | 1 | | 7 | | 12 |
| Reduction of accompanying costs | 30 | 1 | 2 | 2 | | 12 |
| **Scenario 2** | 182 | 9 | 14 | 32 | 39 | 8 |
| *Blocks* | | | | | | |
| Energy efficiency | 45 | 2 | 1 | 4 | | 8 |
| Resource efficiency | 54 | 1 | | 7 | | 12 |
| Reduction of accompanying costs | 30 | 1 | 2 | 2 | | 12 |
| **Scenario 3** | 182 | 9 | 14 | 32 | 39 | 8 |
| *Blocks* | | | | | | |
| Energy efficiency | 45 | 2 | 1 | 4 | | 8 |
| Resource efficiency | 54 | 1 | | 7 | | 12 |
| Reduction of accompanying costs | 30 | 1 | 2 | 2 | | 12 |

**Table 2.13** Objective function value for optimal solution under each scenario

| Scenario | Value of $q^T(E-A)^{-1}\alpha$ | Remarks |
|---|---|---|
| Scenario 1 | 0.41579 | – |
| Scenario 2 | 0.48513 | – |
| Scenario 3 | 0.21178 | – |
| *Note* | | |
| For the 1995 direct input matrix | 0.22669 | Does not satisfy constraints (2.29) |

3–7 % to 15–25 %. The recommended increase is different for different industries, ranging from a high of 10–17 % in agriculture, light, and food industry, and machine building to a low of 3–4 % in chemical and petrochemical industry and in wood and lumber industry. In ferrous and nonferrous metallurgy and in construction, this indicator actually should be decreased.

Note that the proposed model is primarily intended to identify the main directions of change of technological coefficients. It has to be modified to examine specific reform measures, especially on the micro level.

Measures intended to produce technological–structural changes are usually implemented as (industry-specific) innovation programs. The set of all possible projects is usually finite, but the number of projects is so large that it is impossible to choose the appropriate projects by simple enumeration. It is thus necessary to consider a discrete version of the model (2.24)–(2.29).

Let $N$ be the number of projects, and for each project we know what changes $\Delta a_{ij}^t$ of the matrix $A$ and $\Delta q_j^t$ of the vector $q$ it will produce ($i, j = \overline{1,n}$, $t = \overline{1,N}$). It is required to find a list of projects that ensures the maximum increase of real consumer income subject to constraints that avoid cost inflation and subject to bounds on resource availability $z_t$.

Introducing the variables

$$z_i = \begin{cases} 1 & \text{if project } t \text{ is implemented,} \\ 0 & \text{otherwise,} \end{cases}$$

we rewrite the model in the form

$$\frac{q(z)\left(E - A(z)\right)^{-1} h}{1 - q(z)\left(E - A(z)\right)^{-1} \alpha} \to \max,$$

where

$$q(z) = \left(q_1(z), \ldots, q_n(z)\right), \quad A(z) = \left\{a_{ij}(z)\right\}_{ij}^{nn},$$
$$q_j(z) = q_j + \sum_{t=1}^{N} \Delta q_j^t z^t, \quad a_{ij}(z) = a_{ij} + \sum_{t=1}^{N} \Delta a_{ij}^t z^t,$$

subject to

$$\sum_{\substack{i=1 \\ i \neq j}}^{n} \frac{a_{ij}(z)}{1 - a_{ij}(z) - l_j\left(q_j(z) + d_j\right)} \le \beta, \quad j = \overline{1,n},$$

$$0 \le a_{ij}(z) \le 1, \quad 0 \le q_j(z) \le 1, \quad i, j = \overline{1,n},$$

$$\sum_{t=1}^{N} b_{kt} z_t \le B_k, \quad k = \overline{1,K}.$$

Here $K$ is the number of resources, $B_k$ is the volume of resource $k$ allocated for the projects of technological–structural change, and $b_{kt}$ is the consumption of this resource by project $t$.

We can also add constraints specifying mutual exclusion of some projects:

$$\sum_{(t_1,t_2)\in R} z_{t_1} z_{t_2} = 0,$$

where $R$ is the set of mutually exclusive projects.

This model leads to a discrete optimization problem classifiable as a multi-dimensional knapsack problem with a nonlinear nonadditive objective function and nonlinear constraints. The latter complicates solution of the problem by the branch-and-bound method and algorithms based on sequential analysis of partial solutions (such as the Bellman method). Certain difficulties are also attributable to the existence of multiple extrema of the objective function. The problem should thus be considered as a combinatorial optimization problem on the space of combinations and solved by the descent vector method [34], which is a simple iterative algorithm for approximate solution of such problems.

If we define the unit neighborhood in the combination space [34, 36] and restrict the local search to such neighborhoods, then the calculation of the descent vector components requires bounding the objective function changes when project $t_1$ is added to the set of previously accepted projects $t_m$ or when project $t_1$ is excluded from the previous set. As we have noted above, innovation projects are inherently sectoral, and therefore all the changes in the components of the matrix $A(z)$ are concentrated in a particular column $j(t_1)$. This fact can be exploited to bound the objective function. Let

$$A^{T_m} = \left\{a_{ij}^{T_m}\right\}_{ij}^{nn}, \quad a_{ij}^{T_m} = a_{ij} + \sum_{t\in T_m} \Delta a_{ij}^{t},$$
$$q^{T_m} = \left\{q_j^{T_m}\right\}_{j}^{n}, \quad q_j^{T_m} = q_j + \sum_{t\in T_m} \Delta q_j,$$

$$A_j^{t_1} \text{ is the column vector } \begin{pmatrix} \Delta a_{ij(t_1)} \\ \vdots \\ \Delta a_{nj(t_1)} \end{pmatrix}.$$

We have the following theorem.

**Theorem 2.1.** *Let $H_{t_1}^{+}$, $H_{t_1}^{-}$ be the objective function increments following the inclusion (elimination) of project $t_1$ in (from) the set $t_m$, $t_m^{+}$ and $t_m^{-}$ are the resulting new sets of projects. Then*

$$H_{t_1}^{+} = \frac{\overline{H}h\left(1 - q^{T_m}\left(E - A^{T_m}\right)^{-1}\alpha\right) + \overline{H}\alpha q^{T_m}\left(E - A^{T_m}\right)^{-1}h}{\left(1 - q^{T_m}\left(E - A^{T_m}\right)^{-1}\alpha\right)^2 + \overline{H}\alpha\left(1 - q^{T_m}\left(E - A^{T_m}\right)^{-1}\alpha\right)},$$

$$H_{t_1}^{-} = \frac{\hat{H}h\left(1 - q^{T_m}\left(E - A^{T_m}\right)^{-1}\alpha\right) + \hat{H}\alpha q^{T_m}\left(E - A^{T_m}\right)^{-1}h}{\left(1 - q^{T_m}\left(E - A^{T_m}\right)^{-1}\alpha\right)^2 + \hat{H}\alpha\left(1 - q^{T_m}\left(E - A^{T_m}\right)^{-1}\alpha\right)},$$

*where*

$$\overline{H} = \left(\left(q + \Delta q_{j(t_1)e_{j(t_1)}}\right)\left(\frac{1}{1 - g_{j(t_1)}}\overline{G}\right) + \Delta q_{j(t_1)e_{j(t_1)}}\right)\left(E - A^{T_m}\right)^{-1},$$

$$\hat{H} = \left(\left(q + \Delta q_{j(t_1)e_{j(t_1)}}\right)\left(\frac{1}{1 + g_{j(t_1)}}\overline{G}\right) - \Delta q_{j(t_1)e_{j(t_1)}}\right)\left(E - A^{T_m}\right)^{-1},$$

$\overline{G} = \left(E - A^{T_m}\right)^{-1} A_j^{t_1}$, $g_j$ *is the* $j$*-th element of* $\overline{G}$, $e_j$ *is the* $j$*-th unit vector. Here*

$$\left(E - A^{T_m^+}\right)^{-1} = \left(E - A^{T_m}\right)^{-1} + \frac{1}{1 - g_i}\overline{G}\left(E - A^{T_m}\right)^{-1}, \tag{2.34}$$

$$\left(E - A^{T_m^-}\right)^{-1} = \left(E - A^{T_m}\right)^{-1} - \frac{1}{1 + g_i}\overline{G}\left(E - A^{T_m}\right)^{-1}. \tag{2.35}$$

*Proof.* By definition,

$$H_{t_1}^{+} = \frac{\left(q + \overline{\Delta q}\right)\left(E - A^{T_m^+}\right)^{-1}h}{1 - \left(q + \overline{\Delta q}\right)\left(E - A^{T_m^+}\right)^{-1}\alpha} - \frac{q\left(E - A^{T_m}\right)^{-1}h}{1 - q\left(E - A^{T_m}\right)^{-1}\alpha}, \tag{2.36}$$

where $\overline{\Delta q} = \Delta q_{j(t_1)}e_{j(t_1)}$.

Using the estimates from [39] for the inverse matrix when one of the columns of the original matrix is changed, we obtain

$$\left(E - A^{T_m^+}\right)^{-1} - \left(E - A^{T_m}\right)^{-1} = \frac{1}{1 - g_i}G\left(E - A^{T_m}\right)^{-1}, \tag{2.37}$$

where $G = \left(E - A^{T_m}\right)^{-1} \Delta A^{T_m}$, $g_j$ is the $j$th element of $G$, $\Delta A^{T_m}$ is the vector of changes made in column $j$ of the matrix $A^{T_m}$. If the project $A_j^{t_1}$ is added, the column $j(t_1)$ is changed, and the associated changes form the vector $A_j^{t_1}$. We thus have $G = \left(E - A^{T_m}\right)^{-1} A_j^{t_1} = \overline{G}$, and equality (2.34) follows from (2.37). Let us estimate

$$\begin{aligned}\overline{H} &= \left(q + \overline{\Delta q}\right)\left(E - A^{T_m^+}\right)^{-1} - q\left(E - A^{T_m}\right)^{-1} \\ &= q\left(\left(E - A^{T_m^+}\right)^{-1} - \left(E - A^{T_m}\right)^{-1}\right) + \overline{\Delta q}\left(E - A^{T_m^+}\right)^{-1}.\end{aligned}$$

From (2.37) and (2.34) we obtain

$$\overline{H} = \left(\left(q + \overline{\Delta q}\right)\left(\frac{1}{1 - g_{j_{t_1}}}\overline{G}\right) + \overline{\Delta q}\right)\left(E - A^{T_m}\right)^{-1}.$$

Using this estimate, we reduce the right-hand side of (2.19) to a common denominator and simplify the resulting expression. The end result is the sought equality for $H_{t_1}^{+}$.

The equality for $H_{t_1}^{-}$ and relationship (2.35) are derived similarly, noting that when the project $t_1$ is eliminated the changes in the column $j(t_1)$ of the matrix $A^{T_m}$ will have opposite signs and

$$H_{t_1}^{-} = \frac{\left(q - \overline{\Delta q}\right)\left(E - A^{T_m^-}\right)^{-1} h}{1 - \left(q - \overline{\Delta q}\right)\left(E - A^{T_m^-}\right)^{-1} \alpha} - \frac{q\left(E - A^{T_m}\right)^{-1} h}{1 - q\left(E - A^{T_m}\right)^{-1} \alpha}.$$

□

These estimates enable us to calculate the components of the descent vector without inverting the matrices $E - A^{T_m^-}$ and $E - A^{T_m^+}$, which essentially simplifies the construction of an iterative procedure for an integer model. The integer model, combined with the solution method of the continuous problem (2.24)–(2.29) by modeling [13] appropriate information base, and a good user interface could provide a basis for a special-purpose DSS for the advisory and analytical services of the Cabinet of Ministers, whose responsibility would include analysis and screening of proposals for improvement of the technological structure of the economy and development of technical policy recommendations. Solution of these problems would also require analysis of interindustry processes affecting employment.

## 2.5 Models and Information Technologies for Decision Support During Structural and Technological Changes

Structural reforms are one of the basic problems solved during market transformation of the economy. Planning such reforms involves severe difficulties, since it should rest upon the productive forces available and should account for the interests of various social groups and strata formed in the transition economy and turning the existing social and economic structure to their advantage. In contrast to the early stages of transition economy, where there were no alternatives for reforms in the majority of postsocialist countries, structural changes under such circumstances should be chosen from a great number of possible alternatives. The great number of such alternatives and aspects to assess them and the strong effect of the risk and uncertainty on the feasibility of the plan of reforms necessitate modern information decision support technologies to be applied. The development of mathematical models and relevant computational algorithms is an integral part of the development of such technologies. In this book, we consider such models and methods for the decision support of planned structural and technological changes, being one of the most important components of structural reforms.

The current system of industrial technologies was developed during a long time under the conditions of centralized economy. It was assumed at that time that prices were controlled, production was scheduled and unprofitable with losses covered from the state budget, and industrial expenses in some branches were high. Such preconditions contradict the fundamentals of the functioning of market economy and cause a wide use of "soft budgetary constraints" [20], i.e., covering a part of expenses of some economic entities from the financial resources of the whole society. Such a practice hampers structural reforms, it generates corruption and "paternalistic" relationships between the authority and particular economic entities, which objectively makes decision making at the nation-wide level non-transparent and cast some doubt on the necessity of the further democratization of the society. Thus, structural and technological changes to eliminate the above-mentioned disproportions play an important role in the social stabilization. Reducing the industrial expenses, which creates preconditions for the increase of the payment for labor and profitableness on the inflation-free basis and thus decreasing the motivation of applying "soft budgetary constraints," should become the main trend of these transformations.

Two important aspects related to the general decrease of industrial expenses are noteworthy. One is the necessity of intensifying energy- and resource-saving under fast increase in the prices for resources (especially energy ones) in the world markets. Under conditions of artificial underevaluation of fuel and electric energy in the former USSR in the 1970–1980s, the energy intensity of production was several times greater than that in developed countries with market economy. It is only radical reduction of energy consumption that can compensate for the consequences of a quick growth in prices for hydrocarbon fuel. There are two basic ways of energy saving:

(i) direct energy saving, where the introduction of new technologies, modernization of the equipment available, refusal of the most power-consuming industries, and replacements of their production with import reduce the specific industrial expenses of energy resources;
(ii) indirect energy saving, where the general consumption of energy decreases due to replacing power-consuming industrial products with less power-consuming analogs and reducing the consumption of some power-consuming products with preserving (and may be increasing) the total volume of output.

The capabilities of the indirect energy saving are as a rule greater; however, the great variety of alternatives of its implementation requires various, including quantitative, methods of economic studies (including mathematical economic modeling) to be applied for management decision support in choosing this way.

The second important aspect related to the reduction of production costs is the change of the expenditure pattern of the state budget. Rejecting the practice of "soft budgetary constraints" allow concentrating the financial resources on the solution of actual social problems. At the same time, structural and technological changes involve significant budgetary resources. Estimating their efficiency is also of great importance.

In the first decade of the twenty-first century, the authors have developed a series of models to determine structural and technological changes and to solve the above-mentioned problems. In particular, to determine the main trends of reducing production costs, the studies [27, 37] have proposed an intersectoral optimization model with variable coefficients of direct costs. To search for changes in the structure of export and import that reduce the total energy consumption, the linear model considered in [27, 28], and using the equations of intersectoral balance has been developed. The influence of risk and uncertainty factors on the solution of the above problems was studied in [26, 28].

The further development of the models, first of all, of the intersectoral optimization model [27,37] has allowed using them as a basis to create prototype versions of information technologies of management decision support in carrying out structural and technological changes. Let us consider the most important (in our opinion) aspects of these technologies—the basis model and its modifications, numerical methods of computations and their implementation on modern computers, and some aspects of the implementation of model computations.

The section is structured as follows. Section 2.5.1 reviews briefly the results on modeling structural and technological changes [27, 37], considers some problems of carrying out model computations, and analyzes the properties of the intersectoral optimization model that can be used in this case. Section 2.5.2 presents an approach (alternative to that proposed in [27, 37]) to performing computations based on this model. Section 2.5.3 briefly describes the IOMSTC program system, which is a basis of the information technology of management decision support. Section 2.5.4 considers the modifications of the intersectoral optimization model. The final subsection formulates conclusions and proposes fields for further studies.

### *2.5.1 Optimization Model of Planning Structural and Technological Changes*

To determine the structural and technological changes that would reduce the manufacture costs and thus would increase the incomes of ultimate customers and make the economy more dynamic the following optimization model has been proposed in [27, 37].

Let us consider the economy formed by $n$ pure industries manufacturing only one type of products. Let $i$, $j$, be the numbers of these branches ($i,\ j = \overline{1,n}$). Denote by $a_{ij}$ the value of direct production costs of the branch $i$ for manufacturing a unit of production of the branch $j$. This quantity can be expressed in both natural and cost measures depending on the information available. The matrix $A = \{a_{ij}\}$, $i,\ j = \overline{1,n}$ is called the matrix of the coefficients of direct costs (the matrix of technological coefficients, the Leontief matrix) [22].

Assume that the incomes $D$ of ultimate customers are proportional to the volume of production:

$$D = \sum_{i=1}^{n} q_i x_i = (q, x),$$

where $q_i$ is the share of ultimate incomes (payment for labor, social transfers, and profit) in the price of the production of the branch $i$, $x_i$ is the gross output of this branch, $q = (q_1, \ldots, q_n)$, $x = (x_1, \ldots, x_n)$. In the model, we also assume that the final output $y_i$ of each branch is proportional to ultimate incomes

$$y_i = a_i D + h_i, \quad i = \overline{1,n},$$

where the coefficients $a_i$ reflect mainly the structure of individual consumption and internal investments, and $h_i$ is determined from the export/import balance of branches and the structure of public consumption. Let $h = (h_1, \ldots, h_n)$, $a = (a_1, \ldots, a_n)$.

One more assumption of the model is a linear relationship between the share of the added cost $\tilde{q}_i$ in the price of the product of the branch $i$ and the share of ultimate incomes in the price of this product

$$\tilde{q}_i = l_i q_i + d_i, \quad i = \overline{1,n},$$

where $l_i$ is a fiscal multiplier of ultimate incomes [8], and $d_i$ is the share of other components of the added cost in the price of the product of the $i$th branch.

The problem is to determine the structural and technological changes so as to maximize either the ultimate incomes or the multiplier "increase of incomes–increase of output," which would provide the dynamic property of the economy.

In the model from [27, 37], these changes are reflected as the changes $\Delta a_{ij}$ and $\Delta q_i$ of the coefficients $a_{ij}$ and $q_i$, which may be both positive (increase in the corresponding coefficient) and negative (its decrease). A decrease in the technological coefficients requires some limited assets (financial, material, or intellectual). The model presented below implies a linear relationship between the decrease in the coefficients and expenditures of the assets; however, more complicated dependences are possible.

In view of the assumptions above, we obtain the following optimization model.

It is necessary to determine $\Delta a_{ij}$ and $\Delta q_i$ $(i, j = \overline{1, n})$ such that would satisfy the following:

- the constraints that exclude the intensification of the inflation of costs

$$\sum_{\substack{i=1\\ i\neq j}}^{n} \frac{a_{ij} + \Delta a_{ij}}{1 - \left(a_{jj} + \Delta a_{jj}\right) - \left(l_j \left(q_j + \Delta q_j\right) + d_j\right)} \leq \beta, \quad j = \overline{1, n}, \tag{2.38}$$

where $0 < \beta < 1$ is a preset confidential parameter;
- the relationships that follow from the physical meaning of the coefficients $\Delta a_{ij}$ and $\Delta q_j$:

$$0 \leq q_j + \Delta q_j \leq 1, \quad 0 \leq a_{ij} + \Delta a_{ij} \leq 1, \quad i, j = \overline{1, n}; \tag{2.39}$$

- the balance of the expenses and added cost

$$a_{jj} + \Delta a_{jj} + l_j \left(q_j + \Delta q_j\right) + d_j \leq 1, \quad j = \overline{1, n}; \tag{2.40}$$

- constraints for the possible ranges of variation of the coefficients due to specific features of the technologies available

$$\begin{aligned} \underline{\gamma_{ij}} \leq \Delta a_{ij} \leq \overline{\gamma_{ij}}, \quad i, j = \overline{1, n}; \\ \underline{q_i} \leq \Delta q_i \leq \overline{q_i}, \quad i = \overline{1, n}, \end{aligned} \tag{2.41}$$

where $\underline{\gamma_{ij}}$, $\overline{\gamma_{ij}}$ are the lower and upper limits of the possible variation in the technological coefficients, $\underline{q_i}$ and $\overline{q_i}$ are the lower and upper limits of the possible variation in the share of ultimate incomes;
- the resource constraints

$$\sum_{j=1}^{n} \sum_{i=1}^{n} b_{kij} \max\left(0, -\Delta a_{ij}\right) \leq B_k, \quad k = \overline{1, K}, \tag{2.42}$$

where $K$ is the number of resources, $B_k$ is the volume of the $k$th resource intended to carry out structural and technological changes, $b_{kij}$ is the expenditure of this resource in taking the measures that provide a unitary decrease in the expenses of the production of the branch $i$ to produce a unit of production of the branch $j$.

The variations $\Delta a_{ij}$ and $\Delta q_i$ from the set satisfying constraints (2.38)–(2.42) should be chosen so as to maximize the ultimate incomes of consumers

$$D = f_1(\Delta A, \Delta q) = \frac{(q + \Delta q)^T \left(E - (A + \Delta A)\right)^{-1} h}{1 - (q + \Delta q)^T \left(E - (A + \Delta A)\right)^{-1} \alpha} \longrightarrow \max, \tag{2.43}$$

where $\Delta q = (\Delta q_1, \ldots, \Delta q_n)$, $\Delta A = \{\Delta a_{jj}\}_{i,j=1}^{n}$, $\Delta q = (\Delta q_1, \ldots, \Delta q_n)$, $h = (h_1, \ldots, h_n)$, $\alpha = (\alpha_1, \ldots, \alpha_n)$, $E$—is a unit matrix, and $T$ denotes transposition.

Another objective function is the multiplier "increase of incomes–increase of production"

$$f_2(\Delta A, \Delta q) = (q + \Delta q)^T \left(E - (A + \Delta A)\right)^{-1} \alpha^1 \longrightarrow \max, \tag{2.44}$$

where $\alpha^1 = (\alpha_1^1, \ldots, \alpha_n^1)$ is the vector of the structure of ultimate internal consumption.

For a detailed substantiation of constraints (2.38)–(2.42) and objective functions (2.43), (2.44) see [27, 37].

The optimization model may be considered as a one-criterion problem (with one of the functions (2.43) or (2.44)) or in a multicriterion case (with both objective functions considered earlier). Since a limited consumption of some scarce imported energy resources (e.g., natural gas) is important, the list of criteria may be supplemented with their consumption

$$f_3(\Delta A, \Delta q) = g^T \left(E - (A + \Delta A)\right)^{-1} (\alpha D + h) \longrightarrow \min, \tag{2.45}$$

where $g^1 = (g_1, \ldots, g_n)$ is the vector of specific industrial consumption of this resource.

The resultant multicriterion problem can be solved using the convolution method by maximizing the function

$$F(\Delta A, \Delta q) = \tilde{\beta}_1 f_1(\Delta A, \Delta q) + \tilde{\beta}_2 f_2(\Delta A, \Delta q) + \tilde{\beta}_3 f_3(\Delta A, \Delta q), \tag{2.46}$$

where the weight coefficients $\tilde{\beta}_i$, $i = \overline{1, 3}$ can be determined by serial-regression methods based on expert estimates [27].

Since the criteria (2.43) and (2.45) are opposite, it is expedient to complement constraints (2.38)–(2.42) with a constraint for the minimum level of ultimate incomes

$$f_1(\Delta A, \Delta q) \leq \widehat{f}_1, \tag{2.47}$$

where $\widehat{f}_1$ is a desirable level of incomes. Note that the problem (2.38)–(2.42), (2.46), (2.47) has no basic differences from the problems (2.38)–(2.43) and (2.38)–(2.42), (2.44); therefore, in what follows, we will consider the two last problems.

Solving problems (2.38)–(2.43) or (2.38)–(2.42), (2.44) involves some difficulties. First, since the functions $f_1(\Delta A, \Delta q)$ and $f_2(\Delta A, \Delta q)$ are nonconvex, these problems are multiextreme despite the convexity of the set of solutions $X$ specified by (2.38)–(2.42). Second, the objective functions of these problems are not definite for all the $\Delta A$ and $\Delta q$. For the function $f_2(\Delta A, \Delta q)$ to be definite, the matrix $E - (A + \Delta A)$ should be nonsingular. For $f_1(\Delta A, \Delta q)$ to be definite, it is additionally required that the condition $(q + \Delta q)^T \left(E - (A + \Delta A)\right)^{-1} \alpha \neq 1$ be satisfied. Third, the left-hand sides of constraints (2.42) are continuous yet not everywhere differentiable convex functions, and the left-hand sides of constraints (2.38) are discontinuous. Note that the last two problems can be solved. If constraints (2.41) and (2.42) do not reduce all $a_{ij} + \Delta a_{ij}$ up to zero for some $j$ (such a situation is typical for economy since it is impossible to produce anything without material inputs), the denominator of the left-hand side of (2.38) will be non-negative for all the $\Delta A$ and $\Delta q$ satisfying (2.40). The left-hand side of (2.38) increases without limit if this denominator is close to zero; therefore, this constraint will not be satisfied. Thus, for admissible solutions and solutions from some neighborhood of $X$, the denominator of the left-hand side of (2.38) will be positive, and these constraints can be rearranged as

$$\sum_{\substack{i=1 \\ i\neq j}}^{n} \left(a_{ij} + \Delta a_{ij}\right) \leq \beta\left(1-\left(a_{jj} + \Delta a_{jj}\right)-\left(l_j\left(q_j + \Delta q_j\right)+d_j\right)\right), \quad j = \overline{1,n}, \tag{2.48}$$

whence the sufficient existence condition $\left(E-(A+\Delta A)\right)^{-1}$ known as the efficiency condition [22] follows for the matrix $(A + \Delta A)$. It can be shown in a similar way [37] that the condition $(q + \Delta q)^T \left(E - (A + \Delta A)\right)^{-1} \alpha \neq 1$ is satisfied for all the solutions from $X$ and some its neighborhood, and the functions $f_1(\Delta A, \Delta q)$ and $f_2(\Delta A, \Delta q)$ are defined for all the admissible solutions of problems (2.38)–(2.43) and (2.38)–(2.42), (2.44) and also for the inadmissible solutions insignificantly differing from admissible ones. Thus, the objective functions of the problems specified are defined and continuous on some rather small neighborhood $X$.

Note that constraint (2.48) is linear, in contrast to (2.38). To solve the problem thus transformed, well-known methods such as the method of nonsmooth penalty functions [39] can be applied. This does not complicate the problem due to nonsmooth constraints (2.42). The extremum of the penalty function can be found by any non-differentiable optimization method; in particular, the $r$-algorithm [43] was used in [27,37]. Taking the penalty factor sufficiently large always allows localizing the search in a rather small neighborhood of $X$, where the functions $f_1(\Delta A, \Delta q)$ and $f_2(\Delta A, \Delta q)$ are continuous, and the replacement of constraint (2.38) with inequality (2.48) is equivalent.

Implementing this approach involves the dimension problem and of the differentiation of the functions $f_1(\Delta A, \Delta q)$ and $f_2(\Delta A, \Delta q)$, which employ the operation of matrix inversion. The total number of variables equals $n^2 + n$, which is rather large even for $n$ of order 20 $\Longrightarrow$ ..40. To differentiate the functions $f_1(\Delta A, \Delta q)$ and $f_2(\Delta A, \Delta q)$, the authors have proposed [37] formulas that follow from the componentwise differentiation of the matrix function $\overline{F}(Y)$ given implicitly by the relationship $\overline{F}(Y)Y = E$, where $Y$ is an arbitrary $n \times n$ matrix.

The dimension of the problem being solved can be substantially reduced if we exclude from the consideration the positions $(i, j)$, where $\left|\underline{\gamma}_{ij} - \overline{\gamma_{ij}}\right| < \varepsilon$ ($\varepsilon$ is a sufficiently small number) and also $\Delta a_{ij}$ and $\Delta q_i$ such that for any $\Delta A$ and $\Delta q$, the inequalities $\left|\frac{\partial f_1(\Delta A, \Delta q)}{\Delta a_{ij}}\right| < \varepsilon$ and $\left|\frac{\partial f_1(\Delta A, \Delta q)}{\Delta q_i}\right| < \varepsilon$ hold for the problem (2.39)–(2.43), (2.48) and the inequalities $\left|\frac{\partial f_2(\Delta A, \Delta q)}{\Delta a_{ij}}\right| < \varepsilon$ and $\left|\frac{\partial f_2(\Delta A, \Delta q)}{\Delta q_i}\right| < \varepsilon$ hold for the problem (2.39)–(2.42), (2.44), (2.48). In particular, as the computations based on the intersectoral balances composed in Ukraine at the end of the 1990s—beginning of the 2000s show, it was possible to consider no more than 130 variables in the case of $n = 18$, and no more than 250 for $n = 38$.

The fact that the problem being solved is multiextreme can be taken into account by carrying out independent computations by the $r$-algorithm from different initial points. In particular, initial points can be selected from a uniform partition of the diagonal of the hypercube

$$\left\{(\Delta q, \Delta A) : \underline{\gamma_{ij}} \le \Delta a_{ij} \le \overline{\gamma_{ij}}, \quad \underline{q_i} \le \Delta q_i \le \overline{q_i}\right\}$$

i.e. the points $\left(\underline{q}_1, \ldots, \overline{q}_n, \underline{\gamma}_{11}, \ldots, \overline{\gamma}_{nn}\right)$ and $\left(\overline{q_1}, \ldots, \overline{q_n}, \overline{\gamma_{11}}, \ldots, \overline{\gamma_{nn}}\right)$.

Note that the results of computations based on the optimization models (2.38)–(2.43) and (2.38)–(2.42), (2.44) cannot be directive under conditions of transitive economy. They can be used to obtain a desirable structure of industrial technologies that intensify the social and economic development of the country, to reveal the ways of reducing the existing structure to the desirable one, to evaluate the necessary resources, etc. Of importance is the analysis of the variation of the technological coefficients $a_{ij}$ in time during the last several years, revealing the tendencies of the approximation (or removal) of the real values of $a_{ij}$ to the desirable values obtained from the computation using the above-mentioned models. From this point of view, obtaining a series of local extrema of functions (2.43) or (2.44) with the values sufficiently close to their global maxima is more preferable to the subsequent application than searching for the global extremum.

The values of some parameters in the models (2.38)–(2.43) and (2.38)–(2.42), (2.44) may be determined ambiguously. First of all, these are the parameter $\beta$ in constraints (2.38), possible ranges of variation of the technological coefficients and of the share of added value in the price of a production in inequalities (2.41), and some coefficients and the right-hand sides of the resource constraints (2.42). Performing dialogue computations on the models (2.38)–(2.43) and (2.38)–(2.42), (2.44) for different values of the above parameters is therefore of interest. It is expedient that the time interval between the beginning and the end of a current series of computations, i.e., between the formation of a version of the problem with certain values of its parameters and obtaining a series of its locally optimal solutions, does not exceed 35 min. As numerical experiments show, the run time on a Pentium IV computer using the model (2.39)–(2.43), (2.48) and the $r$-algorithm for one initial approximation for $n = 38$ varies from 0.3–0.4 min (if the approaches considered earlier are used to reduce the problem dimension) to 1.3–1.5 min (if the computations are performed for a problem of full dimension). If 10–15 initial approximations are taken, the above-mentioned time constraints may not be satisfied. A way out may be the use of modern multiprocessor devices. A local extremum for each initial approximation is sought using a separate processor. The values of the objective function at the found points are passed to one of the processors, where solutions with insufficiently large values of the objective functions are rejected and other alternatives are ordered and relayed to decision-makers for further analysis. Other formal criteria of rejecting the solutions are also possible. The number of initial approximations for a sufficiently large number of processors can be increased up to several tens or even hundreds, the total run time remaining almost the same as if one extremum is searched for on a one-processor computer, i.e., being quite comprehensible to support the system-user dialogue. The SKIT-2 and SKIT-3 clusters developed recently at the Institute of Cybernetics of the National Academy of Sciences of Ukraine are suitable for computations according the above-stated scheme.

When searching for an extremum by maximizing the penalty function using the $r$-algorithm, additional problems may arise. As noted earlier, the functions $f_1(\Delta A, \Delta q)$ and $f_2(\Delta A, \Delta q)$ are defined on the sets of admissible solutions of problems (2.38)–(2.43) and (2.38)–(2.42), (2.44) and for some neighborhoods of these sets. Nevertheless, when an extremum is searched for, the sequence of approximations obtained using the $r$-algorithm may go beyond the limits of these neighborhoods. This was observed on the average for 6–7 of 10 initial approximations during the numerical experiments. As a result, the computational procedure terminates at an inadmissible point, where the objective function of the problem is not defined. To avoid this, it is proposed to apply a modification of the $r$-algorithm [42], which assumes that the subgradient of the objective function is computed only at admissible points, or [42], which possesses the same properties. From this point of view, of interest is to transform the problems (2.38)–(2.43) and (2.38)–(2.42), (2.44) to a form that avoids the use of functions not defined everywhere, which has underlain the alternative approach to the development of computational algorithms to be considered in the next subsection.

The above-mentioned numerical experiments have shown one more feature of the problems being solved. Different initial approximations have yielded a series of solutions with almost equal values of the objective function (deviations from the maximum value in the series are 0.3–1.0 %), but with several substantially differing components of the solution. The statement below explains this effect to some extent.

**Theorem 2.2.** *Let* $\left(\Delta A^{(1)}, \Delta q^{(1)}\right)$ *and* $\left(\Delta A^{(2)}, \Delta q^{(2)}\right)$ *be arbitrary admissible solutions of the problem* (2.38)–(2.43) *or* (2.38)–(2.42), (2.44) *for which the equality*

$$\left(E - \left(A + \Delta A^{(1)}\right)^T\right)^{-1} \left(q + \Delta q^{(1)}\right) \left(E - \left(A + \Delta A^{(2)}\right)^T\right)^{-1} \left(q + \Delta q^{(2)}\right) \tag{2.49}$$

*holds componentwise. Let also*

$$\left(\Delta A_\lambda, \Delta q_\lambda\right) = \lambda \left(\Delta A^{(1)}, \Delta q^{(1)}\right) + (1 - \lambda) \left(\Delta A^{(2)}, \Delta q^{(2)}\right),$$

*where* $0 < \lambda < 1$ *is an arbitrary number.*

*Then the equality*

$$f_i = \left(\Delta A^{(1)}, \Delta q^{(1)}\right) = f_i \left(\Delta A_\lambda, \Delta q_\lambda\right) = f_i \left(\Delta A^{(2)}, \Delta q^{(2)}\right), \quad i = \overline{1,2},$$

*is true and* $(\Delta A_\lambda, \Delta q_\lambda)$ *is an admissible solution of the above problems.*

The proof of the theorem is based on the statement considered in the next subsection and will be presented after this statement. Let us dwell on some corollaries of the theorem.

Admissible points having identical values of the objective function and belonging to the set satisfying (2.49) form structures similar to a set of intervals of straight lines. If the extreme points of one of such intervals differ in only several components, then all the internal points of the interval also differ from each other in only the values of these components. If the question is values close to the global maximum, the $r$-algorithm for the set of initial points will result in one of such structures, which corresponds to the results of numerical experiments.

Note that the approach proposed to solve problems (2.38)–(2.43) and (2.38)–(2.42), (2.44) allows obtaining their local solutions in a comprehensible time; however, analytic capabilities are rather limited in this case. This necessitates the development of an alternative approach to the solution of these problems.

### 2.5.2 Alternative Approach to Developing a Computational Algorithm

Let us rearrange the problems (2.38)–(2.43) and (2.38)–(2.42), (2.44) by introducing new variables

$$z = \left(E - (A + \Delta A)^T\right)^{-1} (q + \Delta q),$$

whence

$$\left(E - (A + \Delta A)^T\right)z = q + \Delta q. \tag{2.50}$$

Then the problem of maximizing ultimate incomes of consumers can be formulated as the problem of finding $(\Delta A, \Delta q, z)$ that maximize the function

$$\overline{f}_1(z) = \frac{z^T h}{1 - z^T \alpha} \tag{2.51}$$

under the constraints (2.39)–2.42), (2.48), (2.50). The problem of maximizing the multiplier "increase of incomes–increase of output" is formulated as the problem of maximizing the function

$$\overline{f}_2(z) = z^T \alpha^1 \tag{2.52}$$

under the same constraints.

Note that the objective functions and constraints of these problems are defined for any values of $(\Delta A, \Delta q, z)$ (except for $z$ such that $z^T \alpha = 1$ for the problem with the objective function $\overline{f}_1(z)$).

Like (2.39)–(2.43), (2.48) and (2.39)–2.42), (2.44), (2.48), the problems obtained can be solved by the method of nonsmooth penalty functions with the use of the $r$-algorithm. Certain difficulties may arise because of the equality constraints (2.50) that cause the nonconvexity of the set of admissible solutions. Therefore, it is possible to apply here the same approach as that used to solve previous problems—choosing several initial approximations for the $r$-algorithm.

The following statement is true for the problems under study.

**Theorem 2.3.** *Let* $x^{(1)} = \left(\Delta A^{(1)}, \Delta q^{(1)}, z^*\right)$ *and* $x^{(2)} = \left(\Delta A^{(2)}, \Delta q^{(2)}, z^*\right)$ *be arbitrary admissible solutions of problems* (2.39)–(2.42), (2.48), (2.50), (2.51) *and* (2.39)–(2.42), (2.48), (2.50), (2.52) *with the identical values of the variables* $z = z^*$. *Let also* $x(\lambda) = \lambda x^{(1)} + (1 - \lambda)x^{(2)}$, *where* $0 < \lambda < 1$ *is an arbitrary number. Then the values of the functions* $\overline{f}_i(z)$, $i = \overline{1,2}$ *are identical for the solutions* $x^{(1)}$, $x^{(2)}$, *and* $x(\lambda)$, $x(\lambda)$ *being an admissible solution for any* $0 < \lambda < 1$.

*Proof.* Let us show a point $x(\lambda)$ satisfies constraints (2.39)–(2.42), (2.48), and (2.50) for any $0 < \lambda < 1$. Indeed, constraints (2.39)–(2.41), and (2.48) are linear inequalities, and the left-hand side of constraints (2.42) is a convex function. Therefore, the above constraints specify a convex set. If the points $x^{(1)}$ and $x^{(2)}$ belong to this set, then their convex combination $x(\lambda)$ also belongs to it, i.e., it satisfies the relationships (2.39)–(2.42), (2.48). Further, since $x^{(1)}$ and $x^{(2)}$ are admissible, the following equalities are true:

$$\left(E - \left(A + \Delta A^{(1)}\right)^T\right)z^* = \left(q + \Delta q^{(1)}\right),$$

$$\left(E - \left(A + \Delta A^{(2)}\right)^T\right)z^* = \left(q + \Delta q^{(2)}\right).$$

Multiplying the first of these equalities by $\lambda$ and the second by $1 - \lambda$ and adding them yield

$$\left(E - \left(A + \lambda\Delta A^{(1)} + (1-\lambda)\Delta A^{(2)}\right)^T\right)z^* = q + \lambda\Delta q^{(1)} + (1-\lambda)\Delta q^{(2)}. \tag{2.53}$$

Taking into account $z_\lambda = \lambda z^* + (1-\lambda)z^* = z^*$, we can rearrange Eq. (2.53) as

$$\left(E - (A + \Delta A_\lambda)^T\right)z^* = q + \Delta q_\lambda, \tag{2.54}$$

where $\left(\Delta A_\lambda, \Delta q_\lambda, z_\lambda\right)$ are components of the solution $x(\lambda)$. According to (2.54), the point $x(\lambda)$ satisfies Eq. (2.50) and is an admissible solution of the problems (2.39)–(2.42), (2.48), (2.50), (2.51) and (2.39)–(2.42), (2.48), (2.50), (2.52).

Note that the values of the variables $z$ for the solution $x(\lambda)$ coincide with the values of these variables for the solutions $x^{(1)}$ and $x^{(2)}$. Since the objective functions $\overline{f}_1(z)$ and $\overline{f}_2(z)$ of these problems depend only on $z$, their values at the points $x^{(1)}$, $x^{(2)}$, and $x(\lambda)$ coincide too, whence the validity of the theorem follows.

Having proved Theorem 2.3, let us get back to the proof of Theorem 2.2.

Note that the admissibility of the solutions $\left(\Delta A^{(1)}, \Delta q^{(1)}\right)$ and $\left(\Delta A^{(2)}, \Delta q^{(2)}\right)$ yields the existence of matrices $\left(E - \left(A + \Delta A^{(i)}\right)^T\right)^{-1}$, $i = \overline{1,2}$. A unique admissible solution $\left(\Delta A^{(i)}, \Delta q^{(i)}, z^{(i)}\right)$ of the problems (2.39)–(2.42), (2.48), (2.50), (2.51) or (2.39)–(2.42), (2.48), (2.50), (2.52) corresponds to each admissible solution $\left(\Delta A^{(i)}, \Delta q^{(i)}\right)$, $i = \overline{1,2}$, of the problems (2.38)–(2.43) or (2.38)–(2.42), (2.44), the relationship $z^{(1)} = z^{(2)} = z^*$ being true by virtue of (2.49) and (2.50).

The matrix $E - (A + \Delta A_\lambda)$ is also nonsingular, and a unique admissible solution $\left(\Delta A_\lambda, \Delta q_\lambda, z_\lambda\right)$ of the transformed problems also corresponds to the admissible solution $\left(\Delta A_\lambda, \Delta q_\lambda\right)$ of the problems (2.38)–(2.43) or (2.38)–(2.42), (2.44). Therefore, it is possible to apply Theorem 2.3 for $x^{(1)} = \left(\Delta A^{(1)}, \Delta q^{(1)}, z^*\right)$, $x^{(2)} = \left(\Delta A^{(2)}, \Delta q^{(2)}, z^*\right)$ and $x(\lambda) = \left(\Delta A_\lambda, \Delta q_\lambda, z_\lambda\right)$. According to this theorem, $\overline{f}_i(z^*) = \overline{f}_i(z_\lambda)$, $i = \overline{1,2}$. Since the corresponding problems are equivalent in their original and rearranged forms, the equality $f_i\left(\Delta A^{(i)}, \Delta q^{(i)}\right) = \overline{f}_i(z^*)$, $i = \overline{1,2}$ should hold. Herefrom, $f_i\left(\Delta A^{(i)}, \Delta q^{(i)}\right) = f_i\left(\Delta A_\lambda, \Delta q_\lambda\right)$, $i = \overline{1,2}$, which proves the theorem.

According to Theorem 2.3, the set $X_1$ of the values of variables $\Delta A$ and $\Delta q$ admissible for given values of $z$ is convex. This simplifies both constructing this set and searching for extremum points of some additional criteria on it, for example, of the function $f_3 = (\Delta A, \Delta q)$ defined according to (2.45).

Implementing, from several initial points, the procedure of the subgradient method with space dilatation using values of the objective function [39] ($\overline{f}_i(z)$, $i = \overline{1,2}$, are known if $z$ is given), we obtain several points of the above-mentioned set. By virtue of Theorem 2.3, their convex linear combination also belongs to this set. Constructive algorithms of enumerating various points from $X_1$ can also be constructed based on subgradient methods that use external approximation of the set of extrema by ellipsoids, for example, on the method of finding an admissible point of a system of convex inequalities [41], which provides an accelerated convergence to boundary points of the set $X_1$. The approximation of $X_1$ by an ellipsoid, constructed at the previous step of this method, can be used to find the next boundary point of $X_1$.

Note that the components of the vector $z$ reflect the structure of ultimate incomes obtained from various forms of economic activity. Such a structure determines a parity of interests (existing or desirable) among various economic entities. Performing alternative computations (for different $z$) and comparing the resultant values with the solutions of the problems (2.38)–(2.43) and (2.38)–(2.42), (2.44) will make it possible to estimate the influence of partial interests on the efficiency of economic development and to predict whether economic entities accept or oppose a solution (globally optimal or close to it). The models (2.39)–(2.42), (2.48), (2.50), (2.51) and (2.39)–(2.42), (2.48), (2.50), (2.52) considered for given $z$ are models with fixed objectives, where the results are determined but a set of tools (i.e., of measures to achieve these results) should be constructed. Using such models is important for both a preliminary analysis of the situation and final selection of solutions obtained earlier. From this standpoint, generating the set of admissible solutions of the problems (2.39)–(2.42), (2.48), (2.50), (2.51) and (2.39)–(2.42), (2.48), (2.50), (2.52) with the values of objective functions sufficiently close to the global maximum and carrying out interactive computations play an important role. Multiprocessor computers (SKIT-2 and SKIT-3 clusters) and a computation-parallelizing scheme according to which extremum from each initial approximation is searched for at a separate processor can be used.

Rearranging the problems (2.39)–(2.42), (2.48), (2.50), (2.51) and (2.39)–(2.42), (2.48), (2.50), (2.52), it is possible to obtain sufficiently exact estimates of their objective functions at the point of global extremum. Such information is extremely valuable in selecting the solutions obtained.

Changing the variables $\Delta a_{ij} = \Delta a_{ij}^{+} - \Delta a_{ij}^{-}$ and rearranging the objective function (2.51) as $v\left(1 - z^T \alpha\right) = z^T h$, $\overline{f}_1 = v \to \max$ make it possible to consider the problems (2.39)–(2.42), (2.48), (2.50), (2.51) and (2.39)–(2.42), (2.48), (2.50), (2.52) as special cases of the problem of maximizing a linear function

$$Q(x) = (c, x) \to \max \tag{2.55}$$

under quadratic and linear constraints of the form

$$x^T H_1 x + (b_1, x) = q_1, \tag{2.56}$$

$$x^T H_2 x + (b_2, x) = q_2, \tag{2.57}$$

$$x \geq 0, \tag{2.58}$$

where $H_1$ and $H_2$ are some square matrices, $b_1$ and $b_2$ are vectors, and $x$ is the vector of variables.

To obtain upper estimates of the maximum value of the objective function $Q^*$ of the problem (2.55)–(2.58), Lagrange dual estimates $\psi^*$ can be used. Finding them is reduced to minimizing nonsmooth convex functions defined on a family of nonpositive definite symmetric matrices linearly depending on Lagrangian multipliers that correspond to constraints (2.56) and (2.57). The duality may discontinue, i.e., $\Delta^* = \psi^* - Q^*$ will be positive and the estimate $\psi^*$ be insufficiently exact.

One of the ways of reducing $\Delta^*$ involves functionally redundant constraints (which may also increase the number of variables of the problem (2.55)–(2.58)). According to [40] functionally redundant are the constraints that do not change the sets of admissible solutions of the problem (2.55)–(2.58). However, this changes the form of the Lagrangian function, which may decrease $\Delta^*$.

An elementary example of functionally redundant constraints is quadratic corollaries of linear constraints, for example, a quadratic constraint

$$\left(b_1^T x + c_1\right)\left(b_2^T x + c_2\right) \geq 0$$

follows from two linear inequalities $b_1^T x + c_1 \geq 0$ and $b_2^T x + c_2 \geq 0$. Numerous linear constraints present in the problems considered earlier facilitate the application of such an approach to solve them. More complicated functionally redundant constraints considered in [46] may also be applied. In particular, if there are two-sided constraints for three variables of the problem, $0 \leq x_1 \leq 1$, $0 \leq x_2 \leq 1$, and $0 \leq x_3 \leq 1$, functionally redundant constraints can be written as the inequalities

$$\begin{aligned} x_1x_2 + x_1x_3 + x_2x_3 - x_1 - x_2 - x_3 &\geq -1, \\ -x_1x_2 - x_1x_3 + x_2x_3 + x_1 &\geq 0, \\ -x_1x_2 + x_1x_3 - x_2x_3 + x_2 &\geq 0, \\ x_1x_2 - x_1x_3 - x_2x_3 + x_3 &\geq 0. \end{aligned}$$

This approach can easily be generalized to the case of an arbitrary number of two-sided constraints for variables. Note that such constraints are typical for the models earlier considered, which substantially increases the accuracy of the Lagrangian estimates $\psi^*$ for the maximum values of their objective functions. Moreover, introducing new variables by means of quadratic equalities of the form $y_{ij} = x_i x_j$ substantially expands the family of functionally redundant constraints being used and thus increases the accuracy of the estimates $\psi^*$.

When comparing the approaches to performing model computations presented in this and previous subsections, we should note their advantages and shortcomings.

An approach based on solving the problems (2.38)–(2.43) and (2.38)–(2.42), (2.44) without their transformation substantially reduces the problem dimension. Since all the constraints have the form of inequalities, an exact solution of the problem can be obtained for finite values of penalty coefficients, which restricts the ravine property of the function being maximized. The set of admissible solutions of the problem is convex. In some cases, this facilitates finding the first admissible solution. At the same time, the possibilities of the analysis of the solutions obtained with this approach are bounded.

The approach stated in this subsection somewhat increases the dimension of the problem being solved. The set of admissible solutions of the problem becomes nonconvex, and nonlinear (quadratic) equality constraints needs large (theoretically infinite) values of the corresponding penalty coefficients, which increases the ravine property of the penalty function. At the same time, the transformed problem does not use incompletely defined functions, and problems because of abandoning the domain of their definition when solving the problem are excluded. The approach proposed gives ample opportunities for the analysis of solutions and considering models with fixed objectives and allows estimating the values of the objective function at the point of its global maximum.

Thus, both approaches complement each other: therefore, its simultaneous application is expedient in model computations.

### 2.5.3 Software Development

Software is an important component of modern information technologies. The software for decision support in planning structural and technological changes is based on the IOMSTC (Intersectoral Optimization Models of Structural and Technological Changes) software system created at the Institute of Cybernetics of the National Academy of Sciences of Ukraine. The system is implemented in the DELPHI medium, which provides its flexibility and convenience for the user.

IOMSTC consists of the user interface, base of models, and a set of program modules implementing optimization algorithms. The user interface leans upon a developed system of windows and hierarchical menus. It allows reading out initial data of the model from external files (including Excel spreadsheets and Word documents executed in a preconcerted format), forming input files of the model (the matrix $A$, vectors $q$, $h$, $\alpha$, $l$, $d$, coefficients and the right-hand sides of resource constraints, etc.) based on these data, editing initial data of problems composed earlier and newly formulated, selecting parameters of optimization algorithms, visualizing and analyzing results of model computations, including their comparison with actual ones in order to verify the model.

A specific feature of the user interface is the presence of two types of windows in its structure: for a user being an expert in economy and for an expert in optimization methods. Windows of the first type allow forming new models based on data of external files and (or) existing models, renaming the existing models, editing initial data (including during dialogue computations considered in the previous subsections), viewing, analyzing, and saving (as a part of the problem or on external files) the results of model computations. The user can also choose the objective function (the aim of the computations). As alternatives to be chosen, the first version of the system considers the following:

(i) maximization of ultimate incomes of consumers (the objective function $f_1(\Delta A, \Delta q)$ corresponds to it);
(ii) maximization of the multiplier "increase of incomes–increase of output" ($f_2(\Delta A, \Delta q)$ corresponds to it);
(iii) finding an arbitrary point of the set of admissible solutions (arbitrary admissible solution).

Windows of the second type allow viewing the default values of the parameters of the optimization algorithm, changing the values of these parameters if necessary, obtaining information on the problem time, on the value of the objective function, and the residuals of constraints at the extremum point and comparing them with similar parameters for the initial approximation. If searching for extremum is terminated earlier than acceptable accuracy is achieved (e.g., if the objective function appeared not determined for the next current approximation), the user can find out the reason of such actions of the system. The user (an expert in optimization methods) may also change the number of initial approximations and the rule of their choice (by default, initial approximations are chosen similar to that described in Sect. 2.5.1). The structure of windows of the second type also includes a command line where the name of the program module implementing the optimization algorithm is specified. This should be an arbitrary loading module (with the.exe extension) that uses input data from files of the internal (for the system) representations of the model.

As these moduli, the first version of the system used the MULSTR program described in [37] and MULSTR1 implementing the approach to carrying out model computations stated in Sect. 2.5.2. The problems (2.39)–(2.42), (2.48), (2.50), (2.51) and (2.39)–(2.42), (2.48), (2.50), (2.52) are solved by the method of penalty functions followed by the application of the $r$-algorithm. The MULSTR1 program uses three penalty factors (coefficients) for different groups of constraints. The first factor is used for quadratic equality constraints (2.50). It should be considered separately since, as mentioned earlier, the value of this factor should be sufficiently large to obtain an exact solution of the problem. The second factor is used for constraints (2.39), (2.40), and (2.48), and the third for nonsmooth resource constraints(2.42). The two-sided constraints (2.41) imposed on $\Delta A$ and $\Delta q$ are not included in the penalty function and are taken into account by an "even" periodic continuation of the objective function onto the domain of inadmissible values of variables [39]. This approach has proved to be efficient in searching for extremum of a wide class of nonsmooth functions. The MULSTR program intended to search for extremum under inequality constraints uses only the second and third penalty factors; therefore, if the name of this program appears in the command line, the user's attempts to view and change the value of the first factor are blocked.

The controlled parameters of the $r$-algorithm which can be changed by the user for the MULSTR and MULSTR1 programs are: the initial value $h_0$ of the step-by-step factor, the maximum number of iterations of the algorithm, a precision factor $\varepsilon_g$ used in a stopping criterion based on the variation in the argument, a precision factor $\varepsilon_g$ used in a stopping criterion based on the norm of the subgradient of the function being maximized, a space dilation coefficient $\alpha$, and parameters $q_1$, $q_2$, and $n_k$ of adaptive adjustment of a step-by-step factor. The user may vary all these parameters within the limits corresponding to theoretical recommendations for the $r$-algorithm.

A peculiar feature of the IOMSTC system is an internal base of models. Each model is a set of files of standard structure that contain information on the initial data, the method used for the computations, parameters of the computation algorithm, and upon completion of the computations—about their results (sequence of locally optimal solutions). Each model has an individual name given by the user. When a new model is created on the basis of the existing one, all the information contained in files of the base model will be copied to files of the new model. In the case of renaming model, all its files will also be renamed automatically.

The further trends in developing the IOMSTC system are as follows:

(i) extending the list of the applied optimization algorithms due to implementing modern methods developed at the Institute of Cybernetics NASU and in other domestic and foreign scientific centers;
(ii) enhancing service capabilities, including the recognition of handwriting, speech dialogue, etc.

Along with the National Agency on the Efficient Use of Energy Resources, potential users of the system may include divisions of the Ministry of Economic Affairs and Ministry of Industrial Policy of Ukraine involved in preparing structural reforms.

In view of the application domain of the system, the sensitivity analysis of solutions against the variation of ambiguously determined parameters of models becomes of great importance. To this end, numerical experiments have been carried out for a seven-branch model, the results being presented in Table 2.14.

As follows from the results obtained, solutions are most sensitive to variations in the parameters $\underline{\gamma_{ij}}$ and $\overline{\gamma_{ij}}$. The model with the criterion $f_1(\Delta A, \Delta q)$ is a little more sensitive to variations in $\beta$ than to variations in the right-hand sides

**Table 2.14** Variation of optimal values of objective function with the variations in model parameters

| Model parameters | Range of variations | Variation of objective functions of the model (% of the largest value) | |
|---|---|---|---|
| | | $f_1(\Delta A, \Delta q)$ | $f_2(\Delta A, \Delta q)$ |
| $\underline{\gamma_{ij}}$ and $\overline{\gamma_{ij}}$ | 10–50 % of initial values of coefficient | 72.9–100 | 90.4–100 |
| $\beta$ | 0.9–1.0 | 77.3–100 | 91.9–100 |
| $B_k$ | 1.0–3.5 | 80.9–100 | 92.9–100 |

of the resource constraints $B_k$; for the model with the criterion $f_2(\Delta A, \Delta q)$, the sensitivity to variations of parameters taken into account is approximately the same. In general, the influence of variations in parameters of the model on optimal values of the objective functions is relatively small. If all the parameters are varied simultaneously, the range of values of the objective functions is 44–100 % for $f_1(\Delta A, \Delta q)$ and 73.5–100 % for $f_2(\Delta A, \Delta q)$. Similar results have been obtained for models of greater dimensions.

Enhancing the models being used is also an important trend in further development of the IOMSTC. In the next subsection, we consider the models planned to be included in the upgraded version of the system.

### 2.5.4 Modified Intersectoral Models of Structural and Technological Changes

Let us consider the models being the further development of the models (2.38)–(2.43) and (2.38)–(2.42), (2.44) presented in Sect. 2.5.1. The first of these models differs from those considered earlier in the following aspects.

1. Constraints on the available fuel and energy resources are provided, and the possibility of changing unit costs of these resources for the production of branches is taken into account.
2. It is provided that structural and technological changes in some branches (first of all, in electric power industry) can be carried out by two ways:
   (a) by changing the intensity of the use of manufacturing techniques available for the ultimate product of the branch (e.g., increasing the share of thermal engineering due to decreasing the share of atomic power engineering or vice versa);
   (b) by changing the cost structure for available technologies due to introducing technical and technological innovations.

   Introducing new technologies can be considered as a special case of the first of these ways since the intensity of using the corresponding technology was equal to zero at the beginning of its introduction and became a positive number after the introduction.
3. The model takes into account ecological constraints, including constraints for the emission of greenhouse gases.

The model is static; however, its dynamic analogs can be developed in the future. Let us consider the basic assumptions and introduce the notation of the model.

Let the economy consist of $n$ pure branches where structural and technological changes are possible. For a branch $j_0$ (if this is electric power industry, $j_0 = 17$ by the existing nomenclature of intersectoral balance), these changes can be detailed as variations in the intensity of the use of technologies and as variations in the cost structure of forms of production for technologies. Denote by $K_{j_0}$ the number of technologies used in the branch $j_0$ and number these technologies by $k$, $k = \overline{1, K_{j_0}}$. Let $W_k^{(j_0)}$ be the intensity of using the technology $k$ in the branch $j_0$. Then assume that

$$\sum_{k=1}^{K_{j_0}} W_k^{(j_0)} = 1.$$

Denote by $a_{ij}$ the coefficients of direct expenses of production of the branch $i$ for manufacturing a unit product of the branch $j$ (coefficients of the Leontief matrix), $i = \overline{1,n}$, $j = \overline{1,n}$, $j \neq j_0$. Denote by $\overline{a}_{ij_0}^k$ specific expenses of production of the branch $i$ for manufacturing a unit product of the branch $j_0$ using the technology $k$. The corresponding elements of the column $j_0$ of the Leontief matrix are defined as follows:

$$a_{ij_0} = \sum_{k=1}^{K_{j_0}} \overline{a}_{ij_0}^k W_k^{(j_0)}. \tag{2.59}$$

Denote by $q_j$ the share of ultimate incomes in the price of the product of the branch $j$, $j = \overline{1,n}$. Assume that the share of added value $\overline{q}_j$ in the price of the product of the branch $j$ linearly depends on $q_j$:

$$\overline{q}_j = l_j q_j + d_j, \quad j = \overline{1,n}. \tag{2.60}$$

The known coefficients $l_j$ and $d_j$ reflect the value of the fiscal multiplier of ultimate incomes and the share of other components of the added value in the price of the product of the branch $j$. Assume that ultimate consumption $y_i$ of each branch linearly depend on the income of ultimate users inside the system being modeled

$$y_i = \alpha_i D + h_i, \quad i = \overline{1,n}, \tag{2.61}$$

where $\alpha_i$ and $h_i$ are some known coefficients.

The purpose of modeling, similar to the problem (2.38)–(2.43), is to determine the ways of structural and technological changes that increase the efficiency of the economy. In view of this, the model has several objective functions; performing computations can deal with one-criterion optimization, where extremum of one objective function selected from all others is determined, or with a multicriterion problem, where all the objective functions are taken into account simultaneously. To solve such a problem, a combination of the methods of convolutions and constraints considered in [27] can be used.

Denote by $g_j$ the vector of direct unit costs of bounded resources in the branch $j$,$j = \overline{1,n}$, the value of $g_{j_0}$ being determined from unit costs of resources for separate technologies $g_{j_0}^k$ and the intensity of use of technologies $W_k^{(j_0)}$ similar to (2.59):

$$g_{j_0} = \sum_{k=1}^{K_{j_0}} g_{j_0}^k W_k^{(j_0)}. \tag{2.62}$$

Let $G$ be the vector of available resources. In what follows, by $g$ we will denote the matrix consisting of columns $g_j$, $j = \overline{1,n}$.

Denote by $\tilde{g}_j$ the vector of specific emissions of polluting substances for the branch $j$. The vector $\tilde{g}_{j_0}$ is computed similar to (2.62) based on specific emissions of polluting substances for separate technologies $\hat{g}_{j_0}^k$ and the intensity of use of these technologies $W_k^{(j_0)}$:

$$\tilde{g}_{j_0} = \sum_{k=1}^{K_{j_0}} \hat{g}_{j_0}^k W_k^{(j_0)}. \tag{2.63}$$

Denote by $\tilde{G}$ the vector of maximum admissible emissions of polluting substances and by $\tilde{g}$ the matrix created by the column vectors $\tilde{g}_j$.

Along with the resources used during the production, the model takes into account the resources used to change the cost structure of the branch. Such resources are assumed to be spent only to reduce the coefficients of costs of product $a_{ij}$, $\overline{a}_{ij_0}^k$ and costs of resources $g_j$, $\overline{g}_{ij_0}^k$. The relationship between the decrease in the above-mentioned parameters and costs of resources is linear proportionality with coefficients $b_{vij}$, $\overline{b}_{vik}$, $\overline{b}_{vj}$, and $\overline{b}_{vk}$, where $v$ is the resource number. Denote by $B_v$ the amount of the resource $v$.

The task is to determine the changes of the coefficients of direct costs $\Delta a_{ij}$, $i = \overline{1,n}$, $j = \overline{1,n}$, $\Delta_{ij}$ $j \neq j_0$, the value $\Delta\overline{a}_{ij_0}^k$ of unit costs for separate technologies that constitute the branch $j_0$, the intensity of use of technologies $\Delta W_k^{(j_0)}$, $k = \overline{1, K_{j_0}}$, the share of ultimate incomes in the price of the product $\Delta q_j$ and unit costs of the resources $\Delta g_j$, $\Delta\overline{g}_{j_0}^k$, $j = \overline{1,n}$, $j \neq j_0$, $k = \overline{1, K_{j_0}}$, satisfying the following conditions.

1. The equalities that determine $\Delta a_{ij_0}$ and $\Delta g_{j_0}$ and follow from (2.59) and (2.62):

$$\Delta a_{ij_0} = \sum_{k=1}^{K_{j_0}} \left( \left( \overline{a}_{ij_0}^k + \Delta\overline{a}_{ij_0}^k \right) \left( W_k^{(j_0)} + \Delta W_k^{(j_0)} \right) - \overline{a}_{ij_0}^k W_k^{(j_0)} \right),$$

$$\Delta g_{j_0} = \sum_{k=1}^{K_{j_0}} \left( \left( \overline{g}_{j_0}^k + \Delta\overline{g}_{j_0}^k \right) \left( W_k^{(j_0)} + \Delta W_k^{(j_0)} \right) - \overline{g}_{j_0}^k W_k^{(j_0)} \right), \tag{2.64}$$

$$\tilde{g}_{j_0} = \sum_{k=1}^{K_{j_0}} \Delta\hat{g}_{j_0}^k \left( W_k^{(j_0)} + \Delta W_k^{(j_0)} \right), \quad j = \overline{1,n}.$$

2. Constraints for the range of variation of variables of the model determined from technological considerations:

$$\gamma_{ij} \Rightarrow \leq \Delta a_{ij} \leq \overline{\gamma}_{ij}, \quad \underline{\delta}_j \leq \Delta g_j \leq \overline{\delta}_j, \quad i, j = \overline{1, n},$$

$$\begin{aligned} \underline{W}_k^{(j_0)} \leq \Delta W_k^{(j_0)} \leq \overline{W}_k^{(j_0)}, \quad k = \overline{1, K_{j_0}}, \\ \tilde{\gamma}_{ik} \leq \Delta \overline{a}_{ij_0}^k \leq \tilde{\overline{\gamma}}_{ik}, \quad i = \overline{1, n}, \quad k = \overline{1, K_{j_0}}, \\ \underline{\tilde{\delta}}_k \leq \Delta \overline{g}_{j_0}^k \leq \underline{\tilde{\delta}}_k, \quad k = \overline{1, K_{j_0}}, \\ 0 \leq q_j + \Delta q_j \leq 1, \quad i = \overline{1, n}, \\ 0 \leq a_{ij} + \Delta a_{ij} \leq 1, \quad i, j = \overline{1, n}, \end{aligned} \tag{2.65}$$

where $\gamma_{ij} \Rightarrow$, $\overline{\gamma}_{ij}$, $\underline{\delta}_j$, $\overline{\delta}_j$, $\underline{W}_k^{(j_0)}$, $\overline{W}_k^{(j_0)}$, $\tilde{\gamma}_{ik}$, $\tilde{\overline{\gamma}}_{ik}$, $\underline{\tilde{\delta}}_k$, and $\overline{\tilde{\delta}}_k$ are some predetermined quantities.

3. The constraints following from the definition of the concept "the share of added cost in the price of a product":

$$0 \leq a_{jj} + \Delta a_{jj} + l_j \left(q_i + \Delta q_j\right) + d_j \leq 1, \quad j = \overline{1, n}. \tag{2.66}$$

4. The constraints that exclude intensification of the inflation of costs and are similar to those considered in Sect. 2.5.1:

$$\sum_{i=1, i \neq j}^{n} \frac{a_{ij} + \Delta a_{ij}}{1 - \left(a_{jj} + \Delta a_{jj}\right) - \left(l_j \left(q_j + \Delta q_j\right) + d_j\right)} \leq \beta, \quad j = \overline{1, n}, \tag{2.67}$$

where $\beta \leq 1$ is a predetermined quantity.

5. The resource constraints for structural and technological changes

$$\begin{aligned} &\sum_{i=1}^{n} \sum_{j=1}^{n} b_{vij} \max\left(0, -\Delta a_{ij}\right) + \sum_{i=1}^{n} \sum_{k=1}^{K_{j_0}} \tilde{b}_{vik} \max\left(0, -\Delta \overline{a}_{ij_0}^k\right) \\ &+ \sum_{j=1, j \neq j_0}^{n} \overline{b}_{vj} \max\left(0, -\Delta g_j\right) + \sum_{k=1}^{K_{j_0}} \overline{b}_{vk} \max\left(0, -\Delta g_{j_0}^k\right) \leq B_v. \end{aligned} \tag{2.68}$$

6. The equality that determines the profit of ultimate users $D$ after structural and technological changes:

$$D = \frac{(q + \Delta q)^T \left(E - (A + \Delta A)\right)^{-1} h}{1 - (q + \Delta q)^T \left(E - (A + \Delta A)\right)^{-1} \alpha}, \tag{2.69}$$

where $A = \{a_{ij}\}$ is the Leontief matrix, $i, j = \overline{1, n}$; $\Delta A = \{\Delta a_{ij}\}$ is the matrix of changes for $A$, $i, j = \overline{1, n}$; $E$ is a unit matrix; $q = (q_1, \ldots, q_n)$; $\Delta q = (\Delta q_1, \ldots, \Delta q_n)$; $h = (h_1, \ldots, h_n)$, $\alpha = (\alpha_1, \ldots, \alpha_n)$; $T$ is the operation of transposition.

7. The constraints for resources used during the production

$$(g + \Delta g)^T \left(E - (A + \Delta A)\right)^{-1} (\alpha D + h) \le G, \tag{2.70}$$

where $\Delta g$ is a matrix of column vectors $\Delta g_j$, $j = \overline{1, n}$.

8. Constraints for the emission of polluting substances

$$\tilde{g}^T \left(E - (A + \Delta A)\right)^{-1} (\alpha D + h) \le \tilde{G}. \tag{2.71}$$

9. The sum of new values of the intensities of use of technologies should be equal to unity (gross variations of these intensities are zero)

$$\sum_{k=1}^{K_{j0}} \Delta W_k^{(j_0)} = 0. \tag{2.72}$$

Among all the admissible solutions satisfying relationships (2.64)–(2.72), it is necessary to choose such that

(a) maximize ultimate incomes of consumers

$$F_1 = D \to \max; \tag{2.73}$$

(b) maximize the value of the multiplier "increase of incomes–increase of output"

$$F_2 = (q + \Delta q)^T \left(E - (A + \Delta A)\right)^{-1} \alpha^{(1)} \to \max, \tag{2.74}$$

where $\alpha^{(1)} = \left(\alpha_1^{(1)}, \ldots, \alpha_n^{(1)}\right)$ is a predetermined numerical vector;

(c) minimize the costs of especially scarce resource with industrial purpose

$$F_3 = \left(g^{(1)} + \Delta g^{(1)}\right)^T \left(E - (A + \Delta A)\right)^{-1} (\alpha D + h) \to \min, \tag{2.75}$$

where $g^{(1)}$ is a row of the matrix $g$ containing coefficients of unit costs of this resource, $\Delta g^{(1)}$ is the corresponding row of the matrix $\Delta g$.

Note that the resultant model (2.64)–(2.75) is a rather complex optimization problem. The objective functions $F_2$ and $F_3$ and constraints (2.64), (2.67)–(2.71) are nonlinear. A part of these constraints are equalities, which makes the set of admissible solutions nonconvex and additionally complicates the problem; the functions $F_2$ and $F_3$ are nonconvex too. Therefore, it is necessary to develop specialized algorithms to solve such a problem.

Note that the optimization problem (2.64)–(2.75) belongs to the same class as the problems (2.38)–(2.43) and (2.38)–(2.42), (2.44) do. Therefore, the above-mentioned algorithms can be developed based on the approaches stated in the previous subsections.

In the second of the considered models, the prices for the product manufactured are considered to be a tool of varying the coefficients of direct costs.

In spite of the fact that the capabilities of a state to control prices are limited under conditions of transitive (and the more so market) economy, the price regulation remains to be an efficient tool of economic policy. In particular, it can be used for technological and structural changes, stimulation of energy- and resource-saving. In this connection, of interest is to develop models similar to (2.38)–(2.43) and (2.38)–(2.42), (2.44), in which the controlled parameters influencing the coefficients of direct costs are the prices for production of branches. Below, we consider one of such models.

Like in the problems considered earlier, assume that the national economy consists of $n$ branches. Denote by $a_{ij}^0$, $i, j = \overline{1,n}$, the coefficients of direct costs in value terms, computed for the prices valid at the moment of calculating the intersectoral balance. It is necessary to determine expedient changes of these prices that maximize the ultimate incomes of consumers or the multiplier.

Taking into account the formulation of the problem, denote by $p_i$ the changed price index for the production of the branch $i$ ($i = \overline{1,n}$) related to the price valid at the moment of calculating the balance, and by $\tilde{q}_i$ the share of incomes of consumers in the price of the production of the branch $i$. The last quantity is expressed in terms of the original price of this production. Similar to the previous models, assume that the value $q_i$ of the added cost in the cost of a unit of production of the branch $i$ linearly depends on $q_i$:

$$\tilde{q}_i = l_i q_i + d_i, \quad i = \overline{1,n},$$

where $l_i > 0$ and $d_i > 0$ are some given coefficients having the same meaning as for the previous models.

The equations of intersectoral balance for prices [31]

$$p_j = \sum_{i=1}^{n} a_{ij}^0 p_i + \tilde{q}_j, \quad j = \overline{1,n},$$

yield the relationships

$$p_j = \sum_{i=1}^{n} a_{ij}^0 p_i + l_j q_j + d_j, \quad j = \overline{1,n}, \tag{2.76}$$

that form the first group of constraints of the model under study.

The coefficients of direct costs $a_{ij}(p)$ for the established prices, depending on the price indices $p = (p_1, \ldots, p_n)$, are determined from $a_{ij}^0$ as follows:

$$a_{ij}(p) = \frac{a_{ij}^0 p_i}{p_j}, \quad i, j = \overline{1,n}. \tag{2.77}$$

By virtue of the law of value, price variation should not change the sum of price indices. Since the equality $p_i = 1$, $i = \overline{1,n}$, was true for prices of the balance, the current values of $p_i$ should satisfy the constraint

$$\sum_{i=1}^{n} p_i = n. \tag{2.78}$$

The values of the model variables $p_i$ and $q_i$ should be non-negative:

$$p_i \geq 0, \quad q_i \geq 0, \quad i = \overline{1,n}. \tag{2.79}$$

Among $p = (p_1, \ldots, p_n)$ and $q = (q_1, \ldots, q_n)$ satisfying conditions (2.76)–(2.79), it is required to find those maximizing the ultimate income of consumers

$$\tilde{f}_1(p, q) = \frac{q^T \left(E - A(p)\right)^{-1} h}{1 - q^T \left(E - A(p)\right)^{-1} \alpha} \to \max, \tag{2.80}$$

or the value of the multiplier "increase of incomes–increase of output"

$$\tilde{f}_2(p, q) = q^T \left(E - A(p)\right)^{-1} \alpha^1 \to \max. \tag{2.81}$$

In (2.80), (2.81), we use the same notation as in (2.43), (2.44), $A(p) = \{a_{ij}(p)\}$.

For the problems (2.76)–(2.80) and (2.76)–(2.79), (2.81) to be formulated correctly, it is required that the matrix $\left(E - A(p)\right)^{-1}$ exists at any point of the set of admissible solutions. To this end, it is possible to use the necessary and sufficient existence condition for inverse matrix [31]

$$1 > \lambda\left(A(p)\right), \tag{2.82}$$

where $\lambda\Big(A(p)\Big)$ is the Frobenius number of the matrix $A(p)$ (a real positive eigenvalue of this matrix, whose absolute value is maximum among all the eigenvalues).

Taking into account inevitable discrepancy of the initial data and the necessity of creating reserves during management decision-making, it is expedient to consider the following condition instead of (2.82) as the model constraint:

$$\lambda\Big(A(p)\Big) \le \beta_1, \tag{2.83}$$

where $0 < \beta_1 < 1$ is a preset number sufficiently close to unity.

Along with condition (2.83), it is possible to use the sufficient existence conditions for the matrix $\Big(E - A(p)\Big)^{-1}$, for example,

$$\sum_{i=1}^{n} a_{ij}(p) < 1, \quad j = \overline{1,n}$$

or

$$\sum_{i=1}^{n} a_{ij}(p) < \beta_1 < 1, \quad j = \overline{1,n}. \tag{2.84}$$

Conditions (2.84) specify a narrower domain of solutions compared with (2.83); however, being linear inequalities, they simplify the problem.

The model may also include two-sided constraints for possible ranges of variation of $p_i$ and $q_i$:

$$\underline{p_i} \le p_i \le \overline{p_i}, \quad \underline{q_i} \le q_i \le \overline{q_i}, \quad i = \overline{1,n}. \tag{2.85}$$

The optimization problems obtained in this case have nonconvex objective functions, and if condition (2.83) is taken into account, a nonlinear constraint as well. To solve them, the approaches considered in Sects. 2.5.1 and 2.5.2 can be applied. Because of the presence of equality constraints, of interest is the compatibility of constraints (2.76) and (2.78). Basically, these constraints should be satisfied for $p_i = 1$, $i = \overline{1,n}$, and for $q_i$ that correspond to the share of incomes of consumers in the price of production of the branches at the moment of calculating the balance. However, since model parameters, for example, $l_i$ and $d_i$, $i = \overline{1,n}$, vary depending on the fiscal policy, this satisfaction may be violated and there may be necessity to check quickly the constraints for compatibility. This can be executed as follows. As follows from (2.76),

$$p = \Big(E - A^T\Big)^{-1}\Big(\langle l, q\rangle + d\Big), \tag{2.86}$$

where $A = \{a_{ij}^0\}$, $\langle l, q \rangle$ is the componentwise product of the vectors $l = (l_1, \ldots, l_n)$ and $q = (q_1, \ldots, q_n)$, $d = (d_1, \ldots, d_n)$. According to (2.86), the dependence of $p$ on $q$ is continuous. Components of the matrix $\left(E - A^T\right)^{-1}$ (total expenditure coefficients), all the $l_i$ and $d_i$ are non-negative. Therefore, for $q$ such that the indices $p = (p_1, \ldots, p_n)$ defined by (2.86) satisfy (2.78) to exist, it is necessary and sufficient that the inequality

$$\tilde{e}^T \left(E - A^T\right)^{-1} d \leq n$$

holds and there exist admissible values of $q = q^1$ such that the following inequality holds:

$$\tilde{e}^T \left(E - A^T\right)^{-1} \left(\langle l, q \rangle + d\right) \leq n,$$

where $\tilde{e}$ is an $n$-dimensional vector consisting of unities.

The models (2.64)–(2.75) and (2.76)–(2.81), (2.83)–(2.85) are also planned to be included in the next versions of the IOMSTC system considered in Sect. 2.5.3.

### 2.5.5 *Conclusions and Fields of Further Studies*

The obtained results allow making the following conclusions.

1. Intersectoral optimization models with variable coefficients of direct costs are an effective decision support tool for planning structural reforms. They allow determining basic trends of structural and technological changes to increase the efficiency of the economy, to reduce the industrial consumption of energy resources, and to increase the real incomes of ultimate users without significant inflationary effects.
2. Performing computations with the use of the above-mentioned models necessitates searching for a conditional extremum of nonconvex functions defined with the use of inversion of the matrix of variables. Two approaches are proposed to develop solution algorithms for such problems. The first of them is based on using a differentiation formula for implicitly defined functions, the second—on the problem transformation by introducing additional variables. In both cases, the problem reduces to maximizing a nonsmooth penalty function by the $r$-algorithm. Since the resultant penalty functions are nonconvex, it is proposed to search for points of their optimum from several initial approximations. The search can easily be parallelized and implemented with a modern multiprocessor equipment, in particular, on the SKIT-2 and SKIT-3 clusters developed at the Institute of Cybernetics of NASU.

3. Both approaches to the solution of optimization problems have advantages and shortcomings. In particular, within the framework of the first approach, the problem dimension can be substantially reduced due to eliminating "insignificant" components of the matrix of direct cost coefficients, at the same time, the second approach increases the dimension due to additional variables. However, since these variables can be interpreted as parameters of the branch structure of ultimate incomes, the latter approach provides wider opportunities for the analysis of the solution, estimation of the influence of the developed or desirable structure of incomes (i.e., interests of separate economic entities) on the choice of trends of structural and technological changes and revealing the contradictions arising in this case. It is proved for models that for the specified structure of incomes, the set of admissible solutions is convex, which simplifies such an analysis.
4. The models proposed together with the necessary software and dataware and a worked-out application form a basis of the information technologies of management decision support implemented in a specialized DSS. The IOMSTC system has been developed at the Institute of Cybernetics of NASU as a prototype version of such a DSS.
5. An upgraded version of the IOMSTC system is planned to include intersectoral optimization models in view of intrabranch technological structure and the model of influence of price factors. This leads to solving more complex optimization problems, but the approaches considered earlier can be applied to construct computation algorithms for them. In our opinion, further studies in this field will be carried out in the following directions:

   – developing computation algorithms based on advanced models;
   – creating components of software and dataware for these computations;
   – expanding service capabilities of the IOMSTC system and means of its switching to multiprocessor computers;
   – generalizing the experience of applying the above-mentioned models for management decision support.

## 2.6 Expert Assessments Processing by Ordinal Regression Methods

In order to build the fuzzy set it is necessary to determine the universe of elements $X$ and the membership function $\mu(x) \in [0; 1]$. For any $\in$ the value $\mu(x)$ determines the degree of element's $x$ membership in the fuzzy set. The higher values of the membership function correspond with the elements which posses a higher degree of membership with this set. The building of $\mu(x)$ can be based on the expert assessments of the membership degree of separate elements. Generally, these are ordinal assessments of the following type: "the element $x^{(1)}$ has a higher (smaller) degree of membership with the set than the element $x^{(2)}$," that is why the methods of their processing with the purpose of building the function $\mu(x)$ in its analytical form are called ordinal regression methods [24].

The building of $\mu(x)$ by these methods consists of two stages.

1. We build an arbitrary, bounded on $X$ function $F(x)$ which is the membership degree function.
2. $F(x)$ is transformed in such a way that its values belong to the interval [0; 1].

To define the function $F(x)$ we first need to assign a certain analytical form $\overline{F}(x,\gamma)$ for it, which is dependent on some unknown parameters $\gamma$. For example, if $x$ is an $n$-dimensioned numerical vector, this could also be a linear function $\overline{F}(x,\gamma) = \sum_{i=1}^{n} \gamma_i x_i$, where $\gamma_i$ are unknown coefficients. More complex analytical forms are also possible. To identify $\gamma$ we use a trial sample – a certain set of elements $\left(x^{(1)},\ldots,x^{(N)}\right)$ and relations between them of the following type: "the element $x^{(i)}$ has a higher (smaller) degree of membership with the set than element $x^{(j)}$." The experts determine such relations. These relations can be depicted in the form of a matrix $t = \{t_{ij}\}$, where

$$\begin{cases} 1, & \text{if the degree of membership with the set is higher for element } x^{(i)} \\ & \text{than for element } x^{(j)}, \\ t_{ij} = -1 & \text{if the degree of membership for } x^{(i)} \text{ is higher than for } x^{(j)}, \\ 0, & \text{if the expert cannot compare the elements among themselves} \\ & i,\ i,j = \overline{1,N}. \end{cases}$$

It is reasonable to assume that

$$\overline{F}\left(x^{(i)},\gamma\right) \geq \overline{F}\left(x^{(j)},\gamma\right), \text{ if } t_{ij} = 1,$$
$$\overline{F}\left(x^{(i)},\gamma\right) \leq \overline{F}\left(x^{(j)},\gamma\right), \text{ if } t_{ij} = -1.$$

Thus, the values of $\gamma$ can be estimated as the solution of the system of inequalities:

$$\left(\overline{F}\left(x^{(i)},\gamma\right) - \overline{F}\left(x^{(j)},\gamma\right)\right) t_{ij} \geq 0, \quad i,j = \overline{1,N}. \tag{2.87}$$

If the system (2.87) has at least one solution, then the set of its solutions will coincide with the set of points of the function maximum:

$$G(\gamma) = \min\left(0, \min_{i,j=\overline{1,N}}\left(\left(\overline{F}\left(x^{(i)},\gamma\right) - \overline{F}\left(x^{(j)},\gamma\right)\right)t_{ij}\right)\right),$$

which can be found applying numerical methods of non-differential optimization [39]. Under the condition that the solutions of the system (2.87) do not exist, the point of the function $G(\gamma)$ maximum can be viewed as the pseudo-solution of this system.

Optimization algorithms can also be applied for $\gamma$ estimation under more complex structures of expert assessments. For instance, if the degree of membership of elements $\left(x^{(1)},\ldots,x^{(L)}\right)$ with the fuzzy set is estimated by $L$ experts, their estimates form matrixes $t^{(1)},\ldots,T^{(L)}$, similar to $t$, and the sets $M(T^{(k)})$ of the solutions of the system of the type (2.87) provided $t = T^{(k)}$, $k = \overline{1,L}$, are not empty, the value $\gamma$ can be estimated having solved the following problem:

$$G_1(\gamma) = \max_{k=\overline{1,L}} d\left(\gamma, M\left(T^{(k)}\right)\right) \to \min, \tag{2.88}$$

where

$$d\left(\gamma, M\left(T^{(k)}\right)\right) = \min_{\overline{\gamma}\in M\left(T^{(k)}\right)} \|\gamma - \overline{\gamma}\|^2 .$$

This is also a non-differential optimization problem.

Having solved the system (2.87) or the optimization problem (2.88), we derive the estimator $\hat{\gamma}$ for parameters $\gamma$. Let us consider $F(x) = \overline{F}(x,\hat{\gamma})$. After $F(x)$ is built, we define $\mu(x)$ as:

$$\mu(x) = \frac{F(x) - F^{\min}}{F^{\max} - F^{\min}},$$

where

$$F^{\max} = \max_{x\in X} F(x), \quad F^{\min} = \min_{x\in X} F(x).$$

Similar to simply ordered sets, elementary set-theoretical operations, in particular those of unification, intersection, and complement were defined for fuzzy sets. At the same time, it is considered that the universe $X$ is identical for all sets. Then the membership function of sets $A$ and $B$ intersection l equals:

$$\mu(x) = \min\left(\mu_A(x), \mu_B(x)\right),$$

where $\mu_A(x)$, $\mu_B(x)$ are the membership functions for sets $A$ and $B$.

By analogy, the membership function for $A$ and $B$ sets unification is $\mu(x) = \max\left(\mu_A(x), \mu_B(x)\right)$, and the complement function for the set $A$ is $\mu(x) = 1 - \mu_A(x)$. This allows identifying the fuzzy set membership function by the membership functions of the sets that generate it and by all the operations with these sets. The sequence of operations can be shown as a graph, the final nodes of which ("tree leaves") are the initial fuzzy sets, transient nodes are operations performed in a certain sequence. The arches of the graph define those results of the former operations that are being used in the next operation. The initial node of the graph ("tree root") is the result of the performance of a certain operations sequence. After we identify the membership functions $\mu_i(x)$ for the final nodes (these functions can be built by the above-considered methods of ordinal regression), we successively build the membership function $\mu_j(x)$ for the intermediate nodes $j$ according to the following formula:

$$\mu_j(x) = \begin{cases} \min\limits_{i \in I(j)} \left(\mu_i(x)\right), & \text{if the node } j \text{ corresponds with the intersection operation,} \\ \max\limits_{i \in I(j)} \left(\mu_i(x)\right), & \text{if the node } j \text{ corresponds with the unification operation,} \\ 1 - \mu_{I(j)}(x), & \text{if the node } j \text{ corresponds with the complement operation} \end{cases}$$

In relation (2.88) $I(j)$ is the set of nodes adjacent with the node $j$. In case of the complement operation we will consider that this set is always comprised of one node.

The building of the membership function ends in the initial node, which corresponds with the membership function $\mu_0(x)$ of the set comprised of the sequence of operations.

Notably, such approach could be applied for comparing alternatives using complex-hierarchy structure [26] criteria aggregates. In contrast with some other methods designed for such problems solving, as AHP [22], for example, this method has a more strict theoretical validation, as well as allows to explain such phenomena as rank reversion [26] and to interpret the results obtained more distinctly. The above properties of this approach determine its application for solving the problems of relations harmonization around the suburban area.

## 2.7 Example of Ordinal Regression Methods Application for Defining the Limits and Status of the Suburban Area

To define the limits of the suburban area let us make use of the above-considered fuzzy sets theory methods and those of ordinal regression algorithms. Obviously, the suburban area will include the highly urbanized areas, which gravitate toward the city – the center of agglomeration. Thus, if we consider the suburban area as a fuzzy set, it will be the intersection of two fuzzy sets:

1. The area with a sufficient level of urbanization.
2. The area (or a list of towns and other settlements) with strong gravitation toward the agglomeration center.

The membership functions for these sets were constructed by ordinal regression methods.

We have built the membership functions for the areas with a sufficient degree of urbanization for individual regions of Kyiv oblast (for administrative units of the lower level). The following indicators were taken into account:

$f_1^{(1)}$ is share of urban population in the total population of the area (%);

$f_2^{(1)}$ is share of industry products and services in the gross regional

product (%);

$f_3^{(1)}$ is share of those employed in non-agricultural sector in the total number of the region's employees (%).

The data for these indicators for a range of regions in Kyiv oblast is provided in Table 2.15. Along with the areas neighboring Kyiv, the table presents data for some distant regions. This was done with the purpose of trial sample analysis simplification. The trial sample is comprised of the data for Ivankivsky, Kaharlytsky, Baryshevsky, Yahotynsky, and Vasylkivsky regions. Based on formal and normalized arguments, the experts in economical geography placed these regions according to the level of urbanization (from the least urbanized to the most urbanized) in the sequence given above. Having solved the system of the type (2.18) comprised of the indicators for these regions, we derived the following function:

$$F_1(f) = f_1^{(1)} + 1.4 f_2^{(1)} + 1.9 f_3^{(1)},$$

which numerically depicts greater or smaller degree of the object membership in the fuzzy set being considered. Taking into account that the maximum possible value of this function, which can be achieved at 100 % values of indicators, is $F^{\max} = 430$, and the smallest (which is obtained at the zero values of the indicators) is $F^{\min} = 0$, we will define the membership function as

$$\mu_1(f) = \frac{F_1(f)}{430}.$$

The values of the functions $F_1(f)$ and $\mu_1(f)$ are also provided in Table 2.15. It should be emphasized that all the regions adjacent to Kyiv have sufficiently high values of the membership function: for them $\mu_1(f) > 0.6$. Thus, these areas could be considered as sufficiently urbanized. At the same time, the Kaharlytsky region, which is quite close to Kyiv, is not such an area and it is inappropriate to consider it as part of the suburban area.

The model that defines the area's degree of gravitation toward the big city could be considered as the generalization of a well-known "gravitation" model, which is broadly applied for preliminary macroeconomic analysis. The variables of the suggested model are as follows:

$f_1^{(2)} = d_{cc}$ the distance (in km) between the centers of the populated area being considered and the big city;

$f_2^{(2)} = d$ the distance (in km) between the closest points of the populated area and of the big city area;

$f_3^{(2)}$ distance (in km) between the closest points of the populated area considered and of the "compact" agglomeration. It is considered that the "compact" agglomeration around a big city is comprised of the populated areas located at the distance $d$ from each other which is not more than 1 km;

$f_4^{(2)}$ the share, % of industrial products produced in this populated area that are being sold in a big city – the center of agglomeration. This indicator shows the degree of local production orientation toward the big city market;

$f_5^{(2)}$ the share of work-capable population which is employed in a big city. This indicator reflects the participation of local population in pendulum migration.

The values for these indicators calculated for some towns around Kyiv are presented in Table 2.16.

According to these methods of ordinal regression, and utilizing the trial sample consistent of: Irpin, Glevaha, Boryspil, Vasylkiv, Dymer, Yahotyn (these towns are placed in the order of their decreasing level of gravitation toward Kyiv) we built the following function, which depicts this degree:

$$F_2(f) = -f_1^{(2)} - 1.5f_2^{(2)} - 1.87f_3^{(2)} + 2.203f_4^{(2)} + 2.597f_5^{(2)}.$$

The negative coefficients for $f_1^{(2)}$, $f_2^{(2)}$, $f_3^{(2)}$ indicate that an increase in distance has a negative influence on the degree of gravitation toward the agglomeration center (Tables 2.15 and 2.16).

Taking into account that $F_2^{\max} = 190$, $F_2^{\min} = -270$, we derive the following expression for the fuzzy set membership function for the towns that gravitate toward Kyiv:

$$\mu_2(f) = \frac{F_2(f) + 270}{460}.$$

The values of the functions $F_2(f)$ and $\mu_2(f)$ are also presented in Table 2.16.

It was mentioned earlier that the suburban area could be considered as a fuzzy set of populated areas – the intersection of the two previously built sets. Thus, the membership function $\mu(f)$ for this set equals:

**Table 2.15** Membership indicators and membership function values for a fuzzy set of an urbanized area

| Regions | Adjacent to a big city | Indicators $f_1^{(1)}$ | $f_2^{(1)}$ | $f_3^{(1)}$ | $F_1$ | $\mu_1$ |
|---|---|---|---|---|---|---|
| **Vyshgorodsky** | Yes | 53.7 | 67.3 | 77.3 | 294.79 | 0.686 |
| Ivankivsky | No | 31.2 | 49.05 | 45.0 | 186.0 | 0.433 |
| Borodyansky | No | 56.3 | 69.9 | 76.9 | 300.27 | 0.698 |
| Irpin (town and settlements) | Yes | 100.0 | 93.6 | 96.8 | 414.98 | 0.965 |
| Makarivsky | No | 26.2 | 62.4 | 76.7 | 259.39 | 0.603 |
| KyivNoSvyatoshynsky | Yes | 47.6 | 74.7 | 86.3 | 316.15 | 0.735 |
| **Fastivsky** | No | 66.7 | 79.7 | 73.1 | 317.17 | 0.738 |
| Vasylkivsky | No | 56.2 | 70.4 | 79.2 | 305.24 | 0.710 |
| Obuhivsky | Yes | 60.4 | 79.2 | 81.4 | 325.94 | 0.758 |
| Kaharlytsky | No | 32.5 | 48.7 | 49.1 | 193.97 | 0.451 |
| Brovarsky | Yes | 59.1 | 82.9 | 83.2 | 333.24 | 0.775 |
| Boryspilsky | Yes | 46.8 | 78.1 | 80.9 | 309.85 | 0.722 |
| Baryshevsky | No | 42.7 | 71.5 | 75.3 | 285.87 | 0.665 |
| Pereyaslav-Khmelnytsky | No | 42.9 | 66.1 | 69.1 | 266.73 | 0.620 |
| Yahotynsky | No | 56.7 | 69.7 | 71.1 | 289.37 | 0.673 |
| Rokytnyansky | No | 36.6 | 49.1 | 50.1 | 200.53 | 0.466 |

$$\mu(f) = \min(\mu_1(f), \mu_2(f)),$$

where $f$ is the vector of indicators for a corresponding populated area (for the $\mu_2(f)$ function) or for the region where it is located (for $\mu_1(f)$ function). The values of $\mu(f)$ are given in Table 2.17. This data indicate that such towns as Pereyaslav-Khmelnytsky, Berezan, Fastiv and Yahotyn (as well as the corresponding regions) should not be considered as part of a suburban area if the membership criteria is set as the satisfaction of inequality $\mu(f) > 0,6$ (i.e., if the level of trust exceeds 60 %).

Thus, we can assert that the Kyiv suburban area is comprised of Baryshevsky (excluding the eastern part around Berezan town), Boryspilsky, Borodyansky, Brovarsky, Vasylkivsky (with the exception of its southern part), Vyshgorodsky, Kyiv-Svyatoshynsky, Makarivsky, Obuhivsky (with the exception of the southern part) regions and the following towns of oblast subordination: Boryspil, Brovary, Vasylkiv, Irpin (together with the subordinated settlements) and Rzhyshchiv. The suburban area occupies around 5.7 thousand square km. The population is around 820 thousand people.

After we at least approximately define the limits of the suburban area we encounter the problem of appropriate legislation formulation which would regulate the relations between the local and oblast authorities in the area. The various relevant proposals appearing in the press define the main three possible alternative concepts of such legislation:

**Table 2.16** Membership indicators and membership function values for "gravity" fuzzy set

| Towns and settlements | Indicators | | | | | $F_2$ | $\mu_2$ |
|---|---|---|---|---|---|---|---|
| | $f_1^{(2)}$ | $f_2^{(2)}$ | $f_3^{(2)}$ | $f_4^{(2)}$ | $f_5^{(2)}$ | | |
| **Vyshgorod** | 15.5 | 0 | 0 | 45.0 | 32.0 | 166.74 | 0.949 |
| Irpin (town) | 21.5 | 0.5 | 0 | 51.0 | 31.0 | 170.61 | 0.958 |
| Bucha | 25 | 3.5 | 0 | 49.0 | 31.0 | 158.2 | 0.931 |
| **Vorzel** | 29 | 6.5 | 0 | 50.0 | 34.0 | 159.7 | 0.933 |
| Chabany | 15 | 1.8 | 1.8 | 43.0 | 41.0 | 180.1 | 0.978 |
| Boryspil | 35 | 8 | 8 | 47.0 | 35.0 | 135.44 | 0.882 |
| Glevaha | 25.5 | 10.5 | 7.5 | 49.0 | 38.0 | 151.36 | 0.916 |
| Vasylkiv | 34.5 | 16 | 13 | 48.0 | 33.0 | 108.64 | 0.823 |
| Obuhiv | 42 | 11 | 7 | 53.0 | 35.0 | 136.1 | 0.883 |
| Borodyanka | 46 | 29 | 14 | 55.0 | 36.0 | 98.99 | 0.802 |
| Rzhyshchiv | 65 | 38 | 30 | 54.0 | 37.0 | 36.951 | 0.667 |
| Baryshevka | 56 | 34 | 29 | 47.0 | 32.0 | 25.416 | 0.642 |
| Berezan | 66 | 44 | 38 | 44.0 | 33.0 | −20.4 | 0.543 |
| Pereyaslav-Khmelnytsky | 79 | 54 | 49 | 49.0 | 27.0 | −73.56 | 0.427 |
| Makariv | 43 | 28 | 21.5 | 53.0 | 42.0 | 100.63 | 0.806 |
| Yahotyn | 92 | 56 | 53 | 46.0 | 24.0 | −111.4 | 0.345 |
| Fastiv | 79 | 67 | 62 | 44.0 | 27.0 | −128.39 | 0.308 |
| Dymer | 34 | 22 | 20 | 47.0 | 39.0 | 100.4 | 0.805 |
| Piskivka | 65 | 48 | 30 | 52.0 | 32.0 | 22.74 | 0.636 |
| Velyka Dymerka | 33 | 5 | 5 | 59.0 | 37.0 | 176.2 | 0.970 |

1. The subordination of suburban populated areas to the Kyiv city authorities. At the same time, the northern regions of Kyiv oblast which suffered from Chernobyl Nuclear Power Plant Accident the most, as well as some of the regions of the neighboring Zhytomyr oblast, which are in the same condition, should form a separate centrally controlled administrative unit. The Kyiv oblast will include only southern regions to which some regions of the closest oblasts could be added. The oblast authorities will be placed in a local sub-regional center – a city with a 200 thousand population (this is a typical size of an oblast center for agricultural regions of Ukraine).
2. The subordination of Kyiv city authorities to the oblast authorities that is to convert Kyiv to the status of other big cities in Ukraine.
3. The establishment of consultive-advisory body, which would coordinate the activities of city and oblast authorities, related to the suburban area, and would take into account the interest of local governing institutions of the suburban area.

These alternatives were analyzed by means of fuzzy sets theory, taking into consideration various criteria. The hierarchy of criteria was defined in a form of a graph. The various interests of individual participants were reflected as different

**Table 2.17** Values of suburban area membership function for towns and settlements

| Towns and settlements | $\mu_1$ | $\mu_2$ | $\mu$ | Including into suburban area (significance level $\mu_0 = 0.6$) |
|---|---|---|---|---|
| **Vyshgorod** | 0.686 | 0.949 | 0.686 | Yes |
| Irpin (town) | 0.965 | 0.958 | 0.958 | Yes |
| Bucha | 0.965 | 0.913 | 0.931 | Yes |
| **Vorzel** | 0.965 | 0.933 | 0.933 | Yes |
| Chabany | 0.735 | 0.978 | 0.735 | Yes |
| Boryspil | 0.722 | 0.882 | 0.722 | Yes |
| Glevaha | 0.710 | 0.916 | 0.710 | Yes |
| Vasylkiv | 0.710 | 0.823 | 0.710 | Yes |
| Obuhiv | 0.758 | 0.883 | 0.758 | Yes |
| Borodyanka | 0.698 | 0.802 | 0.698 | Yes |
| Rzhyshchiv | 0.758 | 0.667 | 0.667 | Yes |
| Baryshevka | 0.665 | 0.642 | 0.642 | Yes |
| Berezan | 0.665 | 0.543 | 0.543 | No |
| Pereyaslav-Khmelnytsky | 0.62 | 0.427 | 0.427 | No |
| Makariv | 0.603 | 0.806 | 0.603 | Yes |
| Yahotyn | 0.673 | 0.345 | 0.345 | No |
| Fastiv | 0.738 | 0.308 | 0.308 | No |
| Dymer | 0.686 | 0.805 | 0.686 | Yes |
| Piskivka | 0.698 | 0.636 | 0.636 | Yes |
| Velyka Dymerka | 0.775 | 0.970 | 0.775 | Yes |

values of the weight coefficients which were multiplied by membership functions of the previous sets ("the decision is good under a separate criterion") while building the resultive fuzzy set ("the decision is good under the totality of criteria).

The performed analysis demonstrates that the first alternative will be the best, and the second – the worst taking into consideration the interests of Kyiv city authorities and those of local governing institutions of suburban areas. Taking into consideration the interests of oblast authorities – the first alternative is the worst and the second alternative is the best. For the state authorities the third alternative is the best, the second alternative is the worst. Thus, the third alternative should be considered as a reasonable compromise between multi-directed interests of the participants of the processes we studied.

## *Conclusions*

1. In a transition economy, the processes of interaction between a big city and a suburban area are very complex, unsteady as well as multifaceted. A range of interrelated economic, ecological, social, technical as well as legal issues arise during the course of these processes. All this stipulates the rational for Compram method application to research the processes mentioned.
2. Within the framework of the above-mentioned method, a detailed problem description has been made; the major concepts for its solution have been identified. Also, we defined the phenomena which require top-priority attention. In particular, the development of legislation, which would coordinate the actions of city and regional authorities in the suburban area, is considered in this paper as a mental idea for solving the existing problems.
3. The definition of suburban area geographical limits and the comparison of alternatives for legislation development are important tasks, which need to be accomplished in order to implement the above-mentioned idea.
4. While solving these problems we should consider various weakly structured factors. It is recommended to implement the methods of fuzzy sets theory and ordinal regression.
5. On the basis of the suggested approach, a multi-criteria analog of a gravitation model was developed, where the gravitation degree of separate localities toward the big city as well as the degree of urbanization of the surrounding area were taken into account, a comparison procedure for legislation development alternatives was developed. Together, all this allowed to formulate some "knowledge islands" regarding some issues considered. In particular, this allowed defining the geographical limits of the suburban area and to formulate a hypothesis regarding the participants' interests in legislative changes implemented in a particular direction. With regard to the considered example of Kyiv and its interaction with the suburban area, the crucial statement is that about a considerable size of the territory which is occupied by such an area (for a range of populated centers their borders are more than 60–70 km away from the city center), about the limited interest of the procedures participants in the gradualist approaches toward legislation changes, and about the size of the interests conflict case of the radical changes.
6. The results achieved allow passing in the nearest future to the development of semantic and casual models of interaction issues of a big city and a suburban area.
7. Fuzzy sets theory methods, as well as the ordinal regression methods could be applied within the Compram method scheme as the means of individual hypothesis, theories and islands of knowledge formulation. An important advantage of these methods is the possibility to take into account imperfect, weakly structured information, in particular the ordinal expert estimations.

8. The approach suggested is universal enough. It could be applied toward the research of interaction processes of a big city and a suburban area for both transition and developed market economy countries.

The proposed decision support procedure is based on the similar AHP approach, but it has some advantages in comparison with the latter. The procedure has stronger theoretical background from the fuzzy set theory viewpoint; it explains the possible rank reversion as the description of the efficient ranking set. Implementation of different logical operations allows to present different hierarchical relations between criteria and alternatives.

The membership function calculation algorithm allows not only to compare existing alternatives, but also to explain the comparison result and to determine the "weak point" for each alternative. The procedure creates the possibilities to analyze the consequences of the changes in the criteria values and preferences.

The minmax operators are used in the mentioned algorithm. It means that the "better" values of some criterion (or of a criteria group) cannot compensate the "worst" value of other criteria. Such an assumption seems to be quite natural if we deal with incomparable high-level criteria. However, at the lowest hierarchy level, where criteria are interchangeable, the other operators (for instance, weighed sum) are also available.

The features of the hierarchical human expertise procedure allow to use it in applied DSSs.

To develop such systems is also an integral part of our further research.

## References


1. Bellman, R.E., Zadeh, L.A.: Local and fuzzy logics. In: Dunn, J.M., Epstein, G. (eds) Modern Uses of Multiple-Valued Logic, pp. 103–165. D. Reidel, Dordrecht (1977)
2. Bradley, S.: The Micro Economy Today. New York: McGraw-Hill/Irwin (2010)
3. Cagan, P.: The monetary dynamics of hyperinflation. In: Friedman (ed) Studies in the Quantity Theory of Money, pp. 25–117. University of Chicago Press, Chicago (1956)
4. Cardenete, M.A., Guerra, A.-I., Sancho, F.: Applied General Equilibrium: An Introduction. Springer, Berlin (2012)
5. Cavalletti, B.: Concorrenza imperfetta e imprese pubbliche in modelli numerici di equilibrio economico generale. Polit. Econ. **3**, 315–333 (1994)
6. Cobb, C.W., Douglas, P.H.: A theory of production. Am. Econ. Rev. **18**(Suppl.), 139–165 (1928)
7. Consolidated Interindustry Balances (1990–1995). Minstat, Kiev (1996, in Ukrainian)
8. Dolan, E.G., Lindsey, D.E.: Macroeconomics. Dryden Press, Chicago (1991)
9. Eaton, C.B., Lipsey, R.G.: Exit barriers and entry barriers: The durability of capital as a barrier to entry. Bell J. Econ. **11**(2), 721–729 (1980)
10. Ermoliev, Y.M., Gulenko, V.P., Tsarenko, T.I.: Finite Difference Approximation Methods in Optimal Control Theory. Naukova Dumka, Kiev (1978, in Russian)
11. Forsythe, G.E., Michael, A., Malcolm, M.A., Moler, C.B.: Computer Methods for Mathematical Computations. Prentice Hall Professional Technical Reference, Englewood Cliffs, NJ (1977)
12. Gandolfo, G.: Economic Dynamics, 4th ed. Springer, New York (2010)
13. Glushkov, V.M.: Macroeconomic Models and Design Principles of GSAS. Statistics, Moscow (1975, in Russian)

14. Gruber, M.H.J.: Regression Estimators: A Comparative Study. Statistical Modeling and Decision Science. Academic, New York (1990)
15. Hahn, F.H.: Money, Growth and Stability. Oxford, Blackwell (1985)
16. Hetemaki, M.: KESSU IV: An econometric model of the Finnish economy. In: Olsen, L. (ed) Economic Modeling in the Nordic Countries (Contributions to Economic Analysis, vol. 210), pp. 115–137. Emerald Group Publishing Limited, Bingley, UK, (1992)
17. Hetemäki, M., Kaski, E.-L.: Kessu IV: An Econometric Model of the Finnish Economy, 114 pp.. Economics Department, Ministry of Finance (1 january 1992)
18. Interbranch Balance for 1996, 64 pp. Minstat, Kyiv (1998, in Ukrainian)
19. Johansson, R.: System Modeling & Identification. Prentice Hall, Englewood Cliffs (1993)
20. Kornai, J.: Economics of Shortage. North Holland, Amsterdam (1980)
21. Lancaster, K.J.: A new approach to consumer theory. J. Polit. Econ. **74**(2), 132–157 (1966)
22. Leontief, W.: Essays in Economics. Theories and Theorizing. Oxford University Press, New York (1966)
23. Mikhalevich, M.V.: Peculiarities of Some Pricing Processes in the Transition Period. IIASA WP-93-66 (1–24 December 1993)
24. Mikhalevich, M.V., Koshlai, L.B.: Decision-support multi-functional system "Ukrainian Budget". In: Gruber, J., Tangian, A. (eds) Constructing and Applying Objective Functions. Lecture Notes in Economics and Mathematical Systems, vol. 510, pp. 349–366. Springer, Berlin (2001)
25. Mikhalevich, M.V., Podolev, I.V.: Modelling of Selected Aspects of the State's Impact on Pricing in a Transitional Economy, IIASA WP-95-12, pp. 1–22 (1995)
26. Mikhalevich, M.V., Sergienko, I.V.: Applying stochastic optimization methods to analyze transformation processes in economy. Syst. Res. Inform. Technol. **4**, 7–29 (2004, in Ukrainian)
27. Mikhalevich, M.V., Sergienko, I.V.: Modeling Transition Economy, Models, Methods, Information Technologies. Naukova Dumka, Kyiv (2005, in Russian)
28. Mikhalevich, M.V., Sergienko, I.V., Koshlai, L.B.: Simulation of foreign trade activity under transition economy conditions. Cybern. Syst. Anal. **37**(4), 515–532 (2001)
29. Moroney, J., Toevs., A.: Input prices, substitution, and product inflation. In: Pindyck, R.S. (ed) Advances in the Economics of Energy and Resources, vol. I, pp. 27–50. JAI Press, Greenwich (1979)
30. Nikaido, H.: Monopolistic Competition and Effective Demand (Princeton Studies in Mathematical Economics). Princeton University Press, Princeton (1975)
31. Ponomarenko, O.I., Perestyk, M.O., Burim, V.M.: Modern Economic Analysis, Pt. 2. Macroeconomics. Vyshcha Shkola, Kyiv (2004, in Ukrainian)
32. Saaty, T.L.: The Analytic Hierarchy Process. McGraw Hill, New York (1980)
33. Sargent, T.I., Wallance, N.: Rational expectations an the dynamics of hyperinflation. Int. Econ. Rev. **14**, 328–350 (1973)
34. Sergienko, I.V.: Mathematical Models and Methods for Solving Discrete Optimization Problems, 472 pp. Naukova Dumka, Kyiv (1988, in Russian)
35. Sergienko, I.V.: Methods of Optimization and Systems Analysis for Problems of Transcomputational Complexity. Springer, New York (2012)
36. Sergienko, I.V., Kaspshitskya, M.F.: Models and Methods for Solving Combinatorial Optimization Problems on Computers, 81 pp. Naukova Dumka, Kyiv (1988, in Russian)
37. Sergienko, I.V., Mikhalevich, M.V., Stetsyuk, P.I., Koshlai, L.B.: Interindustry model of planned technological-structural changes, Cybern. Syst. Anal. **34**(3), 319–330 (1998)
38. Sergienko, I.V., Mikhalevich, M.V., Stetsyuk, P.I., Koshlai, L.B.: Models and information technologies for decision support during structural and technological changes. Cybern. Syst. Anal. **45**(2), 187–203 (2009)
39. Shor, N.Z.: Methods of Minimization of Nondifferentiable Functions and Their Applications. Naukova Dumka, Kyiv (1979, in Russian)
40. Shor, N.Z.: Nondifferentiable Optimization and Polynomial Problems. Kluwer, Dordrecht (1998)
41. Shor, N.Z., Stetsenko, S.I.: Quadratic Extremum Problems and Nondifferentiable Optimization. Naukova Dumka, Kyiv (1989, in Russian)

42. Shor, N.Z., Stetsyuk, P.I.: Modified $r$-algorithm to find the global minimum of polynomial functions. Cybern. Syst. Anal. **33**(4), 482–497 (1997)
43. Shor, N.Z., Zhurbenko, N.G.: Minimization method using space dilatation toward the difference of two sequential gradients. Kibernetika (3), 51–59 (1971)
44. Shy, O.: The Economics of Network Industries. Cambridge University Press, Cambridge (2001)
45. Statistical Yearbook of Ukraine for 1995, 520 pp. Ministry of Statistics of Ukraine, Tekhnika (1995, in Ukrainian)
46. Stetsyuk, P.I.: New quadratic models for the maximum weighted cut problem. Cybern. Syst. Anal. **42**(1), 54–64 (2006)
47. Ukrainian Economic in Figures - Statistical Yearbook. Minstat, Kiev (1994, in Ukrainian)
48. Ukrainian Economics in Figures. Statistical Yearbooks, 1992–1996. Minstat, Kiev (1997, in Ukrainian)
49. Wallis, K.F., Whitley, J.D.: Macro models and macro policy in the 1980s. Oxf. Rev. Econ. Policy **7**(3), 118–127 (1991)

# Chapter 3
# Modeling of Imperfect Competition on the Labor Market

## 3.1 Imperfect Competition on the Labor Market in a Transition Economy

The social situation of last decades in post-communist countries (especially in countries of the former Soviet Union) has been certainly an extremely difficult one. Various grave problems existed: rising inequalities in income both at downward and upward stages of the business cycle show up in islands of prosperity in the sea of distress and poverty, as well as in capital flights and growing indebtedness. All of these factors lead to the criminalization of society. Extremely low wages are the main reason of such a situation. Theoretical analysis of labor market is necessary to understand essential issues of this situation.

Number of factors defines actuality of theoretical investigations of labor market for transition economy. First, the value of real employees' income essentially influences the amount of final consumption and general economic dynamics. Second, it is in the sphere of employment where the most significant differences between a stable market and transition economy are observed. For example, in several post-communist countries economic decline at the initial stages of reforms was not accompanied by the expected corresponding growth of unemployment (see [54]); the rates of wage increase were sufficiently small in comparison with rates of gross domestic product (GDP) growth when economy was developed, etc. Third, the methods for overcoming the depression that were proposed in classical economics are the least efficient precisely in attempts to improve wage conditions in post-communist world.

It is rather difficult to analyze the peculiarities of labor market in a transition economy by traditional means. A new methodological approach is necessary for this purpose. A lot of phenomena, which have been observed at the labor market in post-communist countries, can be explained if we consider this segment of national market as a market with imperfect competition, as monopsonic market (monopoly position of employers) or as a market with competition between several large



I.V. Sergienko et al., *Optimization Models in a Transition Economy*,
Springer Optimization and Its Applications 101, DOI 10.1007/978-1-4899-7544-7_3

employers. High capital concentration preserved from the centralized economy, absence of powerful, really independent of employers and the state, trade unions, and several other factors support such structure of labor market. It is quite obvious that the development of such economy differs from "pure" market evolution. Fundamental economic investigations including mathematical modeling need to understand such differences. Development, analysis, and verification of economic dynamics models for conditions of imperfect labor market and restricted employers time horizon will be the goal of our monograph. We expect that results of modeling allow us to understand some general peculiarities of economic evolution in such systems and to propose recommendations for social policy improvement aimed the restriction of negative consequences of imperfect competition.

Theoretical investigations of the labor market with imperfect competition started at the third decade of the twentieth century.

Robinson [57, 58] was one of the first economists to systematically study imperfect competition on the labor market (and on the other market segments as well) and analyzed its influence on general economic dynamics. In her publications, she focused on microeconomic analysis of monopolism and monopsony, giving interpretation of several fundamental economic categories from the viewpoint of differences between competitive and monopsonic markets. The progress of trade unions and the establishment of state control over wages were considered the main directions of restricting the consequences of imperfect competition. Rather scarce attention was paid to the **macroeconomic** analysis of the systems with non-competitive labor markets. These problems are not yet solved, probably, because of the fact that the labor market in developed countries is restricted to the standards of a competitive market. (Particular attention to such problems observed in the publications by the of leftist post-Keynesians, for example, in [13], could be explained mainly by traditions). The interest in non-competitive waging analysis increased during the last decade as the result of post-communist studies of the economy. For instance, the peculiarities of the labor market in a transition economy were explained in [52] from the viewpoint of social partnership theory. It is difficult though to believe in such explanations given the Gini coefficient rates of growth in post-communist countries for the last 5 years are approximately equal to the rates of increase of GDP. Another approach, using with the model of a two-side monopoly competition (trade unions against employers) is considered in [20]. Some aspects of the Romanian noncompetitive labor market were also analyzed by Voicu [73]. Papers [15, 21] also contributed to the labor market analysis in post-communist countries. In particular, they paid a great attention to the official statistical information accuracy and analyzed the factors that influenced employers' behavior. It should be noted that several fundamental works have appeared in last years of the twentieth century [28, 39], in which the authors admitted that specific forms of monopsony could be also typical for the market economy. Prof. A. Manning and his scientific school developed this theory; they paid a great attention to the microeconomic analysis of such phenomena.

It is also true that macroeconomic research is an important part of the existing methodology for economic studies. The influence of the imperfect competition on the labor market could be very important for some other market segments and economic processes in general. Different quantitative and qualitative methods, including mathematical modeling, have been applied to study this influence. It should be noted, for instance, the following results received in this direction.

Ridder and van den Berg [56] develop the tool for measuring frictions on the labor market by using information about work duration. The authors propose and apply unconditional inference method that operates with aggregated information relative to uptime.

This method does not require numerical estimation of wages; it is invariant relatively to the wage measuring scale and allows to take into account nonmonetary terms of a work. The empiric analysis for France, USA, Great Britain, Germany and Netherlands was provided with the proposed method. Using the frictions of job search, the authors obtained the estimates of the degree of monopsony on the labor markets in the mentioned countries and analyzed the consequences of lower wage restriction, unemployment benefit change and control for a job search time.

Establishing a low level of wages and changing a size of unemployment benefit, which is financed from an income tax, it is possible to achieve an efficient level of employment and production values in models of a labor market with monopsony.

Strobl and Walsh [68] take account of nonmonetary components of workers' remunerations in models of a monopsonic labor market. Assuming that higher wages increase the value of the labor supply to firms, but less attractive terms of work decrease this value, they prove that the model implies several modifications for working of traditional adjustment instruments on the labor market. For instance, the introduction of minimum wages can worsen nonmonetary working conditions and workers' welfare in contrast to unemployment benefit. Using empirical data from the labor market of Trinidad and Tobago, they show that implementation of minimum wages might indeed worsen working conditions.

Manning [40] uses the simplified model of monopsony on the labor market to develop a unified analysis of different approaches to market adjustment such a market. The conclusion that the averaged wage is squeezed under monopsony is the basic result. In his opinion, to improve the situation it is possible to use such instruments as increasing of minimum wages, strengthening role of trade unions and restriction of labor contracts. Manning proves that, unlike the competitive markets, the tradeoff between *efficiency* and *justice* is not compelling under monopsony. It is not necessary to consider this tradeoff as useful: there is usually an optimal level of regulation beyond which efficiency is lost.

Delfgaauw and Dur [22] analyze the consequences of creating a competitive labor market in an industry formerly controlled by state firm that minimized its costs. Corroborated by the results of empirical investigations, their model demonstrates that under the conditions of competitive labor market producers give large wage incentives to workers, achieve the highest productivity, and hire fewer workers than a state firm does. Consequently, the efficiency of resource allocation increases, but due to higher wages product prices also increase. Since market liberalization has some ambiguous effects, political support may also be restricted.

Flow models of the labor market add further insights into the analysis of a monopsonic labor market. Barth and Dale-Olsen [4] apply this approach to the explanation of a gender wage differentials developed one of such models. The authors show that these differentials could be caused by monopsony discrimination of employers, especially, for different professions and firms. Using the information on the labor market in Norway, the authors analyzed the wage structure both within an organization and at the multi-organization level. They proposed the original proofs of that a surplus of the labor force for several organizations depended more on the bonuses given to male rather female workers. In addition, a labor supply is more wage-elastic for men than for women.

Staiger et al. [67] use the exogenous wage variable at a veteran hospital aiming to estimate the extent of monopsonic power on the labor market for trained nurses. Unlike the previous research, the authors prove that a labor supply in separate hospitals is inelastic enough. They notice that, in response to changes in wages for trained nurses at veteran hospitals, the other hospitals also changed their remuneration level.

Analyzing the liberalization in Russia in the 1990s, Brown and Earle [15] proved the existence of a positive relationship between the degree of competition on local labor markets and efficiency gains for the factors of production. The effect is particularly due to higher labor force mobility and transport infrastructure improvement.

Ioannides and Hsarides [32] propose a labor supply model with senior and junior workers (both as for age and official status), when only a junior workers are affected by a monopsonic power. The result is that senior workers keep on receiving wages according to their marginal product while junior workers' wages are considerably below this level.

Chalkley [17] explores wage determination in a search model under monopsony conditions. He demonstrated the existence of multiple (possibly, inefficient) equilibriums if wages are determined such that monopsony income is maximized. According to the author, temporary state interventions might be required on a labor market of this structure.

Bradfield [14] extends the traditional comparison of the competitive labor market and pure monopsony to the case of long-term monopsony. His results demonstrate that long-term monopsony will show even deeper deviations from a competitive market than short-term one. In a long term, a monopsonistic employer achieves a maximum permanent profit level that is equal to competitive wages. This forces him to reduce employment and wages and to set an upper wage limit during negotiations with trade unions in the case of bilateral competition.

Markusen and Robson [41] propose a two-sector model to analyze the general equilibrium conditions on the market for goods and services with monopsony in one sector. Their results demonstrate that, generally, a monopsonic sector loses efficiency in comparison with a competitive sector, but sometimes the first one can reach an equilibrium state close to the competitive sector and may force the later out from the market.

Haluk and Serdar [30] develop a model, which allows comparing the basic consequences of different forms of labor market organization. In particular, using the model it is proved that employment is the highest under the conditions of a perfect competitive market and for slavery (levels of employment are equal for these two forms). Employment for oligopsony is lower and it will be the least under monopsony. The similar situation emerges as for production of goods per capita. However, an income per capita will be the largest for pure monopsony, then come oligopsony, slavery and pure competition. Equilibrium wages and utility for employed were distributed in accordance with power of employers: they are the highest for a competitive market, then come oligopsony, monopsony and slavery. It should be noted that, under the conditions of complete substitution of productive factors (for instance, during the recession) wages for monopsony and slavery are equal. The ratio of equilibrium margin incomes and wages was the highest for a competitive market, then goes oligopsony. Such an index acquires the least value for slavery and monopsony. All these conclusions demonstrate the low capacity of monopsonic labor market for economic innovations.

Using the geographical monopsony model, Shieh [65] obtains similar results for pricing. He analyzed the consequences of monopsony assuming that the labor supply function is linear. In [66] the author estimates the effect of wage discrimination on employment. According to his conclusions, the combined employment will change, if labor supply functions are linear at two separate markets. Wage discrimination will decrease total employment if a labor supply function is high-elastic and convex, and employment will increase if the same function is low-elastic and concave.

Jones [33] proves that agricultural enterprises, which are monopolists on the labor market and maximize their incomes, have low employment and issues. They do not use possibilities of scale production economy, do not try to reduce their margin costs and, therefore, they are less effective in comparison with enterprises operating on large competitive labor market.

Garibaldi and Wasmer [27] using two-sector model of economy (industry and agriculture are taken into account in it) consider the simple two-sector system with market of wages and a household sector that absorb extra labor supply.

Dong and Putterman [24] prove the presence of monopsonic power on the Chinese industrial labor market that existed in the beginning of the reforms and its gradual diminishing during the reforms.

It should be noted that the mentioned results deal mainly with microeconomic aspects and consequences of monopsony. The insufficient study of macroeconomic aspects of imperfect competition at a labor market, especially its influence onto general economic dynamics, stipulates the actuality of solving these problems. This chapter is devoted to such a study.

The chapter applies dynamic macroeconomic models based on the assumption about monopoly consumption or monopsony and other forms of imperfect competition on the labor market in a transition economy.

The chapter consists of six sections.

The model of economic dynamics in the system with a monopsonic labor market and constant prices is considered in Sect. 3.2. The existence of the equilibrium points in this model is proved and the properties of such points are analyzed.

In Sect. 3.3, this model is generalized for the case when prices are changed according to a pricing mechanism typical for a competitive market of goods. The existence of the cycles similar to the classical business-cycles is proved for this model. The model of oligopsony competition between large employers is analyzed in Sect. 3.4. The existence of Nash equilibrium points is proved with the same properties, which provides the appearance of cycles on monopsony labor market. An empirical analysis of data for Ukrainian economy at the end of the twentieth century is also done there. Static and dynamic models of economy with bilateral monopolistic competition between employer and trade union are considered in Sect. 3.5. The conditions of post-classical business-cycles appearance were analyzed in this section. Some aspects of application of obtained theoretical results to new job-creation problem are discussed in Sect. 3.5.1.

## 3.2 Modeling of Employment and Growth Processes Under Constant Prices

Let us consider a monopsonic labor market with the following structure. There are potentially employed individuals, who supply their labor, and a single monopolist, who is the consumer of labor (an employer). It is assumed that the latter is aware of the labor supply function and the value of demand for goods that are being produced using the labor acquired. The monopolist acts with respect to only his short-time interests striving to maximize his profit obtained from the production of goods. Let us also assume that his additional costs, associated with labor consumption (indirect tax on the wage fund, social payments, etc.), are proportional to the value of the total wage fund $W$ (i.e., the amount of money paid by employer to employees).

Inflation processes, especially in the form of cost inflation, are typical for the early stages of market reforms in post-communist countries. As it was shown in Chap. 1, the main causes of this inflation are the structural and technological imbalances formed in the course of system crisis of a centralized economy and external factors due to a lack of development of the exchange market. Therefore, we can believe that the rate of inflation under such conditions does not depend on real wages, or, more properly, this dependence is so weak and ambiguous that it does not affect the behavior of economic entities over a restricted interval of time. Based on this, the static model presented below uses only real (i.e., corrected for the value of change of the GDP deflator) values of indices expressed in value units.

Let us consider the function $S(W)$–the dependence between $W$ and the maximal amount of labor, which the employer can purchase given the limits of the total wage fund (wage bill) by changing the value of wages $\omega$. In fact, $S(W) = \max\limits_{\omega \in X(W)} L_s(\omega)$, where $L_s(\omega)$ is the labor supply function, $X(W) = \left\{\omega : \omega L_s(\omega) \leq W,\ \omega \geq 0\right\}$. Such properties of $S(W)$ as continuity, concavity and monotony follow directly from the similar assumptions about the properties of the function $L_s(\omega)$.

Let us denote the amount of labor acquired as $T$, $U(T)$ is the utility of labor consumption by the employer, $h$ is the value of the associated costs per wage fund unit.

It should be noted that the employer always acquires the maximum amount of labor within his wage fund limit. Indeed, a lower amount implies less gross profit for the same labor cost, and such a decision will not be the best for the employer. The employer also decreases the size of the wage fund, if the restricted demand for produced goods forces the employer to cut the amount of labor acquired. Then the monopolist–employer will select the value of $W$ so as to maximize the following function:

$$F(W) = U\big(S(W)\big) - W(1+h)$$

provided that $W \geq 0$.

We can assume that the utility of labor consumption for the employer lies in the possibility of obtaining income from the sale of goods produced using this labor; therefore,

$$U(T) = \min(lT, \alpha\nu),$$

where $l$ is the labor capacity per unit of the newly created value added, $\nu$ is the demand for products, and $\alpha$ is the portion of the value added included in the unit value of the products.

Thus, the value of wage fund $W$ established by the employer will be the solution of the optimization problem

$$F(W) = \min\big(lS(W), \alpha\nu\big) - HW \to \max \tag{3.1}$$

for $W \geq 0$, where $H = 1 + h$.

The function $F(W)$ is also continuous and concave under the mentioned assumptions. The point of its maximum will be $W^* = \min\left(W^{(1)}, W^{(2)}\right)$, where $W^{(1)}$ is a solution of the equation $S(W) = \frac{\alpha\nu}{l}$ and $W^{(2)}$ is determined from the relation $\frac{H}{l} \in \hat{\partial}S(W)$. Here $\hat{\partial}S(W)$ is the generalized (by Clark) derivative of function $S(W)$. The point $W = 0$ could also be the solution of the problem (3.1), but this case could be avoided if we assume that the derivative of $L_s(W)$ function in zero point was sufficiently large, thus $\inf \hat{\partial}S(0) > \frac{H}{l}$.

It should be noted that only one component $W^{(1)}$ of the solution $W^*$, out of the two ones, depends on demand $\nu$. Thus, in the case of a deep economic decline when the value of $\nu$ is decreased, the wage fund is also decreased, i.e., an absolute decline of employees' level of life takes place. In the case of economic growth the wage fund will increase up to some limit defined by the point $W^{(2)}$. Further economic growth (an increase of $\nu$) will be accompanied by a constant value of the total wage fund. The relative decline (in comparison with total demand) of employees' incomes will be observed in this case. The dependence between the wage and demand will be nonlinear as a result of the effects mentioned above.

The above-considered model also allows us to account for the low level of unemployment in those transition countries experiencing a deep business depression. It is advantageous for a monopolist–employer to establish the wages, so that the labor supply is equal to the labor demand. A lower price of labor will not allow the monopsonic employer to fully obtain a feasible income, and a higher price will decrease the pure utility of labor consumption due to high labor purchase cost. Thus, there will be no extra labor force (i.e., unemployment in its classical understanding) on the labor market. Redundant labor will simply be ousted from the market due to low wages. Actually, there exists some unemployment that, first, is created by employer to force the employees to agree with low monopoly wages, second, is formed due to disagreement between the expected (used by employer for waging) and actual values of demand, and third, appears as frictional and structural unemployment. However, even taking into account the above-mentioned factors, the total level of unemployment will be considerably lower than the unemployment estimates obtained from the classical labor market models under the condition of a high rate of production decrease.

A further development of the suggested model will be a consideration of a dynamic system consisting of linear auto-correlation equation, which connects the values of demand and the wage fund, and the previously considered equality, which determines the wage fund through the demand.

Let time $t$ be discrete, with a step equal to one. We will assume that the function $S(W)$ does not depend on time, is nonnegative, continuous, and takes zero values for the minimum permissible wage fund $W \leq W_0$ (as a special case, $W_0 = 0$ is possible). Suppose this function is concave, increasing, and differentiable for $W > W_0$, and its derivative from the right of $W_0$ is bounded. Assume also that the coefficients $l$ and $\alpha$ are independent of time. Denote by $\nu_t$ the demand for production in a moment of time $t$ and by $W^t$ the wage fund at the same time. Suppose $W^t$ is determined at each moment of time on the basis of the above-considered model of monopsony waging using $\nu_t$ as the final demand. In other words, we have

$$W^t = \min\left(W^{(1)t}, W^{(2)t}\right), \tag{3.2}$$

where $W^{(1)t} = S^{-1}\left(\frac{\alpha \nu_t}{l}\right)$, $W^{(2)t} = (S')^{-1}\left(Hl^{-1}\right)$, $S^{-1}(\cdot)$ and $(S')^{-1}(\cdot)$ are functions inverse to $S(W)$ and $S'(W)$. By the assumptions made, these functions exist for $W \geq W_0$, the function $S^{-1}(\cdot)$ is increasing and $(S')^{-1}(\cdot)$ is decreasing. Let us redefine the function $S'(\cdot)$ by putting $S^{-1}(0) = W_0$. Note also that for $\nu \geq 0$ the function

$$g(\nu) = \min\left(S^{-1}\left(\frac{\alpha \nu_t}{l}\right),\ \ (S')^{-1}\left(Hl^{-1}\right)\right)$$

will satisfy the Lipschitz condition for some constant $L_1$.

Let us consider the case where the demand for production is determined at each subsequent moment of time as a weighted average value of the demand at the previous time and the current consumer income, i.e.,

$$\nu_{t+1} = (1 - \rho)\nu_t + \rho W^t, \quad t = 0, 1, \ldots, \tag{3.3}$$

where $0 \leq \rho < 1$ is some weighting factor. Due to the above-mentioned reasons, the quantity $\rho$ may be considered as rather small.

We will analyze the properties of the sequence $\{\nu_t\}$ constructed according to (3.3). The following statement is true.

**Statement 3.1.** Let $\rho < \dfrac{1}{2L}$, where $L = 1 + L_1$. Then, the sequence $\{\nu_t\}$, which is determined according to (3.2) and (3.3), is converging, and its limit (depending on the value of $\nu_0$) may be either the solution of the equation

$$S^{-1}\left(\frac{\alpha}{l}\nu\right) = \nu,$$

denoted hereafter by $\nu^{(1)}$, if $\nu^{(1)} < (S')^{-1}\left(Hl^{-1}\right)$, or the point

$$\nu^{(2)} = (S')^{-1}\left(Hl^{-1}\right),$$

if $\nu^{(2)} \geq \nu^{(1)}$.

*Proof.* We will rearrange relation (3.3) as

$$\nu_{t+1} = \nu_t + \rho\left(W^t - \nu_t\right).$$

This equality can be considered as a procedure of search for a maximum of the function $F(\nu) = G(\nu) - 0,5\nu^2$ using the gradient method with a constant step multiplier $\rho$. Here, $G(\nu)$ is the antiderivative of the function $g(\nu)$, this antiderivative existing owing to the continuity and boundedness of $g(\nu)$.

Note that the derivative $g(\nu) - \nu$ of this continuously differentiable function will satisfy the Lipschitz condition with the constant $L = 1 + L_1$. Therefore, by the Lagrange theorem we have

$$\begin{aligned}
F(\nu_{t+1}) &= F(\nu_t) + F'\big(\nu_t + \lambda(\nu_{t+1} - \nu_t)\big)\,(\nu_{t+1} - \nu_t) \\
&= F(\nu_t) + \big(F'\big(\nu_t + \lambda(\nu_{t+1} - \nu_t)\big) - F'(\nu_t)\big) \\
&\quad \times (\nu_{t+1} - \nu_t) + F'(\nu_t)(\nu_{t+1} - \nu_t) \\
&\geq F(\nu_t) + \rho\big(F'(\nu_t)\big)^2 + L\rho^2\big(F'(\nu_t)\big)^2,
\end{aligned}$$

where $0 < \lambda < 1$, and for $\rho < \dfrac{1}{2L}$ we obtain

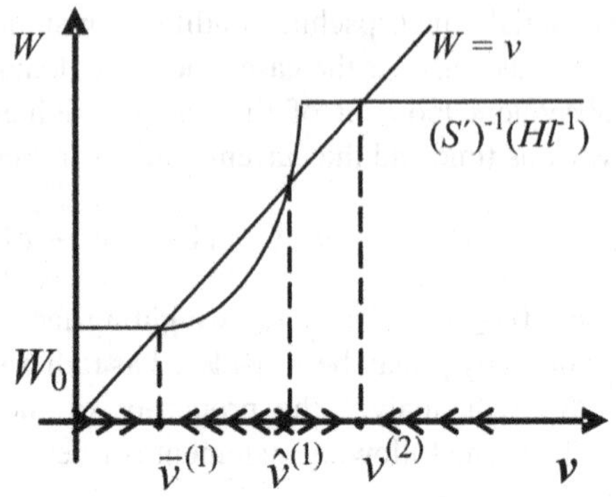

**Fig. 3.1** The case of two roots

$$F(\nu_{t+1}) \geq F(\nu_t) + 1/2\rho\left(F'(\nu_t)\right)^2.$$

The bounded sequence $\left\{F(\nu_t)\right\}$ will be nondecreasing, and $F(\nu_{t+1}) > F(\nu_t)$ if $F'(\nu_t) \neq 0$. Whence, according to the convergence condition for the gradient methods with a constant step multiplier (see [34]), the sequence $\left\{\nu_t\right\}$ will be converging, and its limit $\nu^*$ must satisfy the condition

$$F'(\nu^*) = g(\nu^*) - \nu^* = 0.$$

The validity of the statement directly follows from the latter equality.

Let us now consider the disposition of the points $\nu^{(1)}$ and $\nu^{(2)}$ and the behavior of the convergence of the sequence $\left\{\nu_t\right\}$. The following cases are possible.

1. The equation $S^{-1}\left(\frac{\alpha}{l}\nu\right) = \nu$ has two roots $\overline{\nu}^{(1)}$ and $\hat{\nu}^{(1)}$ on the interval $\left[0;\ S\left((S')^{-1}(Hl^{-1})\right)\right]$, Fig. 3.1, with $\nu^{(2)} = (S')^{-1}(Hl^{-1}) > \hat{\nu}^{(1)}$ existing. In particular, if $W_0 = 0$, then $\overline{\nu}^{(1)} = 0$.

   In all of the figures in this section, the following designations are assumed:

$$\ggg \text{ means that } \nu_t \text{ increases and } \lll \text{ means that } \nu_t \text{ decreases.}$$

The points $\overline{\nu}^{(1)}$ and $\nu^{(2)}$ will be the local maxima of the function $F(\nu)$ and $\hat{\nu}^{(1)}$ is its local minimum. The sequence $\left\{\nu_t\right\}$ will converge to the point $\overline{\nu}^{(1)}$, if $\nu_t$ falls into $\left[0; \hat{\nu}^{(1)}\right)$, and to the point $\nu^{(2)}$ if $\nu_t \in \left(\hat{\nu}^{(1)}, \infty\right)$. In this case, if $\nu_t \in \left(\overline{\nu}^{(1)}, \hat{\nu}^{(1)}\right)$, then a decrease in production (reduction of the demand $\nu$) will be accompanied by a reduction in the real wage fund $W$, which will lead to a new decrease in production. When the sequence $\left\{\nu_t\right\}$ reaches a sufficiently small neighborhood of the point $\overline{\nu}^{(1)}$, the decrease in production will slow down, and, at certain time moments, the demand may increase if $\nu_t < \overline{\nu}^{(1)}$ has been satisfied before. However, the rate of increase will be insignificant, the increase itself will be small, and the demand will be stabilized eventually at the point $\overline{\nu}^{(1)}$.

On the half-interval $\left(\hat{\nu}^{(1)}, \nu^{(2)}\right]$, the convergence of the sequence $\{\nu_t\}$ to the point of the upper equilibrium between the demand and wages $\nu^{(2)}$ will be accompanied by an increase in demand, but if $W^t < W^{(2)t}$, then also by an increase in wages. At a value of demand exceeding $\nu^{(2)}$ it will decrease, but the real wages will remain unchanged. Note that on the strength of the property of the function $(S')^{-1}(\cdot)$ a decrease in $H$ (e.g., due to reduction in tax load on the wages fund) will lead to an upward shift of the horizontal half-line $W = (S')^{-1}\left(Hl^{-1}\right)$, however, in this case, the position of the curve $W = S^{-1}\left(\frac{\alpha}{l}\nu\right)$ will not change, and, consequently, the disposition of the points $\overline{\nu}^{(1)}$ and $\hat{\nu}^{(1)}$ will not change neither. Thus, the above-described mechanism of economy depression on the half-interval $\left[\overline{\nu}^{(1)}, \hat{\nu}^{(1)}\right)$ will remain the same. This measure may change the situation only in the case where the value of demand previously belonged to the interval between the previous and new values of $\nu^{(2)}$. In other cases, in order to eliminate the negative dynamics, it is necessary to shift the curve $W = S^{-1}\left(\frac{\alpha}{l}\nu\right)$ to the left and upward. Such an effect can be achieved either through a decrease in labor supply in the monopolized segment of economy (e.g., due to development of small and medium size enterprises that do not relate to this segment), or through an increase in the value added share $\alpha$ (e.g., by reducing the taxation of income and value added and executing radical structural and technological transformations; see [63]). However, these measures may just narrow the boundary of depression domain $\left[\overline{\nu}^{(1)}, \hat{\nu}^{(1)}\right)$ and raise the lower limit of the demand reduction $\overline{\nu}^{(1)}$, but they do not fully eliminate the associated negative processes.

2. The equation $S^{-1}\left(\frac{\alpha}{l}\nu\right) = \nu$ has a unique positive root within the interval $\left[0; S\left((S')^{-1}\left(Hl^{-1}\right)\right)\right]$. The intersection point of the lines $W = \nu$ and $W = (S')^{-1}\left(Hl^{-1}\right)$ also falls into this interval (Fig. 3.2).

In this case, there exists the unique point $\nu^{(1)}$ satisfying the conditions of Statement 3.1. For a value of demand greater than $\nu^{(1)}$, the production will decrease to the level defined by this point. In this case, the real wages remain the same if $\nu_t > S\left((S')^{-1}\left(Hl^{-1}\right)\right) = \nu'$, and start to decrease as soon as the demand is less than $\nu'$. Since with constant wages the amount of labor force purchased by a monopolist–employer remains the same and the quantity of production decreases with a fall in demand, a drop in labor capacity will occur in this case, and continue more slowly after the reduction in the wages takes place.

A special case of the above-considered situation is when the scope and the rate of depression do not depend up on changes in the additional labor cost $H$. This case corresponds to the deepest and longest-occurring business depression. The only way to reduce the negative consequences of the depression will be a reduction in labor supply accompanied by a growth in the profitability of production. This will cause

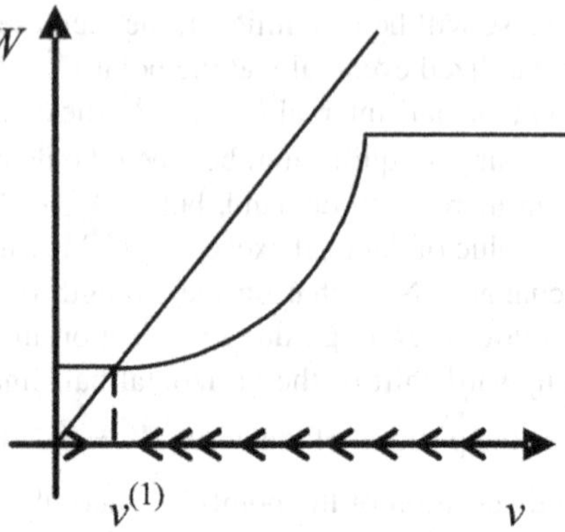

Fig. 3.2 The case of one root

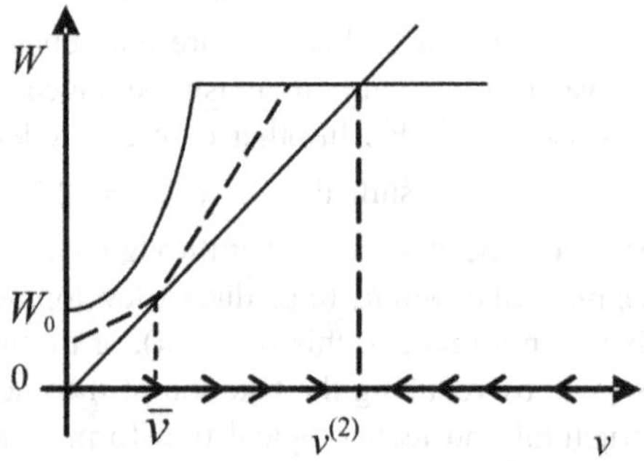

Fig. 3.3 The point of intersection of $v^{(2)}$ and $W = v$

a left-ward and upward shift of the curve $W = S^{-1}\left(\frac{\alpha}{l}\nu\right)$ and will raise the limit $\nu^{(1)}$ up to which the business depression will continue.

3. For $\nu = \nu'$ the inequality $S^{-1}\left(\frac{\alpha}{l}\nu\right) \geq \nu$ is fulfilled, and there exists the point $\nu^{(2)}$ of intersection of the lines $W = \nu$ and $W = (S')^{-1}\left(Hl^{-1}\right)$ lying to the right of $\nu'$ (Fig. 3.3).

In particular, the curve $W = S^{-1}\left(\frac{\alpha}{l}\nu\right)$ may be tangent to the line $W = \nu$ as it was shown in the figure by a dotted line.

In the absence of tangency points $\overline{\nu}$ demand $\nu_t$ will increase up to the point $\nu^{(2)}$ if $\nu_0 < \nu^{(2)}$, and will decrease if $\nu_0 > \nu^{(2)}$. This growth can be stopped at one of the points $\overline{\nu}$, if any. In this case, a left-ward and upward shift of the curve $W = S^{-1}\left(\frac{\alpha}{l}\nu\right)$ will restore the growth. In contrast to the cases considered before, such a shift will not affect the dynamics of depression with constant wages. The depth and rate of depression could be lowered through a reduction of the additional cost $h$ associated with consumption of the labor force. Here, recommendations for overcoming the crisis will be opposite to those that were effective in the previous cases.

Thus, monopsonic waging on the labor market may be the cause of a number of negative economic processes such as economic depression, decrease in real incomes of population, reduction in the labor capacity, etc. There are several possible scenarios of development of these processes. To eliminate their negative effects, it is necessary to determine which scenario takes place, since anti-crisis recommendations will vary for different scenarios.

It should be pointed out that in real economy the agents do not always follow the optimal strategy. Moreover, external and other ill-predicted factors can affect demand. All of these result in the fact that the law of variation of $\nu_t$ can differ from that considered above due to appearance of an additional component $\xi^t$, which in some cases could be considered as a random variable. In other words, instead of (3.3) the following equality holds

$$\nu_{t+1} = (1-\rho)\nu_t + \rho W^t + \xi^t, \quad t = 0, 1, 2, \ldots, \tag{3.4}$$

where $\xi^t$ is some sequence of random quantities. We will assume that for any $t$ the quantity $\xi^t$ is measurable relative to a $\sigma$-algebra, generated by $(\nu_0, \ldots, \nu_t)$, and bounded with probability one. Due to the obvious nonnegativeness of $\nu_t$, it is reasonable to assume that $\xi^t$ satisfies the condition

$$\xi^t \geq (\rho - 1)\nu_t - \rho W^t \tag{3.5}$$

with probability one.

Let us now analyze how the random actions $\xi^t$ will affect the convergence of the sequence $\{\nu_t\}$ to the set of points $\{\nu^{(1)}, \nu^{(2)}\}$, defined above. Here, we will assume that the random actions are not too large on the average as compared to the determinate trend $\rho(W^t - \nu_t)$. The following statement is true.

**Statement 3.2.** Let $M\left(\xi^t/\nu_0, \ldots, \nu_t\right) = 0$, $M\left((\xi^t)^2/\nu_0, \ldots, \nu_t\right) \leq \rho^2\left(W^t - \nu_t\right)^2$ be satisfied for any $t$ with probability one. Let also $\rho < (2L)^{-1}$. Then, for the sequence $\left\{\nu_t, W^t\right\}$ defined according to (3.2) and (3.4) and for any $\varepsilon > 0$

$$P\left\{(W^t - \nu_t)^2 > \varepsilon\right\} \to 0$$

is satisfied as $t \to \infty$.

*Proof.* In view of (3.2), the boundedness of $\xi^t$, and inequality (3.5), the sequences $\{W^t\}$ and $\{\nu_t\}$ will be bounded with probability one. Indeed, according to (3.2) $W_0 \leq W^t \leq (S')^{-1}\left(Hl^{-1}\right)$ is always fulfilled, and due to the boundedness of $\xi^t$ the quantity $\rho W^t + \xi^t$ will also be bounded with probability one. Hence, the boundedness of the sequence $\{\nu_t\}$ from above follows. Its boundedness from below follows from (3.5). Similar to the proof of Statement 3.1, the following relation can be obtained:

$$F(\nu_{t+1}) = F(\nu_t) + F'\left(\nu_t + \lambda(\nu_{t+1} - \nu_t)\right)(\nu_{t+1} - \nu_t)$$

$$= F(\nu_t) + \left(F'\left(\nu_t + \lambda\left(\rho(W^t - \nu_t) + \xi^t\right)\right) - F'(\nu_t)\right)$$
$$\times \left(\rho(W^t - \nu_t) + \xi^t\right) + F'(\nu_t)\left(\rho F'(\nu_t) + \xi^t\right)$$
$$\geq F(\nu_t) + \rho\left(F'(\nu_t)\right)^2 - L\rho^2\left(F'(\nu_t)\right)^2$$
$$+ 2L\rho F'(\nu_t)\xi^t - L(\xi^t)^2 + F'(\nu_t)\xi^t.$$

From here, we have

$$M\left(F\left(\nu_{t+1}\right)/\nu_0, \ldots, \nu_t\right)$$
$$\geq F(\nu_t) + \rho\left(F'(\nu_t)\right)^2 - L\rho^2\left(F'(\nu_t)\right)^2 - LM\left(\left(\xi^t\right)^2/\nu_0, \ldots, \nu_t\right)$$
$$\geq F(\nu_t) + \rho\left(F'(\nu_t)\right)^2 - 2L\rho^2\left(F'(\nu_t)\right)^2,$$

where $F'(\nu_t) = (W^t - \nu_t)$. From latter it follows that

$$M\left(F(\nu_{t+1})\right) \geq M\left(F'(\nu_t)\right) + \left(\rho - 2L\rho^2\right)M\left(F'(\nu_t)\right)^2.$$

Since $\rho < \dfrac{1}{2L}$, there exists $k > 0$ such that $\rho \leq \dfrac{1}{2L} + k$ ; therefore,

$$M\left(F(\nu_{t+1})\right) \geq M\left(F(\nu_t)\right) + k\rho M\left(F'(\nu_t)\right)^2. \tag{3.6}$$

From the latter inequality it follows that the sequence $\{M\left(F(\nu_t)\right)\}$, which is bounded due to the boundedness of $\{\nu_t\}$, will be nondecreasing, and its limit as $t \to \infty$ exists. Therefore, from (3.6) it follows that

$$\lim_{t\to\infty} M\left(F'(\nu_t)\right)^2 = 0,$$

hence, in turn, the validity of Statement 3.2 immediately follows.

Thus, under the previous assumption, the presence of random actions will not change the behavior of convergence of the sequence $\{\nu_t\}$. The limit distribution of this sequence will be concentrated at the points $\left(\nu^{(1)}, \nu^{(2)}\right)$; their properties have been considered above. Therefore, the negative phenomena mentioned above also remain for this case.

## 3.3 Modeling of Employment and Growth Processes Under Variable Prices

The investigation of distinctive features of transition economies in the last decades of the twentieth century shows, on one hand, the prevalence of monopsonic waging (see [45]) and, on the other hand, a material effect of these mechanisms on the development of basic macroeconomic processes. The previously made analysis of models of economic dynamics allows us to draw the conclusion that if the prices of commodities and services are stable, then the monopsony (the monopolism of consumers) on the labor market leads to the stabilization at the point of equilibrium between payment for work and consumer demand [36]. Such stabilization is preceded by an economic recession or growth (depending on the initial state of the corresponding system) whose rate will be slowed down as the equilibrium state is approached. A system with a monopsonic labor market is characterized by low unemployment (theoretically, equal to zero) in its classical understanding (an excess of labor supply over labor demand), a nonuniform (in time) decrease in job with an outpacing decrease in payment for labor during a recession, and a decrease in the wage share (in GDP) under the conditions of economic growth. As it follows from the performed in Sect. 3.2 analysis of the models, a monopsonic employer is not interested in the increasing the productivity of labor, especially if such an increase requires additional expenditures.

It should be noted that these conclusions were drawn under the following two essential assumptions that, fully, are not inherent in an actual economy: the stability of prices and the monopoly of a manufacturer–employer in the market of goods and services. In this section, macroeconomic processes are analyzed without such assumptions. Therefore, we consider a macroeconomic model of a system with a monopsonic labor market and a market of goods and services that is mainly competitive. An analysis of the model being considered is performed and a semantic economic interpretation of the results obtained is given.

We now consider the behavior of a manufacturer of goods and services who is a monopolist–consumer in some isolated segment of the labor market. We denote by $S(W)$ the dependence between real wage fund $W$ and the amount of labor acquired. By analogy with the previous section we assume that $S(W)$ is continuous, monotone increasing, concave, and belongs to a space $C^k$, $k \geq 4$, i.e., is $k$ times continuously differentiable. We assume that $l$ is the productivity of labor, $H = 1+h$ ($h$ represents additional expenditures of the employer that are conditioned by the purchase of a unit of labor by him), $\alpha$ is the value-added per unit of product, and $\nu$ is the demand for the product manufactured by the monopolist–employer. As it was shown in the previous section with a view to obtaining the maximum income, this employer will fix the payment for labor at the level equal to

$$W = \min\Big(S^{-1}(\alpha l^{-1}\nu),\ (S')^{-1}(Hl^{-1})\Big), \tag{3.7}$$

where $(S')^{-1}$ is the function inverse to the derivative of $S(W)$.

We assume that the labor demand and payment for labor depend on the ever continuously varying time $t$ and that the parameters $l$, $\alpha$, $h$ are invariable in time. By analogy with [36] and Sect. 3.2, we assume that, at each succeeding moment $t + \Delta t$, the demand $\nu$ be the linear combination of the demand $\overline{\nu}$ at a moment of time $t$, which is corrected taking into account established prices, and the real value of the payment for the labor $W$ at the same moment of time:

$$\nu(t + \Delta t) = \Big((1 - \rho)\overline{\nu}(t) + \rho W(t)\Big)\Delta t, \tag{3.8}$$

where the coefficient $0 < \rho < 1$ does not depend on time and is sufficiently small. From the economic viewpoint, the first addend in (3.8) reflects the influence of previous savings and incomes that depend on the current sales volume and the second addend reflects the effect of the current employees' incomes. Taking into account the aforesaid, we can assume that the function $\overline{\nu}(t)$ consists of the following two parts and that one of them is inversely proportional to the current prices $p(t)$ and the other does not depend on these prices:

$$\overline{\nu}(t) = \frac{\nu(t)}{p(t)}\alpha + (1 - \alpha)\nu(t), \quad 0 < \alpha < 1. \tag{3.9}$$

Here, the first addend reflects savings in native currency and also the fixed incomes of non-employees (e.g., retirees and social security beneficiaries). The second addend reflects income obtained from commercial activity and foreign currency reserves.

Substituting (3.9) in (3.8) and passing to the limit as $\Delta_t \to 0$, we obtain the differential equation

$$\dot{\nu} = \left((1 - \rho)\alpha\Big(\frac{1}{p} - 1\Big) - \rho\right)\nu + \rho W, \tag{3.10}$$

where $W(t)$ is determined by (3.7) for the given time moment.

Note that the demand dynamics model investigated in [36] and in Sect. 3.2 can be considered as a special case of Eq. (3.10) in which $p = 1$.

In this case, in contrast to [36] and [46], the dynamics of prices is described by the Samuelson equation $\dot{p} = \lambda(\overline{D}(p) - \overline{S}(p))$ (see [62]), where $\overline{D}(p)$ is the current demand, $\overline{S}(p)$ is the commodity market supply and $\lambda > 0$ is some coefficient. We assume that, in addition to the supply created by the monopolist–employer, there exists an alternative supply created by other manufacturers on the commodity market and also by import of commodities (see [23]). Let the alternative supply function $\psi \in C^k$, $k \geq 4$, be a non-decreasing, upper-bounded, and concave one, i.e., it has all the properties of the competitive market supply function. Thus, in the model being considered, the commodity market consists of two parts, namely, the competitive and monopolistic ones. If the monopolistic component of supply exceeds the competitive one, then it is this component which determines the aggregate supply $\overline{S}(p) = \max(\psi(p), lS(W))$. From this, we obtain

$$\dot{p} = \lambda\Big(\nu - \max(\psi(p), lS(W))\Big). \tag{3.11}$$

Equations (3.7), (3.10), and (3.11) form the macroeconomic model being considered. Note that, in an actual economy, prices cannot be equal to zero and unlimited and the demand rate is always non-negative and bounded. Therefore, the following conditions must be added to the equations specified:

$$p^* \leq p \leq p^{**}, \quad 0 \leq \nu \leq \nu^*, \tag{3.12}$$

that must be fulfilled for any moment of time $t \geq 0$. In what follows, we will assume that the minimum allowable price level $p^*$ (corresponding to the prices of abandonment of production in the AS/AD model; see [23]), the maximum allowable price level $p^{**}$, and the maximum allowable demand rate $\nu^*$ are time-independent. We consider the same waging as in the previous section. Under the assumptions made, we will study the dynamics of solutions $p(t)$ and $\nu(t)$ of the equations obtained.

Let us denote $F(\nu) = \min\Big(S^{-1}(\alpha l^{-1}\nu), (S')^{-1}(Hl^{-1})\Big)$.

From (3.10) and (3.11), we obtain the following system of differential equations and will consider it in the domain specified by inequalities (3.12):

$$\begin{aligned} \dot{\nu} &= \left((1-\rho)\alpha\left(\frac{1}{p} - 1\right) - \rho\right)\nu + \rho F(\nu), \\ \dot{p} &= \lambda\Big(\nu - \max\Big(\psi(p), lS\big(F(\nu)\big)\Big)\Big). \end{aligned} \tag{3.13}$$

System (3.13) considered together with initial conditions $\nu(0) = \nu^0$, $p(0) = p^0$ is a non-standard mathematical model since its right-hand side contains the operations of search for the maximum and the minimum. To analyze the dynamics of the system (3.13), we will construct a vector field, namely, the field of rates of changes in $\big(p(t), \nu(t)\big)$. To this end, we find the curves on the plane $(p, \nu)$ that satisfy inequalities (3.12) and on which the right-hand side of the first or second equation of the system (3.13) becomes zero. The points at which the right sides of both equations of this system become zeros are the states of its equilibrium. The mentioned curves and equilibrium states play a key role in studying the dynamics of the system (3.13) in the phase space formed by all the initial conditions satisfying inequalities (3.12).

We denote

$$c = (S')^{-1}\left(Hl^{-1}\right), \quad \tilde{\nu} = lS(c), \quad g(p) = (1-\rho)\alpha\left(\frac{1}{p} - 1\right) - \rho.$$

The function $g(p)$ has the unique zero at the point $\overline{p}$, $0 < \overline{p} < 1$. If $p > \overline{p}$, then the function $g(p) < 0$ and monotonically decreases.

In investigating the dynamics of system (3.13), we will distinguish between the cases where the value of $\rho$ is small and where it is not small.

We start from the second case. We assume that $p^* > \overline{p}$.

Let us consider the curve $\hat{L}$ that is specified by the conditions

$$g(p)\nu + p\tilde{\nu} = 0, \quad p > p^*, \quad \text{and} \quad \tilde{\nu} \leq \nu \leq \nu^*.$$

It is obvious that $\hat{L}$ is the plot of a monotonously decreasing differentiable function defined on an interval $[p^*, \tilde{p}]$. In this case, we assume that $\tilde{p} > 1$. Let us consider the function $S$. It is obvious that its inverse function $S^{-1}$ is convex, differentiable, and unbounded above. The solutions of system (3.13) are further analyzed under the assumptions given below.

1. The set of solutions of the equation

$$g(p)\nu + \rho S^{-1}(\alpha l^{-1}\nu) = 0 \tag{3.14}$$

is not empty. Then, as it is easy to see, there exists $p_m > 0$ such that, for each $p > p_m$, Eq. (3.14) has two roots $0 < V_1(p) < V_2(p)$, and, at the same time, $V_1(p)$ is a monotonously decreasing and $V_2(p)$ is a monotonously increasing function of $p$. The functions $V_1(p)$ and $V_2(p)$ belong to a space $C^k$, $k \geq 4$. These functions assume identical values at the point $p_m$.
2. The inequality $p_m < \tilde{p}$ is true.
3. The inequality $V_1(p_m) < \tilde{\nu}$ is true.
4. The function $\psi(p)$ is bounded and there exists such $p_b > \tilde{p}$ that the conditions $\psi(p) < \tilde{\nu}$ and $\psi(p_b) = \tilde{\nu}$ are true for all $p^* \leq p < p_b$.
5. The equation

$$\psi(p) = \nu^* \tag{3.15}$$

is solvable. In this case $p_b^*$ is a solution of Eq. (3.15) such that $p_b^* < p^{**}$.
6. The set $L_S = \left\{(p, V_1(p)) : p \geq p_m,\ V_1(p) \geq \psi(p)\right\}$ is not empty.

Conditions 1–5 provide for the dissipativity of the system (3.13) on the rectangle (3.12). Hence, the solutions of the system (3.13) with the initial conditions from (3.12) are extendable to the positive semi-axis and are within the rectangle (3.12).

We denote

$$L_U = \left\{(p, V_2(p)) : p \geq p_m,\ V_2(p) \leq \tilde{\nu}\right\}, \quad \psi = \left\{(p, \psi(p)), p > p^*\right\}.$$

By virtue of conditions 1–5, the curves $\hat{L}$ and $L_U$ have a common point $(\tilde{p}, \tilde{\nu})$. Since the function $V_1(p)$ monotonously decreases, the function $\psi(p)$ monotonously decreases, and the set $L_s$ is not empty, the equation $V_1(p) = \psi(p)$ has a unique root $p_{\max} > p_m$.

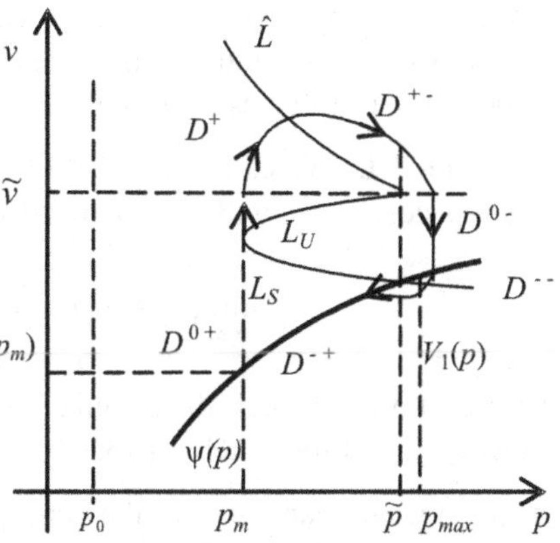


**Fig. 3.4** Phase space. Stabilization for $\lambda > 0$

We now turn back to the above-mentioned analogy with the model with three equilibrium points, which is considered in the previous section. Two of them (the upper one when the quantity $W$ depends on $H$, and the lower one when this quantity is determined by the demand $\nu$ with stable prices) are points of stable equilibrium, and the intermediate point is unstable. In this section, the points of the curve $\hat{L}$ are analogues of the upper equilibrium point, the points from $L_U$ are analogues of the intermediate point, and the points from $L_S$ are analogues of the lower point.

The curves $L_U$, $L_S$, $\hat{L}$, and $\psi$ subdivide the phase space of system (3.13) into domains with different behaviors of the vector field. We denote these domains by $D^{z_1,z_2}$, where $z_1$ is the sign of the right-hand side of the price equation or 0 and $z_2$ is the sign of the right-hand side of the demand equation or 0. The structure of the vector field is presented in Fig. 3.4. We describe the domains obtained.

The set $D^{++}$ is bounded by the segments $\left\{(p^*, \nu),\ \tilde{\nu} \leq \nu \leq \nu^*\right\}$, $\left\{(p, \tilde{\nu}),\ p^* \leq p \leq \tilde{p}\right\}$ and the curve $\hat{L}$. The boundary of the domain $D^{+-}$ is formed by the curve $\hat{L}$, an arc of the curve $\psi(p)$, and the segment $\left\{(p,\ \tilde{\nu}), \tilde{p}\ \leq\ p\ \leq\ p_b^*\right\}$. The domain $D^{0-}$ is bounded by the curve $L_S$, an arc of the curve $L_U$, an arc of the curve $\psi$, and the segment $\left\{(p,\ \tilde{\nu}), \tilde{p} \leq p \leq p_b^*\right\}$. The domain $D^{--}$ is adjacent to $D^{0-}$ and $D^{+-}$ and to the boundary $D^{--}$ belongs an arc of the curve $\psi(p)$ and the plot of the function $V_1(p)$. The curvilinear sector $D^{-+}$ is bounded by a part of the plot of the function $V_1(p)$ and an arc of the curve $\psi(p)$. The domain $D^{0+}$ is a set whose boundary contains an arc of the curve $\psi(p)$, an arc of the curve $L_U$, the curve $L_S$, and the segment $\left\{(p, \tilde{\nu}),\ p^* \leq p \leq \tilde{p}\right\}$.

We denote by $L$ the set of all the stationary points of the system (3.13). It is obvious that we have $L = L_S \bigcup L_U$ and that, for the points of the curve $\hat{L}$, the right–hand side of the price dynamic equation remains positive. Let $V_1(p_m) = V_m$. The set $L_S$ considered without the point $(p_m, V_m)$ is the set of all stable equilibrium states of the system (3.13). The set $L_U$ consists of unstable equilibrium states of this system.

We denote by $p(t) = p(t, p, \nu)$, $\nu(t) = \nu(t, p, \nu)$ the solution of the system (3.13); this solution satisfies the conditions $p(0) = p$ and $\nu(0) = \nu$, where the point $(p, \nu)$ satisfies conditions (3.12).

**Definition 3.1.** We call system (3.13) stabilizable if, for any point $(p, \nu)$ of the phase space of system (3.13), there exists a point $(p_+, \nu_+)$ such that we have

$$\lim_{t\to\infty}\Big(p(t,p,\nu),\ \nu(t,p,\nu)\Big) = (p_+, \nu_+). \tag{3.16}$$

**Definition 3.2.** We call the stabilization set of the system (3.13) the set of all the points $(p_+, \nu_+)$ for which Eq. (3.16) is true.

Note that we can consider the concept of a stabilizable system as an analogy of the convergence concept for a system with discrete time used in [36] and the concept of a stabilization set as an analogy of the limiting points set concept.

Let us find the stabilization conditions of the system (3.13). We assume that $\tilde{p} > p_m$. Semantically, this assumption means that, for different prices, the commodity market supply can be determined by the amounts of products provided both the monopolist–employer and the alternative competitive supply. This segment of the market cannot be considered as completely closed or as completely open to competition.

It is obvious that the system (3.13) belongs to the class of piecewise smooth systems. In what follows, we use the terminology from the theory of dynamic systems, which is commonly accepted in the classical (smooth) case; see [1, 42]. We will not stipulate the choice of some term or another in view of the existing analogy. Let us consider a separatrix going out from the point $(p_m, V_m)$. We denote a phase point moving along the separatrix by $\big(p^s(t), \nu^s(t)\big)$. Let $\big(p^s(0), \nu^s(0)\big) \in D^{0+}$. Then we have $p^s(t) = p^s(0)$, and $\nu^s(t)$ monotonously increases. From the economic viewpoint, this means that the prices remain unchangeable with increasing demand; hence, the expansion in production volume is proportional to demand. We note that such processes are inherent in the growth stage of the classical business cycle. After intersecting the right line $\nu = \tilde{\nu}$, the payment for labor ceases to change proportionally to demand, an imbalance between supply and demand takes place, and this imbalance results in inflation. In the domain $D^{++}$, the functions $p^s(t)$ and $\nu^s(t)$ are monotonously increasing. A growth accompanying by inflation is inherent in the boom stage of the classical business cycle. A rise in prices decelerates demand. After transverse intersection of the curve $\hat{L}$, the phase point passes to the domain $D^{+-}$, where $p^s(t)$ monotonously increases and $\nu^s(t)$ monotonously decreases. By this time, the boom is replaced by a recession, which is aggravated by a depreciation of savings through inflation. If the phase point intersects the right line $\nu = \tilde{\nu}$, then the payment for labor is beginning to change again proportionally to demand. As a consequence, the feedback

"a decrease in demand–a decrease in payment for labor–a decrease in demand," which is considered in the previous section, arises and accelerates the recession. In the domain $D^{0-}$, the function $p^s(t)$ remains constant, $\nu^s(t)$ monotonously decreases, and the recession assumes features of overproduction crisis. In the domain $D^{--}$, to which the phase point passes after intersecting the line $\psi(p)$, the oversupply is beginning to impact on the prices, and $p^s(t)$ and $\nu^s(t)$ will be monotonously decreasing functions. The incipient deflation (a decrease in prices) decelerates the recession and increases the value of retained assets. The combination of deflation and overcooling is typical for the depression stage of the classical business cycle. After intersecting the plot of the function $V_1(p)$, the phase point passes to the domain $D^{-+}$, where $p^s(t)$ monotonously decreases and $\nu^s(t)$ increases. The processes that run in this domain are typical for the beginning of the revival stage of the business cycle. Thus, the order of passage of the specified domains of the phase space by the phase point corresponds to the order of all the basic stages of business activity.

At some moment $t = T^s$, the phase point $(p^s(t), \nu^s(t))$ intersects the curve $\Psi$. It follows from the condition $\tilde{p} > p_m$ that we have $p^s(T^s) < p_{\max}$. The following two cases are possible: $p^s(T^s) \le p_m$ and $p^s(T^s) > p_m$. We first consider the second case. In the domain $D^{0+}$, we have $p^s(t) = p^s(T^s) = \hat{p}$ and $\nu^s(t) \to V_1(\hat{p}) = \hat{V}$ as $t \to \infty$. Then the separatrix enters into a point $(\hat{p}, \hat{V}) \in L_S$. This case is presented in Fig. 3.1. We denote $\lambda = \hat{p} - p_m$. It is obvious that, in this case, we have $\lambda > 0$. In the first case, the system is not stabilizable when $\lambda = p^s(T^s) - p_m < 0$. This case is presented in Fig. 3.5. We note that, for the particular case where $\lambda = 0$, the system (3.13) is also unstabilizable and the separatrix outgoing from the point $(p_m, V_m)$ enters the system. Hereafter, we will call such a curve homoclinic.

It is presented in Fig. 3.6.

To the movement of the point along this curve corresponds the basic business cycle in its "pure" form (without long-term economic growth).

It is obvious that the quantity $\lambda$ does not depend upon the choice of the initial point on the separatrix. We assume that we have $\tilde{p} > p_{\max}$. Then the case is possible where $\left(p^s(t), \nu^s(t)\right)$ intersects the right line $\nu = \tilde{\nu}$ at the point $(p^s(T^s), \nu^s(T^s))$

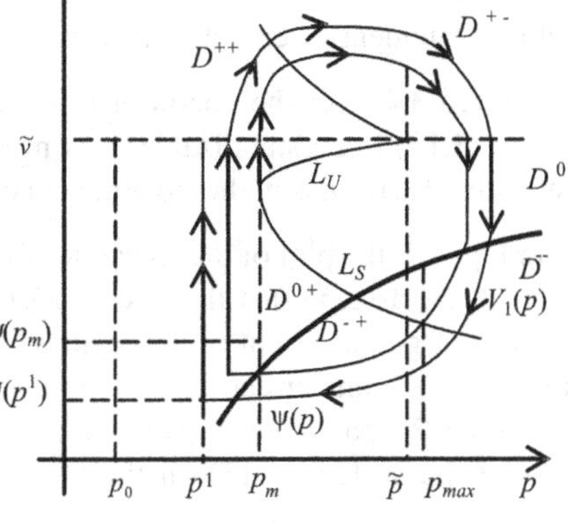


**Fig. 3.5** Cycles for $\lambda < 0$

Fig. 3.6 Homoclinic curve for $\lambda = 0$

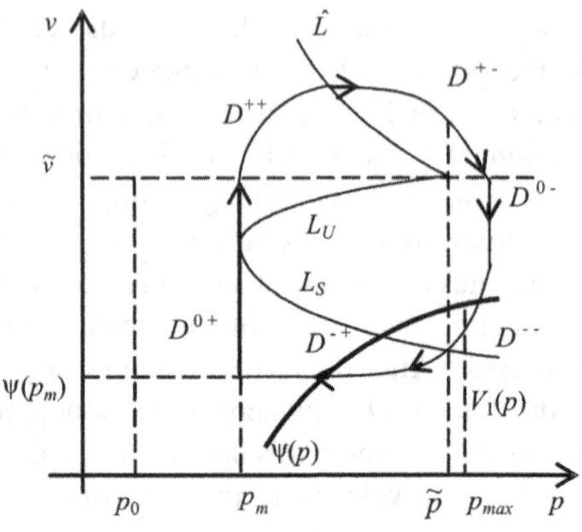


such that $p^s(T^s) < p_{\max}$. In this case, the separatrix also enters a point $\left(\hat{p}, \hat{V}\right) \in L_S$ and we have $\lambda > 0$. From the above reasoning, we obtain the statements formulated below.

**Statement 3.3.** Let assumptions 1–5 be fulfilled. In order that the system (3.13) be stabilizable, it is necessary that the separatrix outgoing from the point $(p_m, V_m)$ enter into a point that belongs to $L_S$ and does not coincide with the point $(p_m, V_m)$.

Let us find the stability criterion of the system (3.13). To this end, we consider an arbitrary point $(p, \psi(p))$ such that $p < p_m$. Denote by $p(t) = p(t, p)$, $v(t) = v(t, p)$ the solution of the system (3.13) that satisfies the condition $p(0) = p$, $v(0) = \psi(p)$.

By analogy with above reasoning, we arrive at the conclusion that there exists the least positive $T^1$ such that the point $\left(p(t), v(t)\right)$ intersects the curve $\Psi$. We denote $p(T^1) = P(p)$.

**Definition 3.3.** We call a function $p \to P(p)$ defined on the interval $[p^*, p_m)$ the successor function of system (3.13).

Under the above conditions, the successor function exists and is differentiable over the interval $[p^*, p_m)$. It follows from the definition of the successor function that the statement given below is true.

**Statement 3.4.** Let the conditions of Statement 3.3 be fulfilled. In order for the system (3.13) to be stabilizable, it is necessary and sufficient that $P(p) > p$ and the system (3.13) has no homoclinic curve.

In Fig. 3.7, the plot of the successor function of the stabilizable system is given.

Next, we assume that the system (3.13) depends on a parameter $\varepsilon$ and study a scenario of getting it out of the regime of stabilization. Note that the choice of the bifurcation parameter can be ambiguous. In particular, the functions $\psi$ and $S$ can depend on the parameter $\varepsilon$. We also assume that the dependence of the right-hand sides of the system (3.13) on the parameter $\varepsilon$ is smooth. We write the obtained

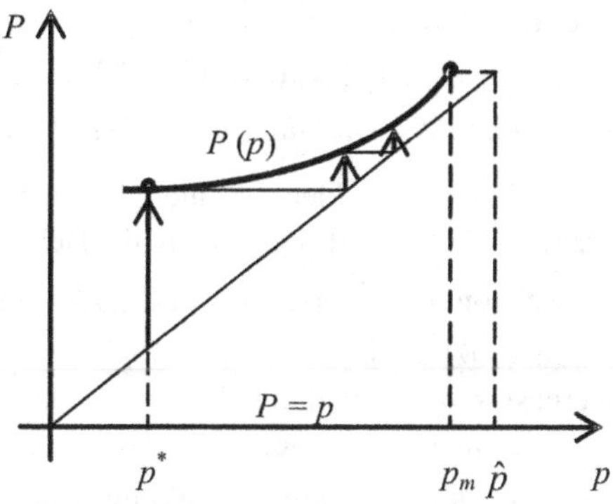


**Fig. 3.7** The successor function of the stabilizable system

one-parametric family of the differential equations (3.13) in the form

$$\dot{x} = f(x, \varepsilon), \quad \varepsilon \in R, \tag{3.17}$$

where $x = (p, \nu)$. We assume that all the functions defined above are differentiable with respect to the parameter $\varepsilon$; hence, the change in the position of the corresponding curves will be sufficiently "smooth."

Let us consider the successor function $P(p, \varepsilon)$ and the quantity $\lambda(\varepsilon)$ that characterize the possibility of stabilization of the system (3.17). Let the system (3.17) be stabilizable when $\varepsilon < \varepsilon_0$. The beginning of getting out of the regime of stabilization is connected with the fulfillment of one of the conditions presented below.

**Condition 1** *Let, when $\varepsilon = \varepsilon_0$, we have*

$$\lambda(\varepsilon_0) = 0, \ \lambda'(\varepsilon_0) < 0, \ P(p, \varepsilon_0) > p$$

*for $p^*(\varepsilon_0) < p < p_m(\varepsilon)$.*

**Condition 2** *There exists a unique $p^0$ such that we have*

$$P(p^0, \varepsilon_0) = p^0, \quad \frac{\partial}{\partial p} P(p^0, \varepsilon_0) = 1, \quad \frac{\partial}{\partial \varepsilon} P(p^0, \varepsilon_0) > 0,$$

$$\frac{\partial^2}{\partial^2 p} P(p^0, \varepsilon_0) < 0, \quad \lambda(\varepsilon_0) > 0.$$

We assume that condition 1 is fulfilled. Consider the system (3.17) when $\varepsilon = \varepsilon_0$. We denote by $G$ the homoclinic curve outgoing from the point $\left(p_m(\varepsilon_0), V_m(\varepsilon_0)\right)$. The statement formulated below is true.

**Theorem 3.1.** *Let condition 1 be true. Then there exist some $\delta > 0$ and a vicinity $U$ of the set $G \bigcup \big(p_m(\varepsilon_0), V_m(\varepsilon_0)\big)$ such that the system* (3.17) *has an orbitally asymptotically stable limiting cycle in $U$ when $\varepsilon_0 < \varepsilon < \varepsilon_0 + \delta$.*

*Proof.* Let us consider a sequence $p_k^s(\varepsilon)$, $k = 1, 2, \ldots$, when $\varepsilon_0 < \varepsilon < \varepsilon_0 + \delta$, where $\delta > 0$ is sufficiently small. Here, $p_k^s(\varepsilon)$ is the first coordinate of the $k$th intersection of the separatrix of Eq. (3.17) that emanates from $\big(p_m(\varepsilon), V_m(\varepsilon)\big)$ with the curve $\psi(p, \varepsilon)$. It is obvious that the sequence $p_k^s(\varepsilon)$, $k = 1, 2, \ldots$, is monotone decreasing one bounded below and, hence, has a limit, which is denoted by $p^+(\varepsilon)$. Based on an unsophisticated analysis, we can conclude that $p^+(\varepsilon) \to p_m(\varepsilon_0)$ when $\varepsilon \to \varepsilon_0$. The phase curve outgoing from the point $\big(p^+(\varepsilon), \psi(\varepsilon, p^+(\varepsilon)\big)$ is closed and orbitally asymptotically stable. Condition 1 implies that, when $\varepsilon_0 < \varepsilon < \varepsilon_0 + \delta$, there exists $p^-(\varepsilon) \le p^+(\varepsilon)$ such that the inequality $P(\varepsilon, p) > p$ is true for all $p^*(\varepsilon) < p < p^-(\varepsilon)$ and, at the same time, we have $P(\varepsilon, p^-(\varepsilon)) = p^-(\varepsilon)$. The phase curve outgoing from the point $\big(p^-(\varepsilon), \psi(\varepsilon, p^-(\varepsilon)\big)$ is closed and orbitally asymptotically stable. Thus, the theorem is proved.

In the case of a general position, a unique limiting cycle apparently bifurcates from the homoclinic curve. Then it is obvious that the number of limiting cycles occurring in the vicinity $U$ is determined by the behavior of $P(\varepsilon, p)$ when $p \to p_m(\varepsilon)$ and $\varepsilon \to \varepsilon_0$. Next, for simplicity, we assume that a unique limiting cycle bifurcates from a homoclinic curve. We note that, when $\varepsilon_0 < \varepsilon < \varepsilon_0 + \delta$ and $\delta$ is small, the period of the limiting cycle increases without bound if $\delta \to 0$, i.e., the duration of each business cycle stage increases with approximating to the crisis value $\varepsilon = \varepsilon_0$.

Using the results obtained above, we describe the distinctive features of getting out of the regime of stabilization when condition 1 is fulfilled. For the system of Eq. (3.17), for a fixed $\varepsilon < \varepsilon_0$, the set of stabilization is $L(\varepsilon) = L_S(\varepsilon) \bigcup L_U(\varepsilon)$. At the "physical level," the set of stabilization of Eq. (3.17) is $L_S(\varepsilon)$ when $\varepsilon < \varepsilon_0$ since the points of the set $L_U(\varepsilon)$ are not realized in view of their instability. System (3.17) is unstabilizable when $\varepsilon = \varepsilon_0$. If $\varepsilon = \varepsilon_0$, then the probabilistically limiting set of Eq. (3.17) is $L_S(\varepsilon_0) \bigcup G$. We note following [2] that the probabilistically limiting set of a dynamic system is understood to be the least closed set that contains $\omega$, i.e., the limiting sets for almost all the points of the phase space. As is obvious, when $\varepsilon < \varepsilon_0$, the probabilistically limiting set of Eq. (3.17) is $L_S(\varepsilon)$. Hence, the value $\varepsilon = \varepsilon_0$ is the crisis bifurcation value; see [2]. In this case, the nature of such a crisis is obvious and is as follows: when $\varepsilon \to \varepsilon_0$, for the points of the phase space of Eq. (3.17) that are outside of the closed curve consisting of the separatrix outgoing from the point $\big(p_m(\varepsilon), V_m(\varepsilon)\big)$ and an arc of $L_S(\varepsilon)$, the measure of the set of points of stabilization tends to zero ($\lambda(\varepsilon) \to 0$). When $\varepsilon < \varepsilon_0$, for the points that are the within the above-mentioned closed curve, the probabilistically limiting set is $L_S(\varepsilon)$. When $\varepsilon = \varepsilon_0$, for the domain $Q(\varepsilon_0)$ that belongs to the phase space and is outside of the closure of the homoclinic curve $\overline{G}$, the probabilistically limiting set of Eq. (3.17) is $\overline{G}$. The semi-trajectories outgoing from $Q(\varepsilon_0)$ approximate to the closed curve $\overline{G}$ and wind around it. A cyclic development of economy corresponds to this.

To investigate the dynamics of the system of Eq. (3.17) when $\varepsilon > \varepsilon_0$, we will consider the separatrix $\Gamma(\varepsilon)$ entering into the point $\left(p_m(\varepsilon), V_m(\varepsilon)\right)$. We denote by $Q(\varepsilon)$ the domain of the phase space that is outside of the closed curve consisting of $\Gamma(\varepsilon)$ and an arc of the curve $L_U(\varepsilon)$. In the domain $Q(\varepsilon)$, the probabilistically limiting set of Eq. (3.17) is the limiting cycle defined in Theorem 3.1. In this domain, the probabilistically limiting set of Eq. (3.17) and maximum attractor coincide. For this case, the maximum attractor is defined below; see [2].

**Definition 3.4.** Let $\{\varphi^t\}$ be the flow generated by a vector field $f$. A domain $B$ is called absorbing if we have $\varphi^t B \subset B,\ t > 0$. The set $A = \bigcap_{t\to\infty} \varphi^t B$ is called the maximum attractor in the absorbing domain $B$. A set is called an attractor if there exists an absorbing domain such that its maximum attractor is this set. The set of all the points through which the trajectories tending to $A$ pass as $t \to \infty$ is called the domain of attraction of the attractor.

In contrast to the case considered above, we assume that the system gets out of the regime of stabilization depending on condition 2. Then, when $\varepsilon = \varepsilon_0$, the system (3.17) contains a semistable cycle. The probabilistically limiting set of the system of Eq. (3.17) consists of the union of the mentioned cycle and $L_S(\varepsilon)$. Hence, as well as in the case considered above, the value $\varepsilon = \varepsilon_0$ is the crisis bifurcation value. It is obvious that, when $\varepsilon_0 < \varepsilon < \varepsilon_0 + \delta$, two limiting cycles bifurcate from the semistable cycle, one of which is orbitally asymptotically stable and the second is unstable. The unstable cycle is within the stable cycle. The probabilistically limiting set of Eq. (3.17) consists of the union of the stable cycle and $L_S(\varepsilon)$. The domain of attraction of the set of stationary points in Eq. (3.17) is the interior of the unstable cycle. All the positive semi-trajectories that belong to the exterior of the unstable cycle approximate to the stable cycle and wind around it.

Thus, business cycles are repeated with a tendency to the total economic growth if the trajectory is within the stable cycle or to a recession if it is outside of it.

When the system (3.17) becomes unstable as a result of the passage through the crisis bifurcation value $\varepsilon = \varepsilon_0$, further possible qualitative changes in this system are of interest. Let us consider a possible variant. We assume that, after getting out of the regime of stabilization, condition 2 is true and, for $\varepsilon > \varepsilon_0$, there are two cycles in the system (3.17), one of which, namely, $\Lambda_S(\varepsilon)$ is orbitally asymptotically stable and the other, namely, $\Lambda_U(\varepsilon)$ is unstable. It is obvious that the exterior of the closed curve $\Lambda_U(\varepsilon)$ is the domain of attraction of the cycle $\Lambda_S(\varepsilon)$. The interior of the cycle $\Lambda_U(\varepsilon)$ is the domain of attraction of $\Lambda_S(\varepsilon)$. Let us consider a scenario of getting the system (3.17) out of the regime of stabilization in the field of attraction of the set $L_S(\varepsilon)$. We denote by $p_S(\varepsilon) < p_m(\varepsilon)$ and $p_U(\varepsilon) < p_m(\varepsilon)$ the first coordinates of the points of intersection of the sets $\Lambda_S(\varepsilon)$ and $\Lambda_U(\varepsilon)$, respectively, with the curve $\psi(p, \varepsilon)$. We assume that there exists $\varepsilon_1 > \varepsilon_0$ such that, for $\varepsilon_0 < \varepsilon < \varepsilon_1$, the system (3.17) has no homoclinic curve and, when $\varepsilon = \varepsilon_1$, its homoclinic curve is $G_1$. We assume that the successor function of the system (3.17) has a unique fixed point on the interval $\left(p_S(\varepsilon), p_m(\varepsilon)\right)$. Then the following two cases are possible:

1. $p_m(\varepsilon) - p_U(\varepsilon) \geq d > 0, \quad \varepsilon_0 < \varepsilon < \varepsilon_1;$
2. $p_U(\varepsilon) \to p_m(\varepsilon), \quad \varepsilon \to \varepsilon_1.$

In the first case, Theorem 3.1 can be used according to which an orbitally asymptotically stable cycle bifurcates from a homoclinic curve.

In the second case, the unstable limiting cycle merges with the homoclinic curve and they annihilate each other. In this case, the limiting cycle in a vicinity of the homoclinic curve is absent. The probabilistically limiting set of Eq. (3.17) is the set $\Lambda_S(\varepsilon) \bigcup L_S(\varepsilon)$. The domain of attraction of $\Lambda_S(\varepsilon)$ is the exterior of the closed invariant curve of Eq. (3.17); this curve is composed of the separatrix $\Gamma(\varepsilon)$ entering into the point $\left(p_m(\varepsilon), V_m(\varepsilon)\right)$ and an arc of the curve $L_U(\varepsilon)$. Within the mentioned closed curve, the probabilistically limiting set of the system (3.17) is the set $L_S(\varepsilon)$. The value $\varepsilon = \varepsilon_1$ is the crisis bifurcation value.

The case is possible where a bifurcation of a homoclinic curve is preceded by several bifurcations producing semistable limiting cycles and their branching. The above scenario of the bifurcation of a homoclinic curve holds true in this case.

We now pass to the bifurcational analysis of the limiting case where $L_S(\varepsilon)$ degenerates into the point $(\tilde{p}, \tilde{v})$ and $L_U(\varepsilon)$ is the empty set (Fig. 3.8).

This case is realized if, in the course of changes, the right line $v = \tilde{v}$ passes through the point of intersection of $\psi(\varepsilon, p)$ and $V_1(\varepsilon, p)$. In what follows, for simplicity, we will consider that $(\tilde{p}, \tilde{v})$ does not depend on the bifurcation parameter $\varepsilon$. The point $(\tilde{p}, \tilde{v})$ is the unique equilibrium position of the family of Eq. (3.17). The structure of the vector field of Eq. (3.17) is retained as a whole. Here, a novelty aspect is the disappearance of the domain $D^{0-}$. Phase points

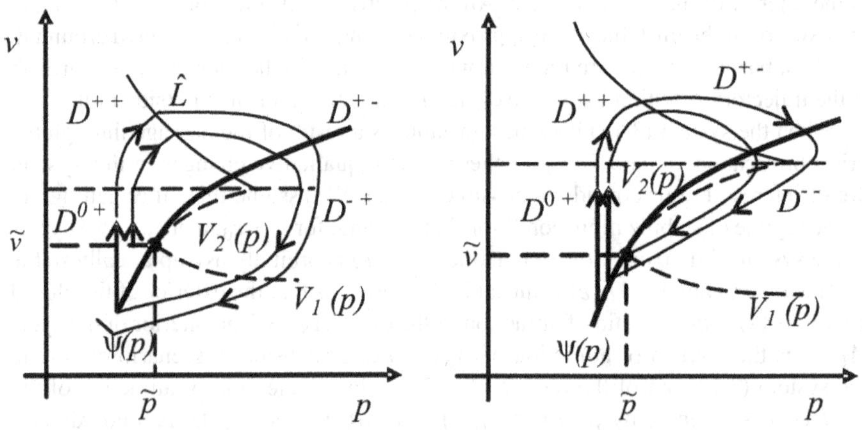


**Fig. 3.8** The case where $L_S(\varepsilon) = (\tilde{p}, \tilde{v})$. The phase curve that folds a spiral (**a**) and case where $L_S(\varepsilon) = (\tilde{p}, \tilde{v})$. The phase curve that unfolds a spiral (**b**)

moving along phase curves describe spirals. In the case of a general position, these spirals are folded into the point $(\tilde{p}, \tilde{\nu})$ (Fig. 3.8a) or are unfolded and move away from it (Fig. 3.8b). In the first case, $(\tilde{p}, \tilde{\nu})$ is the stable focus and, in the second case, this point is an unstable focus. Let us consider bifurcation producing the limiting cycle beginning with the equilibrium position in passing from the stable oscillatory regime to an unstable one. Well-known results on such a bifurcation (e.g., [2, 19, 37]) are connected with smooth systems. In the case considered, the family of Eq. (3.17) is piecewise smooth in a vicinity of $(\tilde{p}, \tilde{\nu})$. This circumstance substantially complicates the bifurcational analysis even at its first stage, namely, in linearizing system (3.13) in $(\tilde{p}, \tilde{\nu})$. System (3.17) linearized in $(\tilde{p}, \tilde{\nu})$ is a system of linear equations with piecewise constant coefficients on a plane. We write this system in the form of systems of linear differential equations on the plane that are pairwise interrelated with one another by common linear boundaries. After the shift of $(\tilde{p}, \tilde{\nu})$ to the origin of coordinates and linearization, we obtain the following equations and domains in which they are defined:

the equations

$$\begin{aligned} \dot{\nu} &= \left((1-\rho)a\left(\frac{1}{\tilde{p}}-1\right)-\rho\right)\nu-(1-\rho)a\,\frac{\tilde{\nu}}{\tilde{p}^2}p, \\ \dot{p} &= \lambda\nu \end{aligned} \tag{3.18}$$

in the domain $D_1 = \Big\{\big(\nu \geq 0,\ p \leq 0\big)\Big\} \bigcap \Big\{\big(\nu \geq \psi'(\tilde{p})\,p,\ p \geq 0\big)\Big\}$;

the equations

$$\begin{aligned} \dot{\nu} &= \left((1-\rho)a\left(\frac{1}{\tilde{p}}-1\right)-\rho\right)\nu-(1-\rho)\,a\,\frac{\tilde{\nu}}{\tilde{p}^2}\,p, \\ \dot{p} &= \lambda\big(\nu-\psi'(\tilde{p})p\big) \end{aligned} \tag{3.19}$$

in the domain $D_2 = \Big\{\big(0 \leq \nu \leq \psi'(\tilde{p})p,\ p \geq 0\big)\Big\}$;

the equations

$$\begin{aligned} \dot{\nu} &= \left(\left((1-\rho)a\left(\frac{1}{\tilde{p}}-1\right)-\rho\right)+\rho l^{-1}(S^{-1})'(\alpha l^{-1}\tilde{\nu})\right)\nu \\ &\quad -(1-\rho)a\frac{\tilde{\nu}}{\tilde{p}^2}p, \\ \dot{p} &= \lambda\big(\nu-\psi'(\tilde{p})p\big) \end{aligned} \tag{3.20}$$

in the domain $D_3 = \Big\{(\nu \leq 0, p \geq 0)\Big\} \bigcap \Big\{(\nu \leq \psi'(\tilde{p})p,\ p \leq 0)\Big\}$;

the equations

$$\begin{aligned}\dot{\nu} &= \left(\left((1-\rho)a\left(\frac{1}{\tilde{p}}-1\right)-\rho\right)+\rho l^{-1}(S^{-1})'(\alpha l^{-1}\tilde{\nu})\right)\nu \\ &\quad -(1-\rho)a\frac{\tilde{\nu}}{\tilde{p}^2}p, \\ \dot{p} &= 0\end{aligned} \tag{3.21}$$

in the domain $D_4 = \Big\{(0 \le \nu \le \psi'(\tilde{p})p,\ p \le 0)\Big\}$.

In what follows, for brevity, we will speak of systems (3.18)–(3.21) without specifying the domain of their definition.

Let us find the boundary of the domain of stability of the zero solution of the systems (3.18)–(3.21) (we recall that the point $(\tilde{p}, \tilde{\nu})$ of the initial model corresponds to zero in the new coordinates). Denote by $p(t, p)$, $\nu(t, p)$ the point that is the solution of the system (3.18) and satisfies the condition $p(0, p) = p$, $\nu(0, p) = 0$, where $p < 0$ (in the new coordinates). Let $T_1(p) > 0$ be the time during which the point $(p(t, p), \nu(t, p))$ intersects the beam $\nu = \psi'(\tilde{p})p$. Let $p(t, p)$, $\nu(t, p)$, be the solution of the system (3.19), and let this solution be a continuous extension of the solution constructed above and be defined on the interval $[T_1(p), T_2(p)]$, where $T_2(p)$ is the time during which the point $\big(p(t, p), \nu(t, p)\big)$ intersects the beam $\nu = 0$, $p > 0$. Similar reasoning allows us to obtain the time $T(p)$ during which the point $\big(p(t, p), \nu(t, p)\big)$ reaches the half-line $\nu = 0$, $p < 0$. It is obvious that the equality $p\big(T(p), p\big) = kp$ is true, where $k > 0$ is a constant that does not depend on $p < 0$ but depends on the parameters of the systems (3.18)–(3.21). The origin of coordinates is the stable focus of the systems (3.18)–(3.21) if $k < 1$ and its unstable focus if $k > 1$. The set of the parameter values for which $k = 1$ determines the boundary of the stability domain of the zero solution of the systems (3.18)–(3.21) in the parameter space. These parameter values are bifurcational for the system (3.17) in a vicinity of the point $(\tilde{p}, \tilde{\nu})$.

The foregoing shows that the bifurcational analysis for the system (3.17) in a vicinity of the point $(\tilde{p}, \tilde{\nu})$ is a complicated problem. We will perform the bifurcational analysis for the system (3.17) in a vicinity of the point $(\tilde{p}, \tilde{\nu})$ if $\rho$ is sufficiently small. We also assume that the function $\psi$ depends on $\rho$, $\psi = \psi(p, \rho)$. Note that the system (3.13) has the equilibrium state $p = 1$, $\nu = \psi(1, 0)$ when $\rho = 0$. Simple reasoning allows one to draw the conclusion that if $\rho > 0$ and is sufficiently small, then there exists the equilibrium state $\big(\tilde{p}(\rho), \tilde{\nu}(\rho)\big)$, of the system (3.13) and this state smoothly depends on $\rho > 0$ and is such that $\tilde{p}(0) = 1$, $\tilde{\nu}(0) = \psi(1, 0)$. We assume that $\psi'_p(1, 0) = 0$. For definiteness, we put $\lambda = 1$. We introduce the following notations:

$$(1-\rho)a\Big(\frac{1}{\tilde{p}(p)}-1\Big)-\rho=-\mu_1(\rho),$$

$$\rho l^{-1}S^{-1}(\alpha l^{-1}\tilde{\nu})=\mu_2(\rho),$$

$$(1-\rho)a\frac{\tilde{\nu}}{\tilde{p}^2}=\omega^2(\rho).$$

Here, we have $\omega(0) > 0$. The bifurcational analysis of the system (3.13) in a vicinity of the point $(\tilde{p}, \tilde{\nu})$ (hereafter, we will not emphasize the dependence of $(\tilde{p}, \tilde{\nu})$ on the parameter $\rho$ with a view to abbreviating notations) is reduced, under the conditions formulated above, to the analysis of a system of the form

$$\begin{aligned}\dot{\nu}&=-\omega^2(\rho)p-\mu_1(\rho)\nu+(p^2-p^3)a_1-(p-p^2)\nu+\cdots,\\ \dot{p}&=\nu,\end{aligned}\tag{3.22}$$

when $\nu \geq 0$ and

$$\begin{aligned}\dot{\nu}&=-\omega^2(\rho)p+\Big(\mu_2(\rho)-\mu_1(\rho)\Big)\nu+(p^2-p^3)a_1-(p-p^2)\nu+\cdots,\\ \dot{p}&=\nu+b_1p^2-b_2p^3+\cdots,\end{aligned}\tag{3.23}$$

when $\nu \leq 0$, here

$a_1=\psi(1,0)=\tilde{\nu}(0), \quad b_1=-\frac{1}{2}\psi''_{pp}(1,0)>0, \quad b_2=\frac{1}{6}\psi'''_{ppp}(1,0).$

Here and in the sequel, omission points denote the terms whose order of smallness is higher than that of the addends shown above.

We denote $\mu(\rho)=\mu_2(\rho)-2\mu_1(\rho)$ and

$$K=\frac{1}{8}-\frac{3b_2}{16}+\frac{(a_1-1)b_1}{8\omega(0)}.\tag{3.24}$$

The following statement is true.

**Theorem 3.2.** *Let we have* $\psi'_p(1,0)=0$, $\psi''_{pp}(1,0)=0$, $\mu'(0)=K\neq 0$, *and* $K<0$. *Then there exists* $\delta_0>0$ *such that we have*

**(i)** *if* $k>0$,

*then, in a* $\delta_0$*–vicinity of* $(\tilde{p}, \tilde{\nu})$, *there exists a periodic solution of system* (3.13); *this solution is unique up to shifts in* $t$, *its period is* $\frac{2\pi}{\hat{\omega}(\rho)}$, $\hat{\omega}(\rho)=\omega(\rho)+o(\rho)$, *and it is of the form*

$$\hat{\nu}(t) = \tilde{\nu} - \frac{1}{2}\left(\frac{k\rho}{-K}\right)^{1/2} \hat{\omega}(\rho) \sin \hat{\omega}(\rho)t + \cdots,$$
$$\hat{p}(t) = \tilde{p} + \frac{1}{2}\left(\frac{k\rho}{-K}\right)^{1/2} \cos \hat{\omega}(\rho)t + \cdots. \qquad (3.25)$$

*This solution is exponentially orbitally stable. The equilibrium state* $(\tilde{p}, \tilde{\nu})$ *of the system* (3.13) *is unstable;*

**(ii)** *if* $k < 0$,

*then, in a* $\delta_0$*–vicinity of* $(\tilde{p}, \tilde{\nu})$*, there exist no nonwandering points distinct from* $(\tilde{p}, \tilde{\nu})$*. The equilibrium state* $(\tilde{p}, \tilde{\nu})$ *of the system* (3.13) *is exponentially stable.*

*Proof.* Following [19], we will use the averaging method (see [12]) to prove the theorem. We introduce $2\pi$-periodic functions $\mu(\theta, \rho)$ and $\eta(\theta)$ as follows:

$$\mu(\theta, \rho) = \begin{cases} -\mu_1(\rho), & 0 \le \theta \le \pi, \\ \mu_2(\rho) - \mu_1(\rho), & \pi < \theta < 2\pi, \end{cases} \qquad \eta(\theta) = \begin{cases} 0, \ 0 \le \theta \le \pi, \\ 1, \ \pi < \theta < 2\pi. \end{cases}$$

In the system (3.22), (3.23), we make the following substitution:

$$\nu = -\omega(\rho)\, r \sin\ \theta,$$
$$p = r \cos\ \theta.$$

As a result, we obtain the following system of equations:

$$\dot{r} = \mu(\theta, \rho)\, r \sin^2 \theta + r^2 C_3(\theta) + r^3 C_4(\theta) + \cdots,$$
$$\dot{\theta} = \omega(\rho) + r D_3(\theta) + \cdots, \qquad (3.26)$$

where

$$C_3(\theta) = -\frac{a_1}{\omega(0)} \cos^2 \theta \sin \theta - \cos \theta \sin^2 \theta + b_1 \eta(\theta) \cos^3 \theta,$$
$$C_4(\theta) = -\frac{a_1}{\omega(0)} \cos^3 \theta \sin \theta + \cos^2 \theta \sin^2 \theta - b_2 \eta(\theta) \cos^4 \theta, \qquad (3.27)$$
$$D_3(\theta) = -\frac{a_1}{\omega(0)} \cos^3 \theta - \cos^2 \theta \sin \theta - b_1 \eta(\theta) \cos \theta \sin^2 \theta.$$

Following [19], we construct the substitution of variables that is smooth in a vicinity of zero:

$$r_1 = r + r u_1(\theta, \rho) + r^2 u_2(\theta) + r^3 u_3(\theta), \qquad (3.28)$$

where $u_1(\theta, \rho)$, $u_2(\theta)$, $u_3(\theta)$ are $2\pi$–periodic piecewise-differentiable functions, $u_1(\theta, 0) = 0$. The conversion (3.28) reduces the system (3.26) to the form

$$\begin{aligned} \dot{r}_1 &= r_1\left(\frac{1}{4}\mu(\rho) + Kr_1^2\right) + \cdots, \\ \dot{\theta} &= \omega(\rho) + \cdots . \end{aligned} \tag{3.29}$$

Here, according to [19], we have

$$K = K(a_1, b_1) = \frac{1}{2\pi}\int_0^{2\pi}\left[C_4(\theta) - \frac{1}{\omega(0)}C_3(\theta)D_3(\theta)\right]d\theta.$$

By virtue of expressions (3.27), Eq. (3.24) is true. The validity of (3.25) follows from the system (3.29) and Eq. (3.28). Now, the validity of the theorem follows from [19].

Reasoning as in proving Theorem 3.2, we arrive at the following theorem.

**Theorem 3.3.** *We assume that* $\psi'_p(1,0) = 0$, $\psi''_{pp}(1,0) = 0$, $\mu'(0) = \hat{k} \neq 0$, *and* $K > 0$. *Then there exists* $\delta_0 > 0$ *such that we have*

**(i)** *if* $\hat{k} < 0$,

*then, in a* $\delta_0$*-vicinity of* $(\tilde{p}, \tilde{\nu})$, *there exists a periodic solution of system* (3.13), *it is unique up to shifts in* $t$, *and its period* $\dfrac{2\pi}{\hat{\omega}(\rho)}$, $\hat{\omega}(\rho) = \omega(\rho) + o(\rho)$, *is of the form*

$$\hat{\nu}(t) = \tilde{\nu} - \frac{1}{2}\left(\frac{-k\rho}{K}\right)^{1/2}\hat{\omega}(\rho)\sin\hat{\omega}(\rho)t + \cdots,$$

$$\hat{p}(t) = \tilde{p} + \frac{1}{2}\left(\frac{-k\rho}{K}\right)^{1/2}\cos\hat{\omega}(\rho)t + \cdots .$$

*This solution is unstable. The equilibrium state* $(\tilde{p}, \tilde{\nu})$ *of systems* (3.13) *is exponentially stable;*

**(ii)** *if* $\hat{k} > 0$,

*then, in a* $\delta_0$*-vicinity of* $(\tilde{p}, \tilde{\nu})$ *there exist no nonwandering points distinct from* $(\tilde{p}, \tilde{\nu})$. *The equilibrium state* $(\tilde{p}, \tilde{\nu})$ *of the system* (3.13) *is unstable.*

By Theorems 3.2 and 3.3, the dynamics of the system (3.13) in vicinity $(\tilde{p}, \tilde{\nu})$ is determined by the quantities $K$ and $k$. Let us find a representation for the constant $k$. To this end, we differentiate the equality

$$\mu_1(\rho)\tilde{\nu}(\rho) = \rho c,$$

which is true by virtue of the definition of $(\tilde{p}, \tilde{\nu})$, with respect to $\rho$. As a result, we obtain

$$\mu_1'(0)\tilde{\nu}(0) + \mu_1(0)\tilde{\nu}'(0) = c.$$

Since $\mu_1(0) = 0$, we have $\mu_1'(0) = \dfrac{c}{\tilde{\nu}(0)}$. By definition, we have

$$k = l^{-1}S^{-1}\left(\alpha l^{-1}\tilde{\nu}(0)\right) - 2\frac{c}{\tilde{\nu}(0)}. \tag{3.30}$$

According to the formula (3.30), the increase in $\tilde{\nu}$ decreases the stability factor of the quiescent state $(\tilde{p}, \tilde{\nu})$.

We pass to the analysis of the first Lyapunov quantity $K$. Note that, for merchandise offering functions such as

$$\psi_1(p) = \gamma_1\left(1 - e^{-\beta_1 p}\, p\right) \text{ and } \psi_2(p) = \gamma_2\left(1 - (p + d)^{-\beta_2}\right),$$

where $\gamma_k > 0$, $\beta_k > 0$, $k = 1, 2$, $d > 0$, the inequality $\dfrac{d^3}{dp^3}\psi_k > 0$, $k = 1, 2$, is true. Hence, for the mentioned functions, we have $b_2 > 0$ and, by virtue of Eq. (3.24), the inequality $K > 0$ if

$$\tilde{\nu}(0) > 1 + \frac{2\omega(0)(2 - 3b_2)}{b_1}$$

is true.

Consider now a corollary from Theorem 3.3, which is important for analyzing the dynamics of system (3.13). It is obvious that in Eq. (3.13), under the conditions formulated in Theorem 3.3 and when $\hat{k} > 0$, a stiff loss in stability (a crisis) of the equilibrium state $(\tilde{p}, \tilde{\nu})$ takes place and, hence, there exists, for small $\rho > 0$, at least one orbitally asymptotically stable limiting cycle whose amplitude is finite (it means that the amplitude is not a small quantity).

Simple reasoning allows us to draw the conclusion that Theorems 3.2 and 3.3 can be generalized to the case where $\psi''_{pp}(1, 0) \neq 0$.

The analysis performed shows that the determining factor influencing the dynamics of the system (3.13) is the relative position of the curves $L_S$ and $\Psi$. For a "low" (relative to $L_S$) position of $\Psi$, the system is stabilizable and its set of stabilization consists of equilibrium points analogous to those for the model (3.2)–(3.3) considered in Sect. 3.2. A change in prices will not exert an appreciable effect on the conclusions drawn in the mentioned section. For a "high" position of $\Psi$, the situation changes. The system (3.13) has solutions that orbitally converge to stable periodic solutions whose characteristics correspond to the basic business cycle. Depending on the initial state of the system, the convergence will be accompanied by the general tendency toward an economic growth or a recession. In this case, the rates of growth (or recession) will decrease and approach the periodic regime.

In the case where the set $L_S$ degenerates into the point $(\tilde{p}, \tilde{v})$, the trajectories of the system that originate not far from $(\tilde{p}, \tilde{v})$ are unfolding or folding spirals. For an economy with the monopsonic labor market and a completely competitive market of goods and services, business cycles are repeated with increasing or decreasing amplitudes and with different duration. Under some additional assumptions, the analytical form of periodic solutions is constructed in a small vicinity of the point $(\tilde{p}, \tilde{v})$ to which the state of simple reproduction in the economy corresponds, the local dynamics of the system is investigated, and the crisis states of the system are specified.

Based on the investigations carried out, we can draw the conclusions listed below.

1. According to the model considered, a macroeconomic system with a monopsonic labor market and predominantly monopolized commodity market passes to an equilibrium state after some transient process.

   Depending on the initial conditions, such a state could be a point of the curve $L_S$. The lower equilibrium point in the model with stable prices (3.2)–(3.3) from the previous section corresponds to this point, in which the rate of manufacturer's monopoly profit is maximal. The points of the curve $L_U$, as well as an intermediate equilibrium point of the mentioned model, are points of unstable equilibrium. In contrast to the model (3.2)–(3.3), the model (3.13) has no upper equilibrium state since the approach of the system toward it will be accompanied by the excess of demand over supply with a constant level of total wage fund. It leads to inflation, depreciates previous savings, lowers demand, and directs the economy to the lower equilibrium point. Thus, if an economy is highly monopolized and external shocks and exogenous technological changes affecting the values of the parameters $l$ and $\beta$ are absent, then, after some period, the economic growth ends. The development comes to the end when prices and the volume of products are such that they provide the greatest superprofit to monopolists.
2. In contrast to a monopolized economy, the cycling of the periods of growth and recession is characteristic of an economy with a monopsonic labor market and a partially competitive (mixed) commodity market. Such dynamics corresponds to the basic business cycle observable in the nineteenth century and in the early twentieth century. According to the results of modeling, the determining role in passing from growth to recession is played by the consumer saving inflationary depreciation and such a role in passing from recession (depression) to growth is played by a deflationary increase in the value of assets. The major cause of cyclicity is the interaction of the competitive commodity market capable (from the viewpoint of the majority of economic theorists) of providing self-regulation under the conditions of economic growth and the monopsonic labor market that is not a self-controllable one (see the previous sections) and allows for only a limited growth level (up to a definite level of production). In the case where the curve $\Psi$ passes through the point $(\tilde{p}, \tilde{v})$, i.e., when the commodity market is competitive with any price level, the cyclic character of development becomes

slack and the economy passes to simple reproduction or becomes stronger. The crisis character of changing the system's behavior when its parameters pass through critical values should be particularly noted.

3. The monopsony in a labor market provides employers with the greatest income over a bounded time interval. However, an economy with such a labor market with different (monopoly, mixed, or competitive) forms of organization of the commodity market demonstrates its incapacity for development and stable growth. In this case (in the long term), the entire society, including employers, suffers losses. Since statistical data [8, 54] indicate that monopsonic relations were (and are) prevalent in some countries with a transition economy, the creation of a competitive labor market should be a key line of further structural reforms.

The consequences of imperfect competition on the labor market require further investigations, including those with the use of mathematical modeling. Such investigations can be realized, in particular, along the following directions:

- analysis of the impact of the labor market with bilateral monopoly competition (monopolists-employers against trade unions) on the economic dynamics at a whole;
- estimation of the effect of external demand shocks (e.g., random changes in export turnover) and subjective employer's expectations of change in labor payment under the conditions of monopsony;
- analysis of models with supply functions that differ from the considered ones and reflect the interaction between the monopolized and competitive segments of the commodity market;
- investigations into an economy with a partially (for some branch segments) monopolized labor market;
- further theoretical economic interpretation of the results achieved.

## 3.4 Model of Interbranch Competition on the Labor Market

In the previous sections of this chapter the labor market of official sectors of economy at the early stages of transition was considered as the typical example of imperfect market with monopoly consumption (or as monopsonic market). The absence of strong trade unions, independent of employers and state and interested in reaching of higher wages and the lack of experience for market behavior of workers result that the level of pay for job is installed by the employers on the base of their own preferences. They have the strong ties into branchial structures which were created in centrally planned economy. Let assume that oligipolean competition on the labor market between such structures is the main kind of competitions which can produce the impact on wages in countries in transition. Can it essentially change the mechanism of wage creation in comparison with the mechanism of monopsony waging? In this section we try to answer this question comparing the results of analysis of the models with monopsony and oligopsony waging. The last one is based on the results of nonantogonistic game theory.

Continued industrial decline was typical for the early stages in market transition. The decrease of effective demand on final production, first of all, at the expense of reduction of expenditures on public account, was its main reason. In these conditions, the Keynesian approach seems more acceptable for employment description [35].

The mentioned decline, as against market economy, is accompanied by strong inflationary processes which are mainly the result of cost inflation [43]. Therefore the main factor defining the value of labor supply will be the value of real wages, i.e., the value of total nominal payments to workers divided to GDP deflator. The value of deflator can be considered as independent from real wages.

The previous models were constructed under the assumption that labor markets controlled by every employer are isolated. In reality this means the separation of employments processes in branches of former state sector in which employers are connected by the previously formed social ties and can realize the monopolean behavior. Therefore every such branch can be considered as monopoly consumer of labor. Actually, this is not absolutely true. Certainly, such factors as necessary professional qualification, regional specialization and conservatism of people hinder the inter-branch labor exchange. Nevertheless, such an exchange has always existed with the economic situation in branches being one of the main reasons. Enterprises with higher real wages are more attractive for workers under the same other conditions. Therefore, if the rate of a wage decreases in branch $A$ is higher than in branch $B$, the value of labor supply in the last one will be greater at the expense of persons wishing to pass between branches. The labor supply function in such situation depends not only on the value of real wages in considered branch but also on payments in other branches. This leads to interbranch competition for labor resources which can be described by methods of the game theory.

Consider a possible model of such competition between $n$ branches. The value of real wage fund for the $i$-th branch is denoted by $W_i$ $\left(i = \overline{1,n}\right)$ and the demand for production of this branch–by $\nu_i$. We assume that the function of acquired labor $S_i(X_i)$ in $i$-th branch depends on the value of related payment for job which is calculated as the ratio of wage fund of $i$-th branch to total wage fund, i.e.,

$$X_i = \frac{W_i}{\sum\limits_{k=1}^{n} W_k}.$$

This function determined within [0; 1] is concave, monotone, and differentiable, $S_i(0) = 0$, $S_i(1) = L$, $S_i'(0) > l_i^{-1}$, where $L$ is the value of the total labor resources. Assume that the value of total payments is chosen by every branch so as to maximize the branch net utility function

$$F_i(W) = \min\Big(l_i S_i(X_i),\ \alpha_i \nu_i\Big) - W_i, \quad i = \overline{1,n},$$

where $l_i$ is the productivity of labor per unit of real gross income for the $i$-th branch, $\alpha_i$ is the real gross income per unit of production of the $i$-th branch. So we consider the nonantagonistic game of $n$ players, where the strategy of the $i$-th player $W_i$ belongs to the set $\{W_i : W_i \geq 0\}$. We assume for a simplicity that the value of associated cost $h = 0$ for all branches. The function $F_i(W)$ will be the payoff function of this player. The following theorem determines the condition of existence of Nash equilibrium in this game and describes its properties.

**Theorem 3.4.** *Suppose that functions $S_i(X_i)(i = \overline{1,n})$ satisfy the above assumptions, $l_i \geq 0$, $\alpha_i \nu_i \geq 0$, and $l_i L > \alpha_i \nu_i$ $\left(i = \overline{1,n}\right)$. Than the Nash equilibrium $\left(W_i^*, \ldots, W_n^*\right)$ exists. The value of $W_i^*$ $\left(i = \overline{1,n}\right)$ can be determined as follows:*

$$W_i^* = \min\left(W_i', W_i''\right), \quad i = \overline{1,n},$$

*where $W_i'$ is the solution of the equation*

$$l_i S_i\left(\frac{W_i}{W_i + \sum\limits_{k=1,k\neq i}^{n} W_k^*}\right) = \alpha_i \nu_i \tag{3.31}$$

*and $W_i''$ is the solution of the equation*

$$l_i S_i'\left(\frac{W_i}{W_i + \sum\limits_{k=1,k\neq i}^{n} W_k^*}\right) = \frac{\left(\sum\limits_{k=1,k\neq i}^{n} W_k^* + W_i\right)^2}{\sum\limits_{k=1,k\neq i}^{n} W_k^*}. \tag{3.32}$$

*Proof.* It should be noted that under the assumption of the theorem the function $S_i\left(\frac{W_i}{W_i + \sum\limits_{k=1,k\neq i}^{n} W_k^*}\right)$ is the concave function of $W_i$.

Really,

$$\frac{\partial^2 S_i\left(\frac{W_i}{W_i + \sum\limits_{k=1,k\neq i}^{n} W_k}\right)}{\partial W_i^2}$$

$$= \frac{\left(\sum\limits_{k=1,k\neq i}^{n} W_k\right)^2}{\left(\sum\limits_{k=1,k\neq i}^{n} W_k + W_i\right)^2} S_i''(X_i) - \frac{2\sum\limits_{k=1,k\neq i}^{n} W_k}{\left(\sum\limits_{k=1,k\neq i}^{n} W_k + W_i\right)^3} S_i'(X_i),$$

where $X_i = \dfrac{W_i}{\sum\limits_{k=1,k\neq i}^{n} W_k + W_i}$. The function $S_i(X_i)$ is concave and monotone increased, therefore $S_i''(X_i) \leq 0$, $S_i'(X_i) \geq 0$ and $\dfrac{\partial^2 S_i(X_i)}{\partial W_i^2} \leq 0$.

The function $F_i(W_1, \ldots, W_i, \ldots, W_n)$ according to properties of concave functions also will be the concave function of $W_i$ specified at the convex set of possible strategies of $i$-th player, so the Nash equilibrium $(W_i^*, \ldots, W_n^*)$ exists in this game. According to the definition of Nash equilibrium

$$F_i\left(W_1^*, \ldots, W_i^*, \ldots, W_n^*\right) = \max_{W_i \geq 0} F_i\left(W_1^*, \ldots, W_i, \ldots, W_n^*\right) = \max_{W_i \geq 0} \varphi_i(W_i).$$

The point $W_i^*$ can be the maximal point of the concave function $\varphi_i(W_i) = F_i\left(W_i^*, \ldots, W_i, \ldots, W_i^*\right)$ in two cases:

1. $\varphi_i'(W_i) > 0$ holds for every $W_i \leq W_i'$, where $W_i'$ is the point in which $\varphi_i(W_i)$ function is nondifferentiable, hence $W_i^* = W_i'$.
2. $\varphi_i'(W_i^*) = 0$ and $W_i^* < W_i'$ holds.

The point $W_i'$ will be the solution of Eq. (3.31), and the Eq. (3.32) implies directly from the equation:

$$\varphi_i'(W_i) = l_i S_i'(X_i) \frac{\sum\limits_{k=1,k\neq i}^{n} W_k^*}{\left(\sum\limits_{k=1,k\neq i}^{n} W_k^* + W_i\right)^2} - 1 = 0.$$

The last one is solvable at [0; 1] if $l_i S_i'(0) > 1$ hods, and Eq. (3.31) is solvable at the same interval due to the condition: $l_i S(1) = l_i L > \alpha_i v_i$. The theorem is proved.

By analyzing Eqs. (3.31) and (3.32) and by comparing them with the model of monopsony waging, the following conclusions can be drawn. There are no principal differences in the structure and properties between employment behavior generated on the basis of monopsony and oligopsony described above. The excess supply crisis resulting from the feedback "decrease of demand for production–decrease of real wages–decrease of demand" which has been analyzed in the previous sections is also possible for the case considered here. Low level of the officially registered unemployment (i.e., excess supply of labor) accompanied by low wages will also be typical in such a situation; hence the competition only between large employers can not create an effectively functioning labor market.

In the considered model we assumed that the value of the demand $v_i$ are independent of the current wage level and can be regarded by employers as a given constant. In a more realistic setting this assumption has to be relaxed. Naturally, the current consumer income influences the value of final demand for goods, but the impact of the past wage level, which created consumers accumulations is much more essential. Such components of final demand as the consumption of non-workers, external consumption, state expenditures, investment demand, etc. are independent of (or only weakly depended on) wages but are time-varying. Thus, the consideration of the dynamic model of multibranch competition is important for the analysis of these aspects.

Let us assume that the time has changed with the step 1, the variable time is denoted as $t$ ( $t = 1, 2, 3, \ldots$). We consider the process of competition for labor between $n$ branches. Let $\nu_i^t$ is the value of the demand for production of the $i$-th branch at time $t$ depending upon the values of total wage bill in the previous time moments $W_i^{t-1}, W_i^{t-2}, \ldots$, but independent of the current values $W_i^t$ of payment. Using the notations from the previous model let us construct the payoff function of the $i$-th employer at the time moment $t$:

$$F_i^t\left(W^t, \nu^t\right) = \min\left(l_i S_i\left(X_i^t\right), \alpha_i \nu_i^t\right) - W_i^t, \tag{3.33}$$

where $X_i^t = \dfrac{W_i^t}{\sum\limits_{k=1}^{n} W_k^t}$.

If we assume that the players make their decisions at every time moment $t$ without any forecasts for the future and any account of past experience, the Nash equilibrium $W^*(t)$, characterized by the Theorem (3.4), should exist for every time moment $t$. However, players cannot quickly change their previous decisions concerning the real wage policy, and so at every moment in time they only make some local steps or changes in the direction toward the maximal point of $F_i^t(W^t, \nu^t)$. It is quite natural to assume that the direction of the maximum local growth, i.e., the gradient of the payoff function at the point $(W^t, \nu^t)$ will lead in such a direction. This leads us to the consideration of the following procedure.

$$W_i^{t+1} = \max(0, W_i^t + \rho_t G_i^t), \quad t = 1, 2, \ldots, i = \overline{1, n}, \tag{3.34}$$

where

$$G_i^t = \begin{cases} l_i S_i'(X_i^t) \dfrac{\sum\limits_{k=1, k \neq i}^{n} W_k^t}{\left(\sum\limits_{k=1}^{n} W_k^t\right)^2} - 1, & \text{if } l_i S_i\left(X_i^t\right) < \alpha_i \nu_i^t \text{ holds}, \\ -1 & \text{otherwise}, \end{cases}$$

$X_i^t = \dfrac{W_i^t}{\sum\limits_{k=1}^{n} W_k^t}$, $\rho_t$ is the step according to which employers change the values of real wage bill. Let us analyze the convergence of such sequence $\left\{W^t\right\}$ in the sense of:

$$\left|F_i\left(W^t, \nu^t\right) - F_i\left(W^*(t), \nu^t\right)\right| \xrightarrow[t\to\infty]{} 0, \quad i = \overline{1,n}. \tag{3.35}$$

The analysis is done in the following theorem.

**Theorem 3.5.** *Let the functions $S_i(X_i)$, $i = \overline{1,n}$ satisfy all previously made assumptions, $\left|\nu_i^{t+1} - \nu_i^t\right| = o(\rho_t)$, $\rho_t \xrightarrow[t\to\infty]{} 0$, $\sum_{t=1}^{\infty} \rho_t = \infty$. Then the sequence $\left\{W^t\right\}$ will converge to $W^*(t)$ in the sense of* (3.35).

*Proof.* The procedure (3.34) can be considered as the procedure of the search of extremal points in nonstationary problem of optimization of the function

$$H\left(W, Z, \nu^t\right) = \sum_{k=1}^{n} F_k\left(W_1, \ldots, W_{k-1}, Z_k, W_{k+1}, \ldots, W_n, \nu^t\right).$$

Its convergence for the case when the function $H(\cdot)$ is concave for $Z$ and the step size $\rho_t$ satisfy the above-mentioned conditions was considered by [53]. The convergence in the sense of (3.35) implies from this result. The procedure (3.34) can be considered as the kind of bargaining procedure for the case when the demand has changed. Its convergence to the solutions of (3.31) or (3.32) equations means that all disadvantages of monopolean pricing also will be saved for this case.

Based on the proposed model, calculations were carried out with statistical data over the period from 1990 to 1998 [50,51,54] for several branches of the Ukrainian economy.

The objective of the calculations was to estimate the compliance of the results of simulation with an actual situation during the mentioned period and to identify the parameters of the models. Here, exponential and logarithmic functions of acquired labor $S(X) = a_0 X^{\nu}$, $0 < \nu < 1$, and $S(X) = a_0 \ln(1 + b_0 X)$ were used. Except for the oligopoly model considered in the current section, a similar model was used with a function of acquired labor depending on the ratio of the wages averaged over the given branch to the same value calculated over all of the branches. This model will hereafter be called a model with relative wages. Each of the mentioned models was considered in two variants: with an annual lag between factors (the arguments of the function $S(X)$) and the value of the labor acquired and without such a lag. The parameters of these functions $a_0$, $\nu$, and $b_0$ were determined using a nonlinear analogue of the least-squares technique separately for each model so as to minimize the total deviation of predicted values of the wages fund from the actual values. We selected the models and functions such that this deviation was minimal. The selection was performed both with no regard for the determination coefficient $R2$ and for the cases where it exceeded certain thresholds.

The results of the calculations are presented in Tables 3.1 and 3.2. The following denotations are used here: $mul$ and $mul^-$ mean models of monopsony waging with an exponential function (without the lag and with it), $log$ and $log^-$ are similar models with a logarithmic function, $rmul$ and $rmul^-$ are models of oligopsony

**Table 3.1** Approximation accuracy and determination coefficient for branch models with monopsony waging

| Branch | Accuracy/determination for models | | | |
|---|---|---|---|---|
| | *log* | $log^-$ | *mul* | $mul^-$ |
| Electric-power industry | 5.6 / 0.122794 | 6.82 / 0.113763 | 6.03 / 0.872188 | 7.19 / 0.085983 |
| Oil-and-gas industry | 0.8 / 0.046438 | 0.57 / 0.191586 | 0.8 / 0.038853 | 0.6 / 0.167225 |
| Coal mining | 31.6 / 0.037036 | 49.0 / 0.494192 | 26.7 / 0.449707 | 45.2 / 0.544654 |
| Other fuel industries | 52.6 / 0.226877 | 66.8 / 0.247339 | 55.6 / 0.185598 | 58.0 / 0.317851 |
| Ferrous met. | 1.0 / 0.750038 | 3.34 / 0.554585 | 1.05 / 0.731624 | 3.38 / 0.540786 |
| Nonferrous metallurgy | 14.20 / 0.010018 | 12.62 / 0.107781 | 14.20 / 0.005758 | 12.03 / 0.128532 |
| Chemistry and petrochemistry | 10.5 / 0.472337 | 6.1 / 0.545591 | 10.35 / 0.478878 | 6.24 / 0.532518 |
| Engineering industry | 12.64 / 0.744955 | 7.08 / 0.826684 | 9.80 / 0.791433 | 6.24 / 0.854646 |
| Forest and woodworking industry | 3.94 / 0.864465 | 10.67 / 0.699132 | 4.40 / 0.851246 | 11.64 / 0.671687 |
| Constructional materials industry | 5.2 / 0.778650 | 22.23 / 0.430934 | 5.69 / 0.771996 | 23.28 / 0.398176 |
| Light industry | 2.43 / 0.926004 | 0.95 / 0.956214 | 1.93 / 0.939958 | 0.93 / 0.955396 |
| Food industry | 5.21 / 0.043753 | 4.67 / 0.297615 | 5.08 / 0.013945 | 5.34 / 0.221931 |
| Other industry branches | 6.81 / 0.738009 | 2.75 / 0.773776 | 7.67 / 0.700039 | 2.94 / 0.741276 |
| Farming | 41.4 / 0.052450 | 59.41 / 0.000208 | 37.38 / 0.010847 | 86.45 / 0.048989 |
| Transport and communication | 26.72 / 0.108930 | 29.92 / 0.117522 | 26.06 / 0.118445 | 30.43 / 0.092859 |
| Construction | 12.7 / 0.737986 | 7.31 / 0.817624 | 11.83 / 0.750070 | 6.52 / 0.831418 |
| Trade and catering | 48.81 / 0.330119 | 0.31 / 0.797073 | 47.88 / 0.340951 | 0.30 / 0.802568 |

waging with an exponential function, $rlog$ and $rlog^-$ are analogous models with a logarithmic function, and $nmul$, $nmul^-$, $nlog$ and $nlog^-$ are models with relative wages (with exponential and logarithmic functions and with and without lag). Relative wages means the ratio of average wages in current branch to the maximal value of this figure taken among all branches.

**Table 3.2** Approximation accuracy and determination coefficient for interbranch models

| Branch | Accuracy/determination for models | | | | | | | |
|---|---|---|---|---|---|---|---|---|
| | $rlog$ | $rlog^-$ | $rmul$ | $rmul^-$ | $nlog$ | $nlog^-$ | $nmul$ | $nmul^-$ |
| Electric-power industry | $\frac{0.46}{0825880}$ | $\frac{12.46}{0.354156}$ | $\frac{0.3}{0.872188}$ | $\frac{11.7}{0.394326}$ | $\frac{0.38}{0.251841}$ | $\frac{12.0}{0.163111}$ | $\frac{0.33}{0.315244}$ | $\frac{11.61}{0.194841}$ |
| Oil-and-gas industry | $\frac{0.29}{0.718341}$ | $\frac{1.81}{0.209983}$ | $\frac{0.25}{0.741142}$ | $\frac{1.69}{0.233179}$ | $\frac{0.23}{0.432278}$ | $\frac{1.65}{0.091839}$ | $\frac{0.21}{0.456635}$ | $\frac{1.58}{0.108365}$ |
| Coal mining | $\frac{19.61}{0.362028}$ | $\frac{32.79}{0.027276}$ | $\frac{18.73}{0.355694}$ | $\frac{33.03}{0.019152}$ | $\frac{17.44}{0.213882}$ | $\frac{30.83}{0.000298}$ | $\frac{17.14}{0.194538}$ | $\frac{31.04}{0.002125}$ |
| Other fuel industries | $\frac{23.38}{0.476567}$ | $\frac{54.05}{0.194081}$ | $\frac{24.32}{0.515474}$ | $\frac{54.53}{0.189547}$ | $\frac{24.19}{0.406021}$ | $\frac{50.26}{0.292276}$ | $\frac{1.0}{0.056029}$ | $\frac{50.9}{0.294516}$ |
| Ferrous met. | $\frac{1.71}{0.276799}$ | $\frac{7.25}{0.086629}$ | $\frac{1.67}{0.288755}$ | $\frac{7.16}{0.096510}$ | $\frac{1.42}{0.058805}$ | $\frac{6.85}{0.036864}$ | $\frac{38.46}{0.056029}$ | $\frac{6.79}{0.057500}$ |
| Nonferrous metallurgy | $\frac{3.81}{0.017589}$ | $\frac{5.55}{0.163217}$ | $\frac{3.91}{0.010852}$ | $\frac{5.77}{0.151600}$ | $\frac{2.62}{0.220258}$ | $\frac{4.77}{0.326865}$ | $\frac{2.71}{0.207896}$ | $\frac{4.98}{0.316951}$ |
| Chemistry and petrochemistry | $\frac{1.32}{0.036299}$ | $\frac{3.56}{0.233004}$ | $\frac{1.32}{0.034408}$ | $\frac{3.50}{0.246154}$ | $\frac{1.21}{0.013111}$ | $\frac{3.55}{0.228543}$ | $\frac{1.21}{0.012366}$ | $\frac{3.49}{0.242193}$ |
| Engineering industry | $\frac{2.50}{0.503467}$ | $\frac{6.94}{0.000178}$ | $\frac{2.57}{0.503786}$ | $\frac{6.98}{0.000788}$ | $\frac{2.54}{0.215119}$ | $\frac{7.19}{0.026692}$ | $\frac{2.57}{0.003079}$ | $\frac{7.24}{0.021974}$ |
| Forest and woodworking industry | $\frac{9.45}{0.003772}$ | $\frac{7.65}{0.181441}$ | $\frac{9.33}{0.007027}$ | $\frac{7.27}{0.211493}$ | $\frac{10.68}{0.009932}$ | $\frac{8.08}{0.218775}$ | $\frac{3.95}{0.014877}$ | $\frac{7.64}{0.251439}$ |
| Constructional materials industry | $\frac{9.90}{0.037469}$ | $\frac{12.31}{0.254814}$ | $\frac{9.65}{0.046193}$ | $\frac{11.58}{0.289379}$ | $\frac{11.13}{0.073076}$ | $\frac{13.11}{0.308846}$ | $\frac{10.74}{0.085673}$ | $\frac{12.21}{0.346209}$ |

(continued)

**Table 3.2** (continued)

| Branch | Accuracy/determination for models | | | | | | | |
|---|---|---|---|---|---|---|---|---|
| | $rlog$ | $rlog^-$ | $rmul$ | $rmul^-$ | $nlog$ | $nlog^-$ | $nmul$ | $nmul^-$ |
| Light industry | $\frac{1.88}{0.553418}$ | $\frac{0.7}{0.070207}$ | $\frac{2.1}{0.518852}$ | $\frac{0.72}{0.061379}$ | $\frac{2.08}{0.090745}$ | $\frac{0.78}{0.310724}$ | $\frac{2.13}{0.072282}$ | $\frac{0.76}{0.325799}$ |
| Food industry | $\frac{0.49}{0.446325}$ | $\frac{1.89}{0.053450}$ | $\frac{0.46}{0.461165}$ | $\frac{1.91}{0.012350}$ | $\frac{0.53}{0.27634}$ | $\frac{1.60}{0.020146}$ | $\frac{0.49}{0.295647}$ | $\frac{1.60}{0.000614}$ |
| Other industry branches | $\frac{2.92}{0.340950}$ | $\frac{3.13}{0.002017}$ | $\frac{2.98}{0.308558}$ | $\frac{3.14}{0.001844}$ | $\frac{2.43}{0.086051}$ | $\frac{2.83}{0.156395}$ | $\frac{2.41}{0.063849}$ | $\frac{2.85}{0.155179}$ |
| Farming | $\frac{4.62}{0.000719}$ | $\frac{17.86}{0.022323}$ | $\frac{4.49}{0.003970}$ | $\frac{18.70}{0.031456}$ | $\frac{4.49}{0.00032}$ | $\frac{17.98}{0.026936}$ | $\frac{4.39}{0.000620}$ | $\frac{18.85}{0.041070}$ |
| Transport and communication | $\frac{3.56}{0.805375}$ | $\frac{33.48}{0.390736}$ | $\frac{3.43}{0.797733}$ | $\frac{31.19}{0.429446}$ | $\frac{3.85}{0.768251}$ | $\frac{34.63}{0.356685}$ | $\frac{3.52}{0.765594}$ | $\frac{32.47}{0.395772}$ |
| Construction | $\frac{2.39}{0.019342}$ | $\frac{13.55}{0.136619}$ | $\frac{2.40}{0.017692}$ | $\frac{13.91}{0.114488}$ | $\frac{2.53}{0.301076}$ | $\frac{13.17}{0.270189}$ | $\frac{2.53}{0.307563}$ | $\frac{13.66}{0.241023}$ |
| Trade and catering | $\frac{54.65}{0.092}$ | $\frac{4.18}{0.737391}$ | $\frac{54.75}{0.096789}$ | $\frac{4.35}{0.728268}$ | $\frac{53.62}{0.097482}$ | $\frac{4.06}{0.748117}$ | $\frac{53.71}{0.090936}$ | $\frac{4.19}{0.790182}$ |

After identification of the parameters, figures of approximation accuracy $\psi = \frac{1}{T}\sum_{t=1}^{T}\frac{|W_t - \overline{W}_t|}{W_t} \times 100\,\%$, where calculated. Here [1; T] is the observation period, $W_t$ is the actual value of the wage fund at the instant of time $t$ and $\overline{W}_t$ is the predicted (obtained by the model) value of the figure. In Tables 3.1 and 3.2, the value of $\psi$ for each of the branches is given as a numerator, and the value of $R2$ as a denominator.

The models with the maximum approximation accuracy for each of the branches are summarized in Table 3.3.

The models were selected both with regard for $R2$, when only models were considered for which $R2 \geq 0.75$ holds, and without regard for this coefficient. The parameters of the models selected are given in Table 3.4.

Analyzing the results obtained, it should be emphasized that models based on the same assumption as for the function of the monopsony and oligopsony waging rather accurately describe the dynamics of wage fund in the majority of the branches in Ukraine (the approximation accuracy is from 0.21 up to 5 %).

**Table 3.3** Branch and inter-branch models with the maximum approximation accuracy

| | Without account *R2* | | | With account *R2* | | |
|---|---|---|---|---|---|---|
| Branch | Model | Value $\psi$ | Value *R2* | Model | Value $\psi$ | Value *R2* |
| Electric-power industry | *rmul* | 0.3 | 0.8722 | *rmul* | 0.3 | 0.8722 |
| Oil-and-gas industry | *nmul* | 0.21 | 0.4566 | *rmul* | 0.25 | 0.7411 |
| Coal mining | *nmul* | 17.14 | 0.02845 | – | – | – |
| Other fuel industries | *nmul* | 1.0 | 0.0560 | – | – | – |
| Ferrous met. | *log* | 1.0 | 0.7500 | *log* | 1.0 | 0.7500 |
| Nonferrous metallurgy | *nlog* | 2.62 | 0.2203 | – | – | – |
| Chemistry and petrochemistry | *nlog* | 1.21 | 0.01311 | – | – | – |
| Engineering industry | *rlog* | 2.5 | 0.5035 | $mul^-$ | 6.24 | 0.8546 |
| Forest and woodworking industry | *log* | 3.94 | 0.8645 | *log* | 3.94 | 0.8645 |
| Constructional materials industry | *log* | 5.2 | 0.7787 | *log* | 5.2 | 0.7787 |
| Light industry | $rlog^-$ | 0.71 | 0.0702 | $mul^-$ | 0.93 | 0.9554 |
| Food industry | *rmul* | 0.46 | 0.4612 | – | – | – |
| Other industry branches | *nmul* | 2.41 | 0.06385 | $log^-$ | 2.75 | 0.7738 |
| Farming | *rlog* | 2.39 | 0.01934 | $mul^-$ | 2.53 | 0.8314 |
| Transport and communication | *nmul* | 4.39 | 0.0006 | – | – | – |
| Construction | *rmul* | 3.43 | 0.7977 | *rmul* | 3.43 | 0.7977 |
| Trade and catering | $mul^-$ | 0.300 | 0.8026 | $mul^-$ | 0.300 | 0.8026 |

**Table 3.4** The parameters of the functions for the models with the maximum accuracy

| Branch | Model | | Parameters |
|---|---|---|---|
| Electric-power industry | *rmul* | $\alpha_0 = 0.08946$ | $\nu = 0.20235$ |
| Oil-and-gas industry | *rmul* | $\alpha_0 = 0.09983$ | $\nu = 0.085808$ |
| Coal mining | *nmul* | $\alpha_0 = l$ | $\nu = 0.3115$ |
| Other fuel industries | *nmul* | $\alpha_0 = 1.093$ | $\nu = 0.2571$ |
| Ferrous metallurgy | *log* | $\alpha_0 = 0.8678$ | $b_0 = 0.1841$ |
| Nonferrous metallurgy | *nlog* | $\alpha_0 = 1.59458$ | $b_0 = 2.2917$ |
| Chemistry and petrochemistry | *nlog* | $\alpha_0 = 0.63906$ | $b_0 = 7.1268$ |
| Engineering industry | *rlog* | $\alpha_0 = 197.203$ | $b_0 = 2.6571$ |
| Forest and woodworking industry | *log* | $\alpha_0 = 2.222$ | $b_0 = 0.2776$ |
| Constructional materials industry | *log* | $\alpha_0 = 2.5053$ | $b_0 = 390$ |
| Light industry | $mul^-$ | $\alpha_0 = 1$ | $\nu = 0.676$ |
| Food industry | *rmul* | $\alpha_0 = l$ | $\nu = 0.33706$ |
| Other industry branches | $log^-$ | $\alpha_0 = 21.058$ | $b_0 = 0.709$ |
| Construction | $mul^-$ | $\alpha_0 = 3.0$ | $\nu = 0.781$ |
| Farming | *nmul* | $\alpha_0 = 0.04031$ | $\nu = 6.8669$ |
| Transport and communication | *rmul* | $\alpha_0 = 1$ | $\nu = 1.1577$ |
| Trade and catering | $mul^-$ | $\alpha_0 = 2.0$ | $\nu = 0.13777$ |

Despite the insufficiency of the statistical data, statistically significant results ($R2 \geq 0.75$) were obtained for 13 out of 18 branches, for which the calculations were performed. They demonstrate that the model of monopsony wage formation with an exponential or logarithmic labor supply function adequately describes the dynamics of employment in such branches as the ferrous metallurgy, constructional materials industry, forest and woodworking industry, trade and catering. In the branches of energy, farming, and transport complexes, the wages are formed mainly by oligopsony competition between employers. Ambiguous results were obtained mainly for branches combining enterprises heterogeneous for the used technologies and requirements on the skill of employees, such as engineering, light industry and others. We can assume that oligopsony competition for labor is possible for some enterprises involved in these branches, while monopsony mechanisms of waging dominate in other branches. Unfortunately, the unavailability of more detailed data did not allow us to check this assumption. It should also be pointed out that the values of the parameters of the functions of acquired labor identified during the calculations were individual for each of the branches. Thus, there are strong grounds to believe that the above-considered monopsony and oligopsony waging are rather typical on the labor market of Ukraine.

The processes of economic growth with monopsony labor market are of contradictory character. An upper limit of wages does not depend on the demand for produced goods. If this limit is achieved, then the growth processes are either suspended or accompanied by a reduction in the portion of employee's income in the total income.

The demand for goods on a rather long time interval stabilizes at a level corresponding to the wages. There may be several equilibrium points. Achieving one of them by the demand depends on its initial value. Some of these points may be the points of unstable equilibrium: in the case of small deviations of the demand from the initial position these cusps do not exists.

The main measure to decrease the depth of depression must be a decrease in labor supply in the monopolized sector of the economy. Due to this, development of alternate forms of employment (in particular, small and medium business) is of particular significance for overcoming the economic crisis. A reduction of the additional costs due to labor consumption will allow slowing down the depression only in the case where the wages have achieved their upper limit. An increase in the labor capacity due to technological factors (e.g., centralized purpose investments) yields an effect inverse to the decrease in the labor supply, i.e., deepens the depression.

Random variations of the demand, for example, due to inaccuracy of data used by the employer in decision making, do not affect the achievement of the above-mentioned equilibrium states under quite realistic assumptions (such as unbiasedness and boundedness of the variance). They may result in only a changeover of the demand to one of such states.

The following conclusions can be elaborated by the results of our research.

1. There are essential differences in mechanisms of wage and employment determination between market and transition economies. The absence of the previous strong trade unions, which are independent of employers and the state, and are interested in increasing wages on one hand as well as the presence of strong corporative ties between employers on the other leads to a situation where the level of wages is determined by employers on the basis of monopsonic pricing mechanisms. As a result, the value of wages corresponds to the level of labor supply that is approximately equal to the value of employers' labor demand, which is, in turn, determined by the expected demand for produced goods. Therefore, in countries with such labor market there is no significant open unemployment in the classical sense because the labor supply does not significantly exceed labor demand. Industrial decline results in a corresponding (and more often, advancing) decrease in the labor supply which is accompanied by wage reduction. The difficulty which the reduced part of labor (i.e., hidden unemployment) gives to registration is swallowed, in main, in the "shadow" economy and criminal structures. The vicious circle described by "industrial decline—wage decrease—decrease of internal effective demand—industrial decline" is a fatal consequence of such labor market structure.
2. Introducing oligopsonistic competition between employers cannot essentially change the mechanism of wage determination compared to monopsony waging. The strategies of employers which belong to Nash equilibrium are of the same structure as optimal strategies under monopsony. All their social disadvantages,

including the possibility of the afore mentioned vicious circle, are preserved. Introducing some forms of bilateral competition on the labor market is necessary to increase the social effectiveness of wage/employment mechanisms.

3. The analysis of the dynamic generalization of the considered processes of oligopsony competition has shown that the step-by-step process of wage adaptation will converge to the previously analyzed static Nash equilibrium under quite natural assumptions.

It is necessary to refer, first, to the following directions of further research development:

- the account for random components of employers' expectations concerning the demand in the dynamical models;
- the development of models for the simulation of inter-branch flows of labor;
- the estimation of the impact of uncertainty, incomplete and non-symmetric information on the labor market.

Certainly, the real data calculations for the considered models also are necessary.

## 3.5 Models of Bilateral Monopolistic Competition on the Labor Market

As it was mentioned in the previous sections, many peculiarities of the labor market in countries with transitional economy may be attributed to imperfect competition in this segment of the national market. In particular, in Sect. 3.2 we considered models based on the assumption of monopsony, i.e., employer's monopoly, and shown that such a labor market may slow down the economic development of the country. One of the reasons is that both the monopolist employer and the employees are not interested in increasing labor productivity. In Sect. 3.3 we analyzed models of economic dynamics and have shown that interaction of a competitive market of goods and services and a monopsonic labor market may generate cycles of economic development, as classical business-cycles with the stages of growth, recession, stagnation, and revival of economy. In this case, there is neither stable nor long-term growth. Consequently, we concluded that overcoming the monopsony on the labor market should become one of the major tasks of economic reforms.

Trade union movement is traditionally considered as an efficient way to accomplish this task. However, one form of imperfect competition on the labor market, monopsony, is substituted here by another one, bilateral monopolistic competition. Estimating the consequences of this replacement should involve complex economic analysis based on different methods, including mathematical modeling. Here, we consider some economic and mathematical models intended for such estimation and the results of their analysis. The main task of the analysis is to estimate the influence of the above-mentioned form of organization of the labor market on general economic dynamics.

This task is achieved in the following way. First a static model of competition of an employer and a union of employees (a trade union) is developed. This model is a game with a nonzero sum. Existence conditions for Nash equilibrium in this game and properties of equilibrium states are presented. Then we consider a dynamic model of competition between a monopolist employer and a trade union under conditions of a varying demand for the employer's product. It is assumed that both parties try to improve their position (to benefit more) by the trial-and-error technique. The model is analyzed using a specially developed dynamic model, which has an independent economic interpretation. The substantial conclusions following from the results obtained and directions of further studies are finished this section.

Let us consider a labor market with the following structure: a unique monopolist employer opposed by a union of employees (a trade union). Assume that this union includes persons who would like to sell their labor and acts as a single player, implementing a coordinated strategy. By analogy [5, 36], we assume that the employer tends to maximize the profit from the production with the use of the acquired labor and that he knows the demand $\nu$ (or an estimate of the demand) for his production. Assume also that the trade union tends to maximize the net profit from the sale of labor. This profit is defined as the difference between the aggregate income of the trade union participants due to sale of labor and the related costs. The latter may include the missed profit due to alternative employment, general losses of workers due to reduction of their spare time, etc.

The model is a nonantagonistic game of two players. The strategy of the first player, the trade union, is an acceptable level of wages $\omega$ determined by it, the strategy of the second one, the employer, is the amount of labor force $G$ he acquires. The payoff function of the first player has the form

$$f_1(\omega, G) = G\omega - \overline{Q}(G, \omega),$$

where $\overline{Q}(G, \omega)$ denotes the costs and losses of the employees due to their sale of labor force in the amount $G$ at the price of $\omega$. Hereafter, we assume that this function is increasing, convex in $\omega$, concave in $G$, and at least three times continuously differentiable.

The payoff function of the second player has the form

$$f_2(\omega, G) = \min(lG, \alpha\nu) - HG\omega,$$

where $l$ is the labor productivity, $\alpha$ is the share of the value added in the price of the production produced by the employer, $H = 1 + h$, $h$ is additional costs due to acquisition of labor force (the cumulative rate of taxes on wage fund, social payments, etc.). Hereafter, we assume that $l > 0$, $1 > \alpha > 0$, $H \geq 1$.

The set of strategies of the first player is $\left\{\omega : \omega \geq 0\right\}$, and of the second player is $\left\{G : 0 \leq G \leq \hat{G}\right\}$, where $\hat{G}$ is the cumulative resource of the labor force. These sets are closed and convex.

Since the function $f_1(\omega, G)$ is concave in $\omega$ and convex in $G$, and the function $f_2(\omega, G)$ is concave in $G$ and linear in $\omega$, it is easy to prove the following statement.

**Statement 3.5.** Let $\omega^{(1)} = lH^{-1}$, $G^{(1)}$ be the solution of the equation $G = g(\omega^{(1)}, G)$, where $g(\omega, G) = \dfrac{\partial \overline{Q}(G, \omega)}{\partial \omega}$. Let also $G^{(1)} < \hat{G}$. Then the Nash equilibrium $(\overline{\omega}, \overline{G})$ exists in the game, $\overline{\omega}$ and $\overline{G}$ being defined either by the relations

$$\overline{G} = g(\overline{\omega}, \overline{G}), \quad \overline{G} = \frac{\alpha \nu}{l}, \quad \text{if} \quad \overline{G} \leq G^{(1)},$$

or by the equalities

$$\overline{G} = g(\omega^{(1)}, \overline{G}), \overline{\omega} = \omega^{(1)}, \quad \text{if} \quad \frac{\alpha \nu}{l} > G^{(1)}.$$

Note that the existence of $(\overline{\omega}, \overline{G})$ under the above assumptions follows from the results of Eichenberg [25]; properties $\overline{G}$ and $\overline{\omega}$ follows from the fact that

$$f_1(\overline{\omega}, \overline{G}) = \max_{\omega \geq 0} f_1(\omega, \overline{G}),$$

$$f_2(\overline{\omega}, \overline{G}) = \max_{0 \leq G \leq \overline{G}} f_2(\overline{\omega}, G).$$

Not everywhere differentiable function $f_2(\overline{\omega}, G)$ may achieve maximum values on $[0; \overline{G}]$ either at the ends of this interval (this case is excluded due to the assumptions that $f_2(\overline{\omega}, G)$ is concave and $G^{(1)} < \hat{G}$) or at the point where the derivative of this function is equal to 0 (this is possible only for $\overline{\omega} = \omega^{(1)}$ and $\dfrac{\alpha \nu}{l} > \overline{G}$) or at the point $\overline{G} = \dfrac{\alpha \nu}{l}$, where this function is nondifferentiable. The function $f_1(\omega, \overline{G})$ is differentiable everywhere; therefore, $\dfrac{\partial f_1(\overline{\omega}, \overline{G})}{\partial \omega} = 0$. Herefrom, we obtain the relations for $\overline{G}$ and $\overline{\omega}$.

Note that under real competition, employers and trade unions cannot frequently determine Nash optimal strategies. In particular, they have no necessary complete information: employers do not know the form of the function $\overline{Q}(G, \omega)$ and employees—the profitability of production $\alpha$. Therefore, searching for a compromise, both sides act by trial and error, purposefully changing their strategies toward the greatest increase of their payoff functions. Since the incomes of employees change, the aggregate demand $\nu$ varies. One of the models of such a dynamic is considered further.

In describing this model, we will use the previously made notations. Let us analyze the "employer–trade union" system functioning in continuous time. Assume that at each instant of time the players change their strategies toward the derivatives (i.e., toward the greatest local increase) of their payoff functions. Let also the rates of change in the demand be proportional to the change in the incomes of the employees. On these assumptions, considering a payment $\omega$, the amount of labor force $G$ bought by the employer, and the aggregate demand $\nu$ as functions of time $t$, we obtain the system of differential equations

$$\begin{aligned} \dot{\omega} &= \lambda_1\Big(G - g(\omega, G)\Big), \\ \dot{G} &= \lambda_2 F(\omega, G, \nu), \\ \dot{\nu} &= \rho(k\omega G - \nu), \end{aligned} \tag{3.36}$$

where $\lambda_1 > 0$, $\lambda_2 > 0$, $\rho > 0$, and $k > 0$ are some constants,

$$F(\omega, G, \nu) = \begin{cases} l - H\omega, \text{ if } lG \le \alpha\nu, \\ \quad -H\omega, \text{ if } lG > \alpha\nu. \end{cases}$$

This system and the initial conditions

$$\omega(0) = \omega^0, \quad G(0) = G^0, \quad \nu(0) = \nu^0 \tag{3.37}$$

constitute the model being considered.

Let the parameter $\rho$ be large compared with the other system parameters. This assumption means that the value of demand $\nu$ should react fast enough to change the revenue $\omega G$ of employees. Thus, in the economic system under study, the production will be intended to meet the demand of domestic customers.

The phase space of the system (3.36) is the set

$$E = \Big\{(\omega, G, \nu) : \; 0 \le \omega \le \hat{\omega}, \; 0 \le G \le \hat{G}, \; 0 \le \nu \le \hat{\nu}\Big\},$$

where $\hat{\omega} > 0$ and $\hat{\nu} > 0$ are some sufficiently large constants and $\hat{G}$ is the cumulative resource of labor force. We assume that once the variables of the system $(\omega, G, \nu)$ achieve their lower or upper bounds, they remain constant, i.e., the right-hand sides of the corresponding equations of the system (3.36) take zero values. We also assume that there are no spasmodic changes in the phase variables. In this connection, the right-hand sides of the system (3.36) have a first-kind discontinuity on a part of the plane $\sigma = \Big\{(\omega, G, \nu) \in E : lG = \alpha\nu\Big\}$, which divides the phase space into two domains, $D_1$ and $D_2$, where $D_1 = \Big\{(\omega, G, \nu) \in E : lG > \alpha\nu\Big\}$ and $D_2 = \Big\{(\omega, G, \nu) \in E : lG < \alpha\nu\Big\}$. In each of these domains, the right-hand sides of the system (3.36) are smooth functions on the interior of $E$. Continuous transition of a phase point following a phase curve, with time increasing, through the surface of discontinuity from one smoothness domain to the other corresponds to the so-called sewing together of solutions by continuity (see [16, 48]).

Existence of unstable sliding motion determined by [26, 48] in the system necessitates, under some conditions, for the surface of discontinuity. Let us determine the solution of the system (3.36) on $\sigma$ according to [26].

Thus, by the solution of the system (3.36) on the interval $[0, T]$ with the initial condition $\left(\omega_0, G_0, \nu_0\right) \in E$, we mean a piecewise differentiable function $\left(\omega(t), G(t), \nu(t)\right)$ that is defined on $[0, T]$, takes values in $E$, satisfies the system at all points except for the boundary $E$ and the points for which $lG = \alpha\nu$, and such that conditions (3.37) are satisfied.

Denote $\rho^{-1} = \mu$. System (3.36) where, according to the assumptions, $\mu$ is a small parameter, is singularly perturbed. Let us consider the following degenerate system defined by [70, 72], which corresponds to the system (3.36):

$$\begin{aligned} \dot{\omega} &= \lambda_1\Big(G - g(\omega, G)\Big), \\ \dot{G} &= \lambda_2 F(\omega, G, v), \\ \nu &= k\omega G. \end{aligned}$$

We may also consider this system as a mathematical model of competition between the employer and trade union provided that the demand for goods is proportional to payment. Therefore, its analysis is of independent importance. Obviously, it is reduced to an analysis of the system

$$\begin{aligned} \dot{\omega} &= \lambda_1(G - g(\omega, G)), \\ \dot{G} &= \lambda_2 Q(\omega, G), \end{aligned} \tag{3.38}$$

where

$$Q(\omega, G) = \begin{cases} l - H\omega, & \text{if } l \le \alpha k\omega, \\ -H\omega, & \text{if } l > \alpha k\omega. \end{cases}$$

We will consider the latter in the phase space

$$\hat{E} = \Big\{(\omega, G) : 0 \le \omega \le \hat{\omega},\ 0 \le G \le \hat{G}\Big\}.$$

Initial conditions (3.37) correspond to the conditions

$$\omega(0) = \omega_0, \quad G(0) = G_0 \tag{3.39}$$

of this system.

Denote the solution of the problem (3.38), (3.39) by $x^0(t) = \left(\omega^0(t), G^0(t)\right)$, and the solution of the system (3.36), (3.37), where $\mu = \rho^{-1}$, by $\Big(\omega(t, \mu), G(t, \mu), \nu(t, \mu)\Big)$. Let $\nu^0(t) = k\omega^0(t)G^0(t)$. For the problems under study, conditions of the Tikhonov theorem [70] are satisfied, except for special initial conditions, which is insignificant in this case. Obviously, the unique solution $x^0(t)$ of the problem (3.38), (3.39) exists on the interval $[0; T]$ for any $T > 0$ and, by virtue of the Tikhonov theorem, the unique solution $x(t,\mu) = \Big(\omega(t,\mu), G(t,\mu), \nu(t,\mu)\Big)$ of the problem (3.36), (3.37) exists for sufficiently small $\mu (\mu < \mu_0)$. If the trajectory $x^0(t)$ belongs to the interior of the set $\hat{E}$, we can pass to the limit for any $t$:

$$\omega(t,\mu) \underset{\mu\to 0}{\longrightarrow} \omega^0(t), \quad G(t,\mu) \underset{\mu\to 0}{\longrightarrow} G^0(t), \quad 0 \le t \le T,$$

$$\nu(t,\mu) \underset{\mu\to 0}{\longrightarrow} \nu^0(t), \quad 0 < t \le T.$$

Using the method of invariant manifolds for singularly perturbed systems of differential equations [69] and applying it to the case under study; we conclude that the results of the qualitative analysis of the system (3.38) can also be applied to the system (3.36). It is clear that the economic interpretation of the conclusions obtained in the analysis of the system (3.38) is equally valid for the model (3.36) for sufficiently small $\mu$.

Let us begin the qualitative analysis of the system (3.38) with determining its stationary points. Denote $\omega^1 = lH^{-1}$ and $\omega^* = l(\alpha k)^{-1}$. It is obvious that the quantity $\beta = \alpha k H^{-1}$ determines the mutual position of the straight lines $\omega = \omega^1$ and $\omega = \omega^*$. If $\beta > 1$, then $\omega^1 > \omega^*$, if $\beta < 1$, then $\omega^1 < \omega^*$. The interval $S = \Big\{(\omega, G) \in \hat{E} : \omega = \omega^*\Big\}$ is the discontinuity line of the system (3.38). If $\omega^1 > \omega^*$, then the interval $S^{(1)} = \Big\{(\omega, G) \in \hat{E} : \omega = \omega^{(1)}\Big\}$ is the set of points at which the right-hand side of the second equation of the system takes zero values. If $\omega^1 < \omega^*$, then the right-hand side of this equation is of constant sign and the system has no stationary points. Let us consider the equation $G = g(\omega^1, G)$ for $\omega^* \le \omega^1 \le \hat{\omega}$. Since $g(\omega^1, G)$ is a monotonically decreasing smooth function $G$ solution $G^1$, $0 < G^1 < \hat{G}$. Thus, when $\beta \ge 1$, there is a unique equilibrium state $P^1 = (\omega^1, G^1)$ of the system (3.38).

Denote $\nu^1 = k\omega^1 G^1$. Since $\beta \ge 1$ yields the inequality $lG^1 \le \alpha\nu^1$, the point $(\omega^1, G^1, \nu^1)$ is the equilibrium state of the system (3.36). It is obvious that the equilibrium state of the system (3.36) is stable if and only if the equilibrium state of the system (3.38) is stable.

To analyze the stationary point $P^1$ of this system for stability, let us use the Lyapunov theorem of first-order stability. As a result, we obtain that stability of the point $P^1$ is determined by the characteristic equation

$$\lambda^2 + \lambda_1 g_\omega\left(\omega^1, G^1\right)\lambda + \lambda_1\lambda_2\Big(1 - g_G\left(\omega^1, G^1\right)\Big) = 0. \qquad (3.40)$$

Assumed that the condition

$$g_\omega\left(\omega^1, G^1\right) > 0 \qquad (3.41)$$

holds, which agrees with the condition of monotonic increase of the function $g(\omega, G)$, with respect to the variable $\omega$. Since $g(\omega, G)$ monotonically decreases

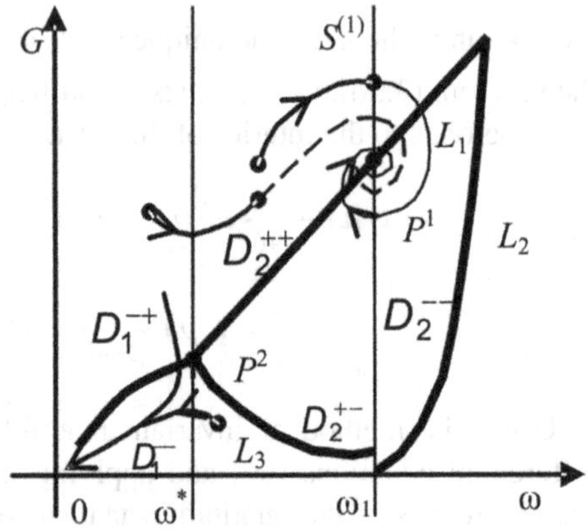


**Fig. 3.9** Formation of postclassical cycles ($\beta > 1$)

in $G$, the inequality $g_G\left(\omega^1, G^1\right) \leq 0$ is true and hence Eq. (3.40) is stable, i.e., its roots are in the left complex half-plane. We will assume below that the determinant of Eq. (3.40) is negative. Then the point $P^1$ by virtue (3.41) is a stable focal point of the system (3.38).

Let us now construct the phase-plane portrait of system (3.38). Denote $L_1 = \left\{(\omega, G) \in \hat{E} : G = g(\omega, G)\right\}$. Let us first consider the case $\beta > 1$ (Fig. 3.9). To the right of the straight line $S$, in the domain $D_2$, the system (3.38) takes the form

$$\begin{aligned} \dot{\omega} &= \lambda_1\Big(G - g(\omega, G)\Big), \\ \dot{G} &= \lambda_2(l - H\omega). \end{aligned} \tag{3.42}$$

The lines $L_1$ and $S^{(1)}$ divide $D_2$ into the sets $D_2^{+-}$ (to the left of the line $S^{(1)}$ and downwards from $L_1$), $D_2^{++}$ (to the left of $S^{(1)}$ and above $L_1$), $D_2^{-+}$ (to the right of $S^{(1)}$ and above $L_1$), and $D_2^{--}$ (to the right of $S^{(1)}$ and below $L_1$). The notation $D_2^{\gamma_1,\gamma_2}$ stands for the domain in which the right-hand side of the equation determining $G$ has the sign of $\gamma_1$, and the right-hand side of the equation determining $\omega$, the sign of $\gamma_2$.

In the domain $D_1$ lying to the left of the straight line $S$, the system (3.38) can be represented as

$$\begin{aligned} \dot{\omega} &= \lambda_1\Big(G - g(\omega, G)\Big), \\ \dot{G} &= -\lambda_2 H\omega. \end{aligned}$$

The right-hand side of the second equation of the system is of constant sign; therefore, $D_1$ consists of the domains $D_1^{-+}$ (above $L_1$) and $D_1^{--}$ (below $L_1$). The notation of signs is the same as in the partition of $D_2$.

Let us now consider the motion of the phase point $\big(\omega(t), G(t)\big)$ of the system (3.38) if it stays within the domain $D_2$. For the sake of determinacy, assume that the motion begins at the time $t = 0$ from a point that belongs to the set $D_2^{++}$. As the phase point passes through the domain, both employment $G$ and wages $\omega$ increase. Such a situation is typical for the stage of growth of an economic cycle. Once the level of wages exceeds its equilibrium value $\omega^{(1)}$, it becomes inexpedient for the employer to hire a new labor force thus increasing employment. For constant labor productivity, this reduces production volumes; the growth is replaced by a decline. As the phase point passes through the domain $D_2^{-+}$, $G$ decreases, and $\omega$ increases. At early stages of decline, the trade union continues realizing the strategy of increasing payment, which was generated for growth. This may deepen the decline. Once the phase point intersects the straight line $L_1$, it falls into the domain $D_2^{--}$. Here, both employment $G$ and payment for work $\omega$ decrease. Such a situation is typical of the completion stage of decline and for depression—the next stage of an economic cycle. Once the phase point intersects the straight line $S^{(1)}$, the employer is interested in increasing employment under conditions of the continuing price reduction of labor force. Production and aggregate demand start increasing. The motion of the phase point on the set $D_2^{+-}$ corresponds to the stage of revival of the economy. Once the phase point intersects the curve $L_1$, it falls again into the domain $D_2^{++}$. Then the process of its motion described above repeats. Since $P^1$ is a stable focal point of the system (3.38), amplitude of the cycle decreases if they are sufficiently close to this point. Processes with damping cyclicity are typical of postclassical cycle of economic development; see, for instance, [23]. Thus, while a classical business cycle under certain conditions may be treated as a consequence of monopsony on labor market (see Sect. 3.3), a postclassical business cycle under similar assumptions may be treated as a result of bilateral monopolistic competition in this segment of the national market.

Note that the point $P^1$ to which the above process converges will be the Nash equilibrium in the above considered "employer–trade union" game. As labor productivity $l$ increases, the point moves to the right along the curve $L_1$, the values of payoff functions of the players increasing. As a result, both labor marketers are interested in innovational development of economy. Such development is accelerated when suitable technological prerequisites appear. If the situation $(G, \omega)$ is close to equilibrium at the time of acceleration of innovational development, this equilibrium will be then violated and the cyclicity of economic development will be strengthened. This conclusion agrees with the basic statements of Shumpeter's theory of innovation cycles developed for postclassical cycles [49].

Beginning the motion in the domain $D_2$, the phase point may fall, except for the equilibrium state $P^1$, on a point of the axis $0\omega$ (in the domain $D_2^{--}$) or to cross the straight line $S$ and fall into the domain $D_1$. In the former case, $G = 0$ and hence, the process arrives at the boundary of the phase space $\hat{E}$ and then stops. This situation corresponds to zero employment and zero volume of production, i.e., the decline of economy happens. To prevent such a case, the trajectory of the phase point should be above the line $L_2$, i.e., the trajectory $x^0(t) = \left(G^0(t), \omega^0(t)\right)$ running through the point $(\omega^1, 0)$. This can be promoted, in particular, by a struggle of trade unions against reduction of employment under conditions of a recession.

Let us consider the motion of a phase point in the domain $D_1$ provided that it started on the set $D_1^{-+}$. In such motion, payment $\omega$ increases and employment $G$ decreases. As a result, the phase point either crosses the straight line $S$ and enters the domain $D_2$ crosses the curve $L_1$ and enters the domain $D_1^{--}$. In the latter domain, both $G$ and $\omega$ monotonically decrease. As a result, either $G(t)$=0 or $\omega(t)$=0 at time $t$ and the process stops. Thus, reaching the domain $D_1^{--}$ by the phase point inevitably causes collapse of economy.

In the analysis of variable-structure systems, it is important to analyze the motion of phase points near the switching surface $S$. The phase point can pass from the domain $D_2$ to the domain $D_1$ only through the interval on the straight line $S$ between the point $(\omega^*, 0)$ and the point of intersection of $S$ with the line $L_1$ (denote this point by $P^2$). In this case, it reaches the domain $D_1^{--}$, and the process ends in collapse of economy. The phase point can pass from the domain $D_1$ to the domain $D_2$ through the part of the straight line $S$ lying above the point $P^2$. Here, it reaches the domain $D_2^{++}$ and continues to move along the trajectory considered earlier. At the point $P^2$, the rate of change of $G$ is positive and the phase point passes to the domain $D_2^{++}$. To avoid collapse of economy due to reaching the domain $D_1^{--}$, the trajectory of the phase point should pass above the line $L_3$, which is a part of the trajectory $\left(G^0(t), \omega^0(t)\right)$, beginning at one of the points of the line $S_1$ and passing through the point $P^2$.

Let us now consider the behavior of the system (3.38) for $\beta < 1$ (Fig. 3.10a). In this case the system has no stationary points, and the curve $L_1$ divides the phase space into the domains $D^{-+}$ and $D^{--}$. Starting the motion in the domain $D^{-+}$, the phase point $\left(G(t), \omega(t)\right)$ will cross (sooner or later) $L_1$ and will reach $D^{--}$. Motion in this domain is similar to the motion in the domain $D_1^{--}$ considered earlier and

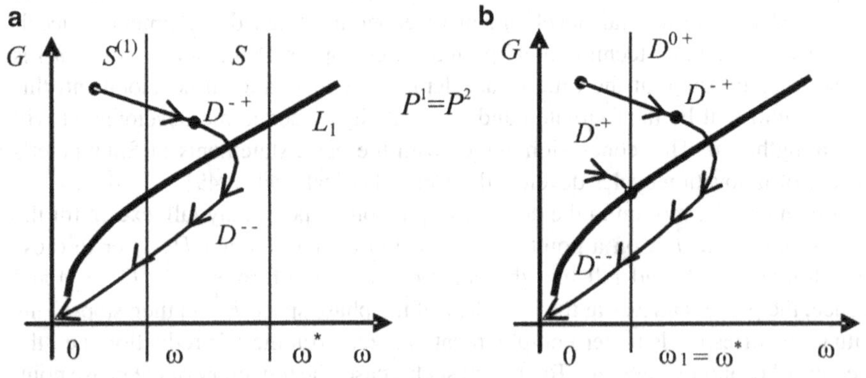


**Fig. 3.10** Collapse of production $\beta < 1$ and collapse of production ($\beta = 1$)

will be finished at one of the points for which either $G = 0$ or $\omega = 0$. Thus, collapse of economy is inevitable in this case.

It remains to consider the case $\beta = 1$ (Fig. 3.10b). For it $\omega^* = \omega^1$, and the points $P^1$ and $P^2$ coincide. Here $P^1$ is a singular point (see [26]), and the motion of phase points is similar to the case $\beta < 1$. The right-hand side of the equation determining $G$ takes zero values at points of the straight line $S = D^{0+}$, but this does not affect the form of the trajectory of motion of the phase point: the employment $G$ decreases and the wages $\omega$ increase above the line $L_1$, and both $G$ and $\omega$ decrease below this curve. Apparently, a unique trajectory $\left(G_0(t), \omega_0(t)\right)$ of the motion of a phase point finishing at the Nash equilibrium state $P^1$ exists here. For all other trajectories, the result is collapse of economy. Note that the stability domain of the focal equilibrium state $P^1$ concentrates near this point as $\beta \to 1^+$. In this case, loss of stability is rigid–catastrophic.

The analysis of the models (3.36), (3.37) and (3.38), (3.39) shows that the result of competition between the employer and trade union may be both a compromise (the Nash equilibrium) and collapse of economy. The latter will be preceded by a rather long downswing. It is possible to avoid such undesirable scenario by establishing the minimum payment at a level greater than $\omega^*$ (this excludes passage to the domain $D_1$ for $\beta > 1$) and not admitting a significant decrease in employment (when the trajectory falls into the domain $D^{--}$, the both marketers are interested in this). However, when $\beta \leq 1$ these measures are insufficiently effective.

To reach a compromise, the condition $\beta = \alpha k H^{-1} > 1$ should be met. For these conclusions to be true also for a more general case, i.e., for the model (3.36), $\beta$ should not be close to unity. As already mentioned, $H = 1 + h$, where $h > 0$ are additional costs due to acquisition of labor force. Mainly, $h$ coincides with the cumulative rate of indirect taxes for wages fund and other social payments. A significant part of inflows from such payments is then redistributed among customers and promote increase of their demand. The coefficient $k$ is greater than 1 due to, first, the indicated redistribution of incomes, and second, the multiplicative effect that strengthens the initial increase of demand owing to the incomes obtained during production, and additional consumption of industrial goods. Thus, we may say that $k = (1 + \gamma h)$, where $\gamma$ is a part of inflows due to social payments, which is redistributed among ultimate consumers, and $m$ is the Keynes multiplier [62]. The condition $\beta > 1$ can be written as $\alpha m(1 + \gamma h)(1 + h)^{-1} > 1$. The last inequality holds if $\gamma$ is close to unity (the fiscal system is socially oriented), yield of production $\alpha$ is not much less than 1 (the economy is efficient), and $m$ is much greater than 1 (the existing macroeconomic proportions create premises for a considerable increase when an aggregate demand increases). Of importance is also the orientation of social productive forces toward satisfaction of mainly domestic demand that is reflected by a large value of the parameter $\rho$ of model (3.36). Cycles of postclassical type, which are superimposed on the general tendency of innovational growth of economy, may appear only when these conditions are realized. In other cases, the two-sided monopolistic competition on the labor market has no less destructive consequences than monopsony.

In addition to the case of using large values of $\rho$, of interest is the analysis of the model (3.36)–(3.37) for small values of this parameter. This takes place, for example, in the case when some production is mainly exported. This situation is sufficiently typical for many branches of production in countries with transitive economy.

We first consider the extreme case, when $\rho = 0$. In this case, the model (3.36)–(3.37) assumes the form

$$\dot{\omega} = \lambda_1\Big(G - g(\omega, G)\Big),\qquad \dot{G} = \lambda_2 F\Big(\omega, G, \nu^0\Big), \tag{3.43}$$

$$F\Big(\omega, G, \nu^0\Big) = \begin{cases} l - H\omega, \text{ if } G \le \dfrac{\alpha\nu^0}{l}, \\ -H\omega, \text{ if } G > \dfrac{\alpha\nu^0}{l}, \end{cases} \tag{3.44}$$

$$\omega(0) = \omega^0, \quad G(0) = G^0. \tag{3.45}$$

This model describes the situation, where the incomes of employed persons do not influence on the demand for the production produced by the employer. This occurs, in particular, in the case of purely capitalized products and (or) when the production is realized only in foreign markets.

The phase space of the system (3.43)–(3.45) is the set

$$E^0 = \Big\{(\omega, G) : 0 \le \omega \le \hat{\omega},\ 0 \le G \le \hat{G}\Big\},$$

where $\hat{\omega}$ and $\hat{G}$ are defined in the same way as in constructing the set $E$ for the system (3.36)–(3.37).

As it was shown in [6], the dependence $G(\omega)$, obtained as a result of solution of the equation $G = g(\omega, G)$, with respect to $G$ for a fixed $\omega$ is a continuous monotonically increasing function. We use this fact in performing a qualitative analysis of the system (3.43)–(3.45). We denote by $L_1$ the line that reflects the mentioned dependence on the plane $G_0\omega$ (Fig. 3.11).

The stationary point of this system is the point of intersection of $L_1$ and the vertical straight line $S^{(1)} = \Big\{(G, \omega) : \omega = l \;/\; H\Big\}$, provided that the coordinate $G$ of this point satisfies the inequality $G^1 \le \dfrac{\alpha\nu^0}{l}$. It may be noted that a Nash equilibrium corresponds to the mentioned point in the static nonantagonistic game of the trade union and employer that is considered at the beginning of this section. The sign of the right sides of both differential equations of the system (3.43)–(3.45) also varies at the point of intersection of $L_1$ and the horizontal straight line

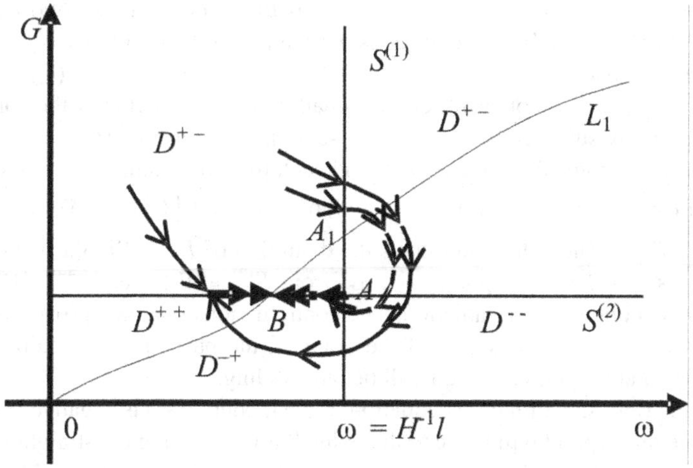


**Fig. 3.11** The intersection point of $L_1$ and $S^{(1)}$ is above the straight line $S^{(2)}$ the motion of the phase point and a sliding motion

$S^{(2)} = \left\{(G, \omega) : G = \frac{\alpha \nu^0}{l}\right\}$. Therefore, in constructing the phase portrait of the system, it is necessary to independently consider the cases when the intersection points $L_1$ and $S^{(1)}$, respectively, are above and below the straight line $S^{(2)}$.

We begin with the first case. We denote by $D^{\gamma_1 \gamma_2}$ the set from $E^0$ for which the sign of the right side of the first equation of model (3.42) is $\gamma_1$ and the sign of the right side of the second equation is $\gamma_2$ ($\gamma_1, \gamma_2 = -, +$). This case is presented at Fig. 3.11.

The set $D^{++}$ is located between the lines $L_1$ and $S^{(2)}$, and the set $D^{+-}$ is above the lines $S^{(2)}$ and $L_1$. The set $D^{--}$ is located between the lines $S^{(2)}$ (lower) and $L_1$ (upper), and it also contains points located to the right of the straight line $S^{(1)}$ and below the line $S^{(2)}$. The set $D^{-+}$ is bounded by the line $L_1$, by the straight line $S^{(2)}$ (from above), and by the straight line $S^{(1)}$ (from the right).

Let us consider the motion of the phase point of the system (3.43)–(3.45) (see Fig. 3.11). On the set $D^{+-}$, wage $\omega$ increases and the employment $G$ decreases. After the intersection of the line $L_1$ by the phase point, both $\omega$ and $G$ decrease. When the phase point intersects $S^{(1)}$, $G$ begins to increase, but $\omega$ decreases in this case. After the intersection of the line $L_1$, both $\omega$ and $G$ increase. Of interest is the behavior of the phase point in the case when it intersects the straight line $S^{(2)}$ to the left of the point $A$. Let us show that, in this case, it remains on this straight line.

Let us assume that this is not the case, i.e., that, at some moment of time $t_0$, the phase point $\left(\omega^0(t), G^0(t)\right)$ has been located on the straight line $S^{(2)}$ to the left of the point $A$, and that, at the moment of time $t_1 > t_2$, it does not belong to this straight line. Since the right side of the first equation of the system (3.43) is negative at the point $A$, the phase point can deviate from $S^{(2)}$ only to the left of the point $A$.

In this case, at the moment of time $t_1$, it can belong to one of the following sets: $D^{+-}$, $D^{-+}$, $D^{++}$ and $D^{--}$. Let us consider the first of these four cases. By virtue of the relative position of $D^{+-}$ and $S^{(2)}$, the condition $G^0(t_1) > G^0(t_0) + \varepsilon$ must be true, where $\varepsilon > 0$ is some sufficiently small number. By virtue of the continuity of $G^0(t)$, for any sufficiently small $\delta > 0$ such that we have $G^0(t_1 - \delta) > G^0(t_0)$ and $t_1 - \delta > t_0$, some $\varepsilon_1 > 0$ can be found such that we obtain $G^0(t_1) > G^0(t_1 - \delta) + \varepsilon_1$. From this we have $G^0(t_1 - \delta) > 0$, but the point $\left(G^0(t_1 - \delta), \omega^0(t_1 - \delta)\right)$ belongs to $D^{+-}$. The latter contradicts the definition of $D^{+-}$. Similarly, under the assumptions that the phase point can pass from $S^{(2)}$ to the sets $D^{-+}$, $D^{++}$, and $D^{--}$, we also obtain a contradiction. The contradictions obtained prove that, after the attainment of $S^{(2)}$ to the left of the point $A$, the phase point remains on this straight line and its further motion will be only sliding.

Since the right side of the first equation of the model (3.43) is negative below the line $L_1$, the phase point will move to the point $B$ after attaining the straight line $S^{(2)}$ in the interval $AB$. In this case, the employment $G$ remains unchangeable and the labor remuneration $\omega$ decreases. If the phase point attains $S^{(2)}$ to the left of the point $B$, then, at the corresponding points of this straight line, the right side of the first equation of the system (3.43)–(3.45) is positive and the phase point moves to $B$. The value of $G$ is also constant and that of $\omega$ increases.

Thus, after attaining $S^{(2)}$, the phase trajectory will terminate at the point $B$. The points $A$ and $B$ will not be stationary points of the system being investigated since they belong to the discontinuity surface $S^{(2)}$. However, these points can be considered as equilibrium points on the basis of the following definition accepted in economic-mathematical modeling: if the phase point is in one of these states, then the values of the variables of the model remain unchangeable later on. In this case, $B$ will be a point of stable equilibrium since the phase point that moves from a sufficiently small vicinity of $B$ will inevitably intersect $S^{(2)}$.

In contrast to $B$, the point $A$ is an unstable equilibrium point. Any small decrease in $\omega$ and (or)$G$ will generate a trajectory that will terminate at the point $B$. This point corresponds a balanced labor remuneration for a given demand (Fig. 3.12).

We note the following distinctive feature of motion of the phase point. If it is not above the straight line $S^{(2)}$ at some moment of time $\bar{t}$, then, when $t > \bar{t}$, it will also be not above this straight line, i.e., the following inequality is true:

$$G \leq \frac{\alpha \nu^0}{l}. \tag{3.46}$$

We continue our analysis of the system. Let us now consider the case when the point of intersection of the line $L_1$ and the straight line $S^{(1)}$ is below the straight line $S^{(2)}$ (Fig. 3.13). By analogy with the previous case, the set $D^{++}$ is bounded by the lines $L_1$, $S^{(2)}$, and $S^{(1)}$ and the set $D^{+-}$ is located above the straight line $S^{(2)}$ and also between the lines $S^{(1)}$ and $L_1$. The set $D^{--}$ is located below the line $L_1$ and to the right of $S^{(1)}$. The set $D^{-+}$ is located to the left of $S^{(1)}$ and below the line $L_1$. Let us consider the motion of the phase point in this case. Let it begin to move from the domain $D^{--}$. The wage $\omega$ and employment $G$ will first decrease. This situation is typical for the stage of a recession in production in a business cycle. When the

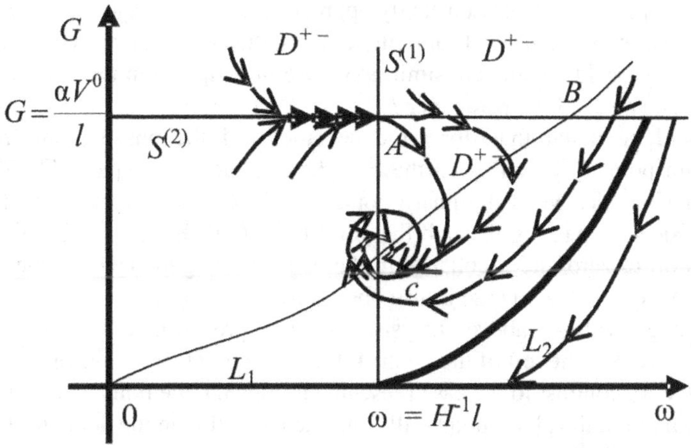


**Fig. 3.12** The point of intersection of $L_1$ and $S^{(1)}$ is below the straight line $S^{(2)}$, the motion of the phase point, and a sliding motion

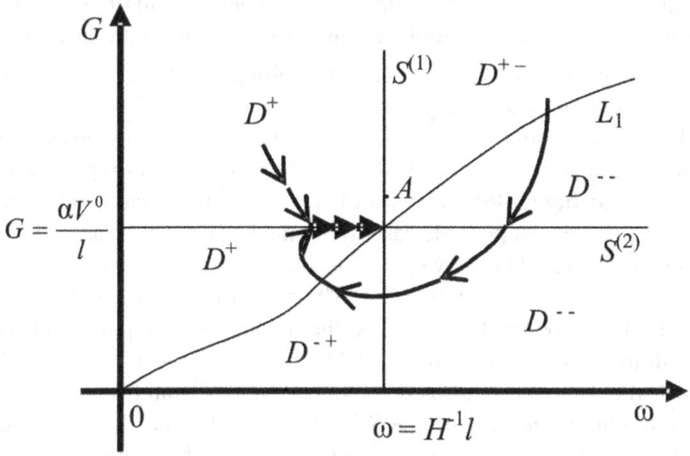


**Fig. 3.13** The lines $L_1$, $S^{(1)}$ and $S^{(2)}$ intersect one point, the motion of the phase point, and its sliding motion

phase point will intersect the straight line $S^{(1)}$ and attain the domain $D^{-+}$, wage $\omega$ will continue to decrease, but the employment $G$ will begin to increase. It is typical for the stage of revival of economy. After the point of intersection of the line $L_1$ by the phase point, a job growth will be accompanied by an increase in the labor remuneration, which is inherent in the stage of economic growth.

When the phase point intersects the straight line $S^{(1)}$, employment begins to decrease and, at the same time, wage increases. This creates preconditions for a renewal of recession that increases as a result of decreasing labor remuneration after the intersection of the line $L_1$ by the phase point. Note that the intersection point of the line $L_1$ in each cycle constantly approaches the equilibrium point $c$, which stipulates a decrease in the cycle amplitude. Thus, the business cycles observed here are of postclassical type and are similar to those investigated in this section for the case of large values of the parameter $\rho$.

It should be noted that, for large deviations of the phase point from the equilibrium point $c$, it can reach the boundary of the phase space $E^0$. This can occur, for example, when the phase point is located to the right of the line $L_2$, i.e., the trajectory passing through the point $(H^{-1}l, 0)$. In this case, employment decreases up to zero and a collapse in the economy begins. Here, some analogy with the above-considered case is also observed.

By analogy with the above analysis, it may be proved that, after attaining the straight line $S^{(2)}$ to the left of the point $A$, the phase point can move only along the straight line. In contrast to the case presented in Fig. 3.9, the point $A$ will not be an equilibrium point, the phase point will continue its motion in the domain $D^{+-}$, and both $\omega$ and $G$ will change.

As in the model (3.38), (3.39), the amplitudes of postclassical cycles in model (3.43)–(3.45) decrease with constant labor productivity $l$, motion terminates at the point $\overline{M}^1$, and a Nash equilibrium corresponds to this point in the static nonantagonistic game of the employer and trade union. With increasing labor productivity, the point $\overline{M}^1$ is shifted to the right along the line $L_1$, which leads to an increase in the wage $\omega$ and employment $G$ and is accompanied by a casual increase in the cyclicity of economic dynamics. As it was noted in the previous section, such a development is inherent in cycles of innovative Schumpeter-type development. We note that inequality (3.46) is also fulfilled for $t > \bar{t}$ in the case considered.

Let us consider the last possible case, when the lines $L_1$, $S^{(1)}$, and $S^{(2)}$ intersect at one point (Fig. 3.13). On the whole, it is similar to the case when the lines $L_1$ and $S^{(1)}$ intersect above the straight line $S^{(2)}$ but, in this case, the interval $AB$ is transformed into the point $A$ that will be the unique stable equilibrium point. By analogy with the case presented in Fig. 3.11, the set $D^{++}$ is bounded by the lines $L_1$ and $S^{(2)}$, the set $D^{+-}$ is located above the lines $S^{(2)}$ and $L_1$, the set $D^{--}$ is located to the right of the straight line $S^{(1)}$ and below the line $L_1$, and the set $D^{-+}$ is located between the lines $L_1$ and $S^{(1)}$. After reaching the straight line $S^{(2)}$ to the left of the point $A$, the phase point moves along this straight line, which is proved by analogy with the previous cases. The motion of the phase point along the sets $D^{+-}$ and $D^{--}$ is accompanied by a decrease in employment, i.e., a setback in production. At the same time, wage $\omega$ increases in the first case and decreases in the second case. The motion of the phase point along the sets $D^{-+}$ and $D^{++}$ is accompanied by a job growth, i.e., a revival of manufacturing. In this case, the value of $\omega$ decreases within the set $D^{-+}$ and increases within $D^{++}$.

Let us generalize the results of our analysis. The considered model illustrates two variants of economic dynamics. According to the first of them (Fig. 3.11) and (Fig. 3.13), after some recession and (or) revival of the economy (whose scope and course are determined by its initial state), the stabilization at an equilibrium point takes place and such a stabilization can be considered as an analogue of a situation of full employment in a system with a competitive labor market [31]. According to the second variant (Fig. 3.12), cycles of postclassical type with decreasing amplitudes arise and their repetition leads to a gradual stabilization at an equilibrium point that is a compromise (a Nash equilibrium) attained in static nonantagonistic game between the employer and the trade union. In both variants, the process can attain the boundary of the phase space $E^0$(an economic collapse) in the case of significant deviations of the state of the system being modeled from equilibrium points.

To realize the second variant (a cyclic development), a high demand for the production produced by the employer and a high profitability of manufacturing are necessary (a significant specific weight of added cost in the price of the production). The productivity gain that shifts the straight lines $S^{(1)}$ and $S^{(2)}$ to the right and downward, respectively, is the factor of the passage to the first variant of development (stabilization at an equilibriums point).

It may be noted that the results of analysis of the model (3.43)–(3.45) can be extended to the model (3.36)–(3.37) in the case when the parameter $\rho$ is so small and the phase point attains an equilibrium state so quickly that the quantity demand $\nu$ cannot essentially change. In the general case, the system (3.36)–(3.37) requires a special analysis when it functions with nonzero values of $\rho$ in a time interval $[0, T]$, where $T$ can be arbitrary large.

We perform such an analysis, assuming that $\rho$ is sufficiently small in comparison with $\lambda_1$, $\lambda_2$, and the other parameters. In other words, the quantity demanded $\nu$ changes more slowly than other variables of the system. We have earlier noted that, when $\beta = \alpha k H^{-1} \le 1$, the motions of phase points along phase curves of equations terminate at the boundary of the domain $E$, i.e., we have an economic collapse. Therefore, in what follows, we will consider only the case when $\beta > 1$. Then, according to the above analysis, the system (3.36) has the unique exponentially stable stationary point $M^1$. In a sufficiently small vicinity of this point, phase curves are of the form of compressed spirals. Of interest is the investigation of the dynamics of this system on the entire phase space, in particular, the estimation of the domain of influence of the point $M^1$. To this end, we lean upon the assumption that follows from the conditions being considered and consists of the fact that the phase point that has attained an arbitrary plane of the form $\nu = C$ remains close to it for some time.

Let us consider a two-parametric family of phase trajectories outgoing from the plane $\nu = \nu^1$, where $\nu^1$ is the corresponding component of the equilibrium point $M^1$. Its behavior in a finite time interval is described by the following system of the differential equations:

$$\begin{aligned} \dot{\omega} &= \lambda_1\Big(G - g(\omega, G)\Big), \\ \dot{G} &= \lambda_2 F\Big(\omega, G, \nu^1\Big), \end{aligned} \tag{3.47}$$

where $F(\omega, G, \nu^1)$ is defined similarly to (3.44) with the replacement of $\nu^0$ by $\nu^1$. The system (3.47) is considered in the domain $E^0$ under initial conditions (3.45).

The phase portrait of the system (3.47) is presented in Fig. 3.14a and is similar to that presented in Fig. 3.12. Here, the straight line $G = G_1 = l^{-1}\alpha\nu^1$ plays the role of the line $S^{(2)}$. We note that, according to the definition of the stationary point $M^1 = \left(\omega^1, G^1, \nu^1\right)$, we have $\nu^1 = k\omega^1 G^1$, but $\omega^1 = lH^{-1}$ and, hence, the truth of the inequality $\beta > 1$ implies the inequality $G_1 > G^1$. The correspondence between the families of phase curves of the systems (3.38) and (3.47) means that, during the time providing the passage of phase points of the system (3.47) from the branch of the curve $L_1$ located to the right of the point $\overline{P}^1$, to its left branch, the projections of trajectories of the system (3.36) onto the plane $\omega 0\, G$ parallel to the axis $\nu$ are approximated by system trajectories (3.47) up to $\rho$. In this case, to the trajectories of the system (3.47) that terminate at the boundary of the phase space also correspond similar trajectories of the system (3.36). Now, taking into account that the dynamics of the system (3.47) is fundamentally the same as the dynamics of system (3.43), we draw the conclusion that the projection of the two-parametric family of phase curves of the system (3.36) mainly coincides with that presented in Fig. 3.14a. We note a distinctive feature connected with phase curves of the family that approach the plane $\alpha$. We choose one of them. The motion of the phase curve after attaining the plane $\sigma$ is sliding, and the projection of this sliding mode onto the plane $\omega 0 G$ is close (is in the vicinity equal approximately to $\rho$) to a fragment of the sliding mode in Fig. 3.14a. As is obvious, the mentioned distinctive feature does not play any significant role in this case.

We now consider the two-parametric family of phase curves of the system (3.36) whose beginnings are on the plane $\nu = \nu^2$, where $\nu^2 > \nu^1$. In this case, a system similar to the system (3.47) can also be constructed. The phase portrait of this system is presented in Fig. 3.14b. Here, the line $S^{(2)}$ is the straight line $G = G_2 = \alpha l^{-1}\nu^2$. In this case, all the distinctive features of the motion of the projection of the family are preserved that have been mentioned in the previous case, including the occurrence of postclassical cycles close to the equilibrium point $\overline{P}^1$ and sliding motion.

The case when the motion begins with the plane $\nu = \nu^3$, $\nu^3 < \nu^1$, and $G_3 = \alpha l^{-1}\nu^3 > G^1$, is similar to the previous one.

We note that, in the three considered cases, the phase trajectory can attain the boundary of the domain $E$ (an economic collapse) as a result of significant deviations of the initial position of the phase point from the equilibrium state $M^1$.

We now consider the two-parametric family of phase curves of the system (3.36) whose beginnings belong to the plane $\nu = \nu^4$, $\nu^4 < \nu^1$, and, at the same time, we have $G_4 = \alpha l^{-1}\nu^4 < G^1$. Next, we prove that all the mentioned curves terminate at the boundary of the phase space and, hence, the set of initial conditions does not belong to the domain of influence of the stationary point $M^1$. By analogy with the previous reasoning, we obtain a system of the form (3.47) in which $\nu^1$ is replaced by $\nu^4$. In this case, the point $\left(G^1, \omega^1\right)$ is above the line $S^{(2)}$ and the phase portrait of the system in Fig. 3.14c) is similar to that presented in Fig. 3.11. Let $\left(\hat{\omega}(t), \hat{G}(t)\right)$ be a phase trajectory that changes into sliding motion when $t = \hat{T}$.

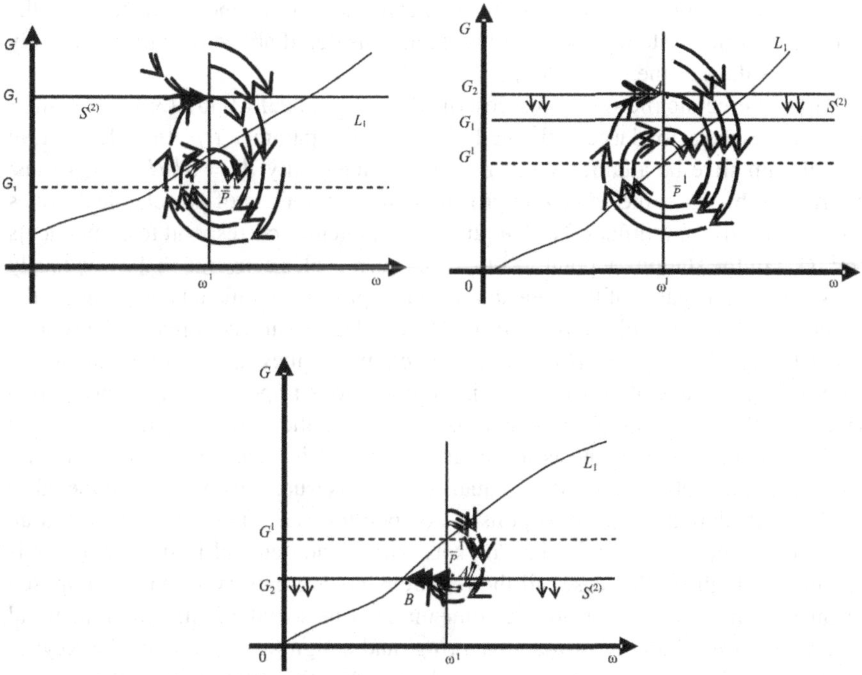


**Fig. 3.14** Two-parametric family of phase trajectories when $v = v^1$ (**a**), $v = v^2 > v^1$ (**b**), and $v = v^2 < v^1$ (**c**); the motion of the phase point, its sliding motion, and a shift of the line $S^{(2)}$

We now consider the trajectory $\left(\omega(t,\rho), G(t,\rho), \nu(t,\rho)\right)$ that is generated by the system (3.36) and satisfies the conditions $\omega(0,\rho) = \hat{\omega}(0)$, $G(0,\rho) = \hat{G}(0)$, and $\nu(0,\rho) = \nu^4$. The theorem on smooth dependence on the parameter of solutions of systems of differential equations implies the existence of $\rho_0 > 0$, and $T(\rho)$ (where $T(\rho)$ is a function smooth in the interval $[0,\rho_0]$ and $T(0) = \hat{T}$ such that $\left(\omega(t,\rho), G(t,\rho), \nu(t,\rho)\right)$ belongs to the fracture plane $\sigma$ when $t = T(\rho)$. It is obvious that the motion along the trajectory $\left(\omega(t,\rho), G(t,\rho), \nu(t,\rho)\right)$ assumes sliding character from the moment of time $t = T(\rho)$. This implies the existence of a moment of time $\tilde{T}$ for which the equality $\left(\hat{\omega}(\tilde{T}), \hat{G}(\tilde{T})\right) = B$ is fulfilled. Using again the theorem on the smooth dependence of the solution on the parameter, we conclude that there exists a moment of time $T_1(\rho)$, i.e., a function smooth in $[0,\rho_0]$ and $T_1(0) = \tilde{T}$, such that, when $t = T_1(\rho)$, the point $\left(\omega(t,\rho), G(t,\rho), \nu(t,\rho)\right)$ belongs to a curve (on the fracture plane $\sigma$) whose projection is the curve $L_1$. In the sliding mode $\left(\omega(t,\rho), G(t,\rho), \nu(t,\rho)\right)$ moves along the mentioned curve to the special stable point $\omega = 0$, $G = 0$. Thus, the above statement on the termination of phase curves of the system (3.36) at the boundary of the domain $E$ is proved for small $\rho$. There is a good reason to believe that the distinctive features of the dynamics of the system (3.36) that has been considered above take place not only for small values of the parameter $\rho$.

Thus, the economic dynamics described by the model (3.36)–(3.37) is similar in many respects for large and small values of the parameter $\rho$. In order that an equilibrium state be attained when $\rho \neq 0$, the inequality $\beta = \alpha k H^{-1} > 1$ must be true. In this case, the economic growth assumes the form of postclassical cycles with decreasing amplitudes. Nash equilibrium strategies correspond to components $(\omega^1, G^1)$ of the stationary point $M^1$ to which such cycles converge in the previously considered static game of the trade union and employer. Significant deviations of the initial position of the phase point from $M^1$ can lead to an economic collapse and, under the condition $\beta = \alpha k H^{-1} \leq 1$, a collapse is inevitable. In contrast to the case of large values of $\rho$, when the initial conditions imposed on $\nu$ do not play a vital part, the dynamics of the system is sensitive to the initial quantity demanded when $\rho$ is small. If demand is less than the critical level $\nu^* = \alpha^{-1} l\, G^1$, then a collapse is inevitable even if the inequality $\beta > 1$ is true. This means that the labor market with double-sided monopolistic competition can efficiently function in an economy oriented toward foreign markets only if the demand in these markets is sufficiently high. In this case, all the requirements are preserved that are imposed on an economy oriented toward the home market (the social orientation of the fiscal system, high profitability of manufacturing, and a significant value of the Keynes multiplier [35]). One can draw the conclusion that the development of the labor market with bilateral monopolistic competition in an open economy is connected (all other factors being equal) with a high risk in contrast to a closed economy.

We also note that, for small values of $\rho$ and initial values $\nu > \nu^*$, both employers and employed persons are interested in a local increase in the labor productivity $l$. In this case, an equilibrium point $\overline{P}^1$ is shifted to the right and upward along the line $L_1$. Thus, both the labor remuneration $\omega^1$ and employment $G^1$ are increased. At the same time, a labor productivity growth can increase the deviation of the current state $(\omega, G)$ from the equilibrium state $\left(\omega^1, G^1\right)$, which increases cyclicity.

Based on the analysis of static and dynamic (the systems (3.36)–(3.39) and (3.43)–(3.45)) models of a bilateral monopolistic competition on the labor market, we may draw the following conclusions.

1. For sufficiently general assumptions about properties of payoff functions in a game model of a competition between an employer and a trade union, there is a pair of compromise strategies–a Nash equilibrium point. However, participants of the competition cannot realize these strategies since there is no necessary information, and have to act by trial and error.
2. An analysis of dynamic models of the process of searching for an equilibrium point by the marketers has shown that when the producers (employers) are assumed to be oriented mainly toward satisfying domestic demand, such a process may result in either reaching an equilibrium point or a collapse of economy.

3. Under conditions of constant labor productivity, the attainment of an equilibrium point happens in the form of repeating economic cycles with decreasing amplitudes. The concept of a Post-classical cycle corresponds to this. The degree of orientation of an economy toward home consumption does not render essential influence on the form of such cycles.
4. For the process of searching for a compromise to converge to an equilibrium point, it is required that the fiscal system of a country is transparent and socially oriented and the technologies used provide high yield of production and create premises for dynamic development of economy. In this case, the risk of collapse also remains and may be reduced by establishing minimum levels of payment and employment.
5. As labor productivity increases, both wage and employment corresponding to the equilibrium point increase. Thus, in reaching a compromise, all the marketers are interested in innovational development of economy. When labor productivity changes fast (which is typical of large technological innovations), the equilibrium is violated and the cyclicity of economic development is temporary increased.
6. In addition to the nonfulfillment of the above-mentioned necessary conditions, the causes of an economic collapse in the model being considered can be significant deviations of wage and employment from their equilibrium values and an excessively fast labor productivity growth.
7. Replacement of the monopsony (or interbranch oligopsony, which, as it was shown in Sect. 3.4, is close to it) by bilateral competition on the labor market can eliminate some negative phenomena, but there are new risks in this case. A transformation to standards of a competitive market through accelerated development of small and average business independent on corporate giants is the most efficient way of reforming a monopsonic labor market.

Given these conclusions, it is of interest to analyze a labor market whose structure is intermediate between a competitive and oligopsonic one. Several large players act on such a market, possibilities of their influence on the prices are bounded, and there is many small-sized participants that make decisions with allowance for the formed expectations. This can be the subject of further investigation.

### 3.5.1 Defining of the Branch and Regional Criteria During the New Job Creation

Let us consider the peculiarities of employment stimulation under the condition of imperfect competition on the labor market. The serious problems in employment that transition countries (for instance, Ukraine) faces during the last decade require forming and realization of considered state policy of job-creating. It is important not only to determine financial and other resources that could be used for programs implementation in this sphere, but also to substantiate the ways of using these resources. The existing of imperfect competition on the labor market can substantially influence the results of job-creating measures if these measures were based on the standard recommendations formed first for a competitive labor market. Let us analyze the effectiveness of some recommendations using the model of the monopsonic labor market, which was described in Sect. 3.2. At the same time we will consider the results of Sect. 3.4, where it was shown that the characteristics of the oligopsonic labor market with competition between different branch structures are similar to the case of monopsony. So, the following conclusions can be generally used also while studying the market with oligopsonic competition between large industrial structures, which appear in many post-communist countries in transition period and are over the control of oligarchic groups.

The most widespread ways of job-creation stimulation are:

- implementation of privilege (reduced) rates of social taxes on the wage fund for job-creating enterprises;
- financing (from the state budget or from the special funds) of the part of the wage fund for persons, who works on recently created places;
- stimulation of entrepreneurs to make technological innovations in order to improve working conditions, implement new technologies, etc.

Let us analyze the effectiveness of these measures using the assumptions of the model (3.2). We remember that according to this model the wage fund $W$ will be determined by employer in the size

$$W = \min\left(S^{-1}\left(\frac{\alpha\nu}{l}\right), \left(S'\right)^{-1}\left(Hl^{-1}\right)\right)$$

and employment $z$ is equal to $S(W)$. Here $\alpha$ is the production profitableness, $\nu$ is the demand on production made by the employer, $H = 1 + h$, $h$ is the extra expenses of the employer while buying labor force (in particular, they include social taxes on the wage fund), $l$ is the labor productivity. The function $S(W)$ shows the dependence between the quantity of labor to be acquired by employer and the wage fund. This function is determined by the aggregate labor supply function on the market segment analyzed. In Sect. 3.2 it is indicated that the function $S(W)$ is concave and increasing. The inverse function to $S(W)$ is denoted as $S^{-1}(\cdot)$, and the inverse function to its derivative is denoted as $\left(S'\right)^{-1}(\cdot)$. At such assumptions these functions will be only positive-valued; $S^{-1}(\cdot)$ is increasing, and $\left(S'\right)^{-1}(\cdot)$—decreasing.

Let us analyze the results of job-creating stimulation measures using the above-described model.

The implementation of privilege rates of social taxes will reduce the value of $H$. This will increase the wage fund $W$ and employment $z$ only in a case when $W = \left(S'\right)^{-1}\left(Hl^{-1}\right)$ that is when the wage fund will not depend on demand and the values of demand are sufficiently large (so that the inequality $S^{-1}\left(\frac{\alpha\nu}{l}\right) > \left(S'\right)^{-1}\left(Hl^{-1}\right)$ is satisfied). Otherwise, the change of $H$ will not influence the employment and the wage. As the implementation of privilege rates will cause the decrease of deductions to the special funds, there will be a need to compensate this decrease. In a case of increase of other taxes the quantity $\alpha$ will decrease, in a case of reduce of the government expenditures the aggregate demand $\nu$ will decrease. In a case when the wage fund is defined only by the quantity $H$, this won't have negative influence on employment. Otherwise, the quantity $W$ will decrease. Therefore, the first way of job-creating stimulation will be effective only in certain conditions, which are characteristic mostly to the stage of economy growth with favorable state of the market. In other conditions, the consequences of the implementation of this approach will be unsatisfying.

In a case when the government will compensate the employer the part $W_0$ of his/her expenditures on the wage of the workers who are employed on new jobs, the model similar to (3.1) will take the form:

$$F(W) = \min(lS(W + W_0), \alpha\nu) - HW \to \min, \quad W \geq 0,$$

therefore

$$W = \min\left(S^{-1}\left(\frac{\alpha\nu}{l}\right), (S')^{-1}\left(Hl^{-1}\right)\right) - W_0.$$

Thus, the employer will reduce own expenditures on the quantity $W_0$, which will reduce the value of wage for persons employed in other jobs. The aggregate wage fund (the employer's expenditures plus government compensations) will not change then. The value of net profit will not change, too. Therefore, this approach will lead only to the increase in number of workers, but the wage of every one of them will reduce. In most cases, such a consequence is not desirable from the social point of view, but in some conditions (e.g., when it is needed to considerably increase the quantity of workers in the region with hard social conditions) it is quite appropriate. Let us also pay attention to the absence of employer's interest to get such financing. This, in particular, can explain little results while implementing this approach in Ukraine for creating new jobs for young people, disabled people and other socially unprotected population groups [38].

Now let us describe the influence of technological innovations on employment in conditions of monopsonic labor market.

The first result of innovations development will be the increase of labor productivity $l$. This will increase the quantity $(S')^{-1}\left(Hl^{-1}\right)$, but decrease $S^{-1}\left(\frac{\alpha\nu}{l}\right)$. As a result, the wage fund $W$ and the employment $z$ can either increase or decrease. The situation in this case is similar to those of implementation of privilege social rates deduce. The extra effects can emerge if the innovations lead to increase of employer's need in highly qualified labor force, which is not in enough quantities now. Then the wage of such workers can increase even if the general wage fund reduces; in such a situation, the employer will reduce the wage of other workers categories. However, if the implementation of new technologies will not set extra requirements to the employee's qualification, the interprofessional ratios in wages will be practically the same.

The most effective way to increase employment and wage in these conditions is to overcome imperfect competition (monopsony) on the labor market. It is possible to estimate extra employment which will emerge in that case with the help of certain microeconomic models described by [22, 30, 59] and in series of other papers. However the creation of competitive market will require taking not only economic, but also radical social-political measures; it contradicts the interests of employers who get superprofits due to their monopsony power and will take much time. In these conditions even material and money expenses, which are only a part of general expenses, cannot be exactly estimated. However, this way of employment stimulation in conditions of transition economy is the most attractive one from the strategic point of view. It should be considered that all the other described ways of employment stimulation are insufficiently effective in conditions of economic decline when the value of demand $\nu$ is not large and is decreasing with time.

At the same time the transition to the positive economic dynamics (even when the transition is local and short-term) can stimulate the monopolist–employer to create new jobs in a certain economic sector (industry and region). In such conditions there will be a problem of distribution of restricted resources (financial, material, etc.), which can be used for stimulation of such actions of the employer that is the problem of determining regional and industry priorities in creating new jobs.

It is important to mention that direct determining of such priorities is an extremely difficult task. While solving such a task one should consider a large number of factors, many of which are not formalized. So, this needs involving expert estimations and every decision about priorities will be partially subjective. Considering this, it is important to concentrate the efforts in the defined industry on the development of methodology of determining the priorities and on determining the ways of economic mechanisms creation; these mechanisms should help (first, indirectly) to form certain regional and industry priorities. These investigations are described in the current chapter.

It is also important to stress that it is expedient to solve simultaneously the problems of formation of regional and industry priorities. It is conditioned by definite industry specialization of most regions and considerable territorial differentiation of values of most criteria which should be considered during the priorities defining. Later under priorities we will understand namely such ones, regional-industry.

Practically it is impossible to define one certain integral index (or a limited group of numeral indices) which would reflect the priorities during job-creating in specific conditions of a transition economy. Therefore, a typical problem of multi-criteria analysis emerges. One of the most popular approaches is the method of Analytic of Hierarchy Processes (AHP); see [60, 61]. This method is the uniting of separate numeral and ordinal criteria in sense groups built on experts' estimates of numeral models of priorities for groups and, with their help, for a large number of criteria in general. Later these models are used for estimation of different possible decisions (alternatives) acceptability and for forming the rational plan of actions. Taking into account the peculiarities of the problem of setting priorities, it is expedient to use this method to solve it.

**Criteria Used During Priorities Estimation**. The analysis of possible alternatives, according to the method we are examining, begins with the defining of criteria according to which these alternatives are estimated. For the schemes of priority development of certain industries in certain regions, forming of which is the aim of problem solution, it is possible to single out the following criteria groups:

**Economic Criteria** These criteria intended to reflect the influence of certain industry development in certain region on regional and general economic situation. Such criteria, from authors' point of view, can be:

**A.** ***Multipliers of influence of production and employment increase in a certain industry on the general growth of production profits and gross domestic product.***

It is known that the lack of consumers' demand was one of the main reasons of economic decline in the 90th. Thus, the increase of production volumes and employment at the constant wage norm must lead to increase of worker's incomes. This in its turn will increase their demand on consumed goods, which will favor the new growth of production. Economic growth will be also promoted by the increase of demand on goods of intermediate consumption, such as raw materials, half-finished products, and knots and details [35] based on the theoretical basics for these processes analysis. One can estimate the influence of certain industries on their development using the equations of inter-industry balance:

$$x_i = \sum_{j=1}^{n} a_{ij} x_j + y_j, \quad i = \overline{1,n}, \tag{3.48}$$

where

$n$ is the number of industries forming the economy;
$a_{ij}$ is the norm of direct product expenses of the $i$-th industry on the production of unit production of $j$-th industry;
$y_j$ is the volume of final consumption (not connected with production processes) of production of the $i$-th industry;
$x_j$ is the aggregate product of the industry $i$.

Let us denote the part of wage in the cost of product of the $i$-th industry as $q_i$; the average wage in this industry–as $Q_i$. Let us assume that the final consumption volumes are proportional to the changes of the population's incomes; at the same time the industry structure of consumption does not change. Therefore, there should be the ratio

$$\Delta y_i = y_i^0 k \Delta D, \tag{3.49}$$

where

$y_i^0$ is the initial value of final consumption (which is not connected with production) of the $i$-th industry,

$\Delta y_i$ is the change of the volumes of final consumption,
$\Delta D$ is the change of aggregate consumers' incomes,
$k$ is the coefficient of proportionality.

Let us assume that in some industry $i_0$ (for simplicity we will further consider that $i_0 = 1$) a new job is created. This will increase the production volume in this industry on the quantity $\Delta x_1 = Q_1/q_1$. The increase of production volumes in other industries caused by the increase of intermediate consumption in industry $i_0 = 1$ without changes of final consumption of all types of products can be found from the system of equations:

$$\begin{aligned} \Delta x_2^{(1)} &= \sum_{j=2}^{n} a_{2j}\,\Delta x_j^{(1)} + a_{21}\Delta x_1, \\ &\ldots\ldots\ldots\ldots\ldots\ldots\ldots\ldots\ldots, \\ \Delta x_n^{(1)} &= \sum_{j=2}^{n} a_{nj}\,\Delta x_j^{(1)} + a_{n1}\Delta x_1, \end{aligned} \tag{3.50}$$

which directly follows from Eq. (3.48). Here $\Delta x_i^{(1)}$, $i = \overline{2,n}$ are the increase of production in industries $2,\ldots,n$ caused by increase of production demand of the industry $i_0 = 1$. These quantities can be defined as

$$\Delta x^{(1)} = (E - \overline{A})^{-1} A_1 \Delta x_1, \tag{3.51}$$

where matrix $\overline{A} = \begin{pmatrix} a_{22}\ldots a_{2n} \\ \ldots\ldots \\ a_{n2}\ldots a_{nn} \end{pmatrix}$, vector $A_1 = \begin{pmatrix} a_{21} \\ \ldots \\ a_{n1} \end{pmatrix}$, $E$ is unit matrix,

$$\Delta x^{(1)} = \left(\Delta x_2^{(1)},\ldots,\Delta x_n^{(1)}\right).$$

The general increase of consumers' incomes caused by the change of production volumes in industries will be equal to

$$\begin{aligned} \Delta D^{(1)} &= q_1\Delta x_1 + q_2\Delta x_2^{(1)} + \cdots + q_n\Delta x_n^{(1)} \\ &= q_1\Delta x_1 + \overline{q}\left(E - \overline{A}\right)^{-1} A_1\Delta x_1, \end{aligned} \tag{3.52}$$

where $\overline{q} = (q_2,\ldots,q_n)$.

The growth of consumers' incomes will increase the final demand (according to Eq. (3.49)) on the quantity

$$\Delta y_i = y_i^0 k\left(q_1 + \overline{q}\left(E - \overline{A}\right)^{-1} A_1\right)\Delta x_1.$$

According to Eq. (3.48), there will be an increase of production volumes in size

$$\Delta x^{(2)} = (E - A)^{-1} y^0 k \left(q_1 + \overline{q}\left(E - \overline{A}\right)^{-1} A_1\right) \Delta x_1,$$

where matrix $A = \begin{pmatrix} a_{11} & \dots & a_{1n} \\ \dots & \dots & \dots \\ a_{n1} & \dots & a_{nn} \end{pmatrix}$, $y^0 = \left(y_1^0, \dots, y_n^0\right)$.

Such production growth will cause the increase of incomes in size

$$\Delta D^{(2)} = \sum_{i=1}^{n} q_i \Delta x_i^{(2)} = q(E - A)^{-1} y^0 k\left(q_1 + \overline{q}(E - \overline{A})^{-1} A_1\right)\Delta x_1. \tag{3.53}$$

The growth of incomes in size $\Delta D^{(2)}$ will lead to the growth of the final demand and production $\Delta x^{(3)}$, which is calculated similar to $\Delta x^{(2)}$. The values of $\Delta D^{(3)}$, $\Delta D^{(4)} \dots$ are calculated in the same way.

The general increase of demand will be equal to

$$\Delta\overline{D} = \Delta D^{(1)} + \Delta D^{(2)} + \dots = \frac{\left(q_1 + \overline{q}(E - \overline{A})^{-1} A_1\right)\Delta x_1}{1 - q\,(E - A)^{-1}\, y^0 k}.$$

As the criterion of industry $i_0 = 1$ priority estimation the quantity of the multiplier can be used:

$$F_1 = \frac{\Delta\overline{D}}{\Delta x_1} = \frac{\left(q_1 + \overline{q}\left(E - \overline{A}\right)^{-1} A_1\right)}{1 - q\,(E - A)^{-1}\, y^0 k}. \tag{3.54}$$

Similarly the value of the multiplier $F_1$ for other industries can be calculated. The more the value of $F_1$, the bigger the general increase of consumers' incomes and production volumes will be acquired due to the increase of employment in the industry $i_0$.

The estimations made using the data of interindustry balance 1998 [50] demonstrated that the largest values of this multiplier in Ukraine in general were in such industries as building, ferrous metallurgy, electric power engineering and building of materials production. This is explained by substantial volumes of intermediate consumption in these industries, considerable product costs in other industries and substantial part of wage in their production's prices. The analysis of balances at the first years of the twenty-first century demonstrates that the mentioned trends still exists.

Till now we assumed that the increase of employment is accompanied by the increase of production. In the condition of its absence the job-creating in the industry $i_0$ will cause the increase of the aggregate consumers' incomes on the quantity $Q_{i_0}$. This will lead to the increase of the final demand on the quantity $\Delta y = y^0 k Q_{i_0}$, which will increase production and consumption, and the general increase of the related incomes will be the following:

$$\Delta\tilde{D} = \frac{Q_{i_0}}{1 - q(E - A)^{-1}y^0k}.$$

The quantity

$$F_2 = \frac{Q_{i_0}}{1 - q(E - A)^{-1}y^0k} \tag{3.55}$$

can be considered the second criterion of industry priorities.

**B.** ***The estimation the change of export–import balance.***

The openness of the economy and the presence of foreign trade operations were not directly considered in the previously described criteria. However these factors can substantially influence the estimation of the industries priority. It is worth mentioning that it is rather difficult to give a precise forecast about which part of extra demand caused by the employment increase will be satisfied due to home production and which one—due to import. This will depend not only on production powers of the country, but also on the situation on the world markets, on carrying out of the import-substitution industry policy, on norms of currency, customs-tariff and fiscal state policy, etc. Therefore nothing remains, but to use partial estimations of changes of net export (export–import balance), built at the certain assumptions. During defining the first of such estimations we assume that the extra import compensates the difference between the growth of production in industries that was reached by the general demand increase and production powers of the country. During this forecast estimations $\Delta H_i$ of the export change in industries ($i = \overline{1,n}$) are also used.

Let us assume that we know $\overline{x_i}$–capacities of industries $i = \overline{1,n}$. Then the import increase will be the difference between the expected need (i.e., the production increase without considering the capacities) and the capacities. The need $\Delta x$ is found from (3.49) through the general incomes increase $\Delta\overline{D}$ as

$$\Delta x = (E - A)^{-1}\left(y^0k\Delta\overline{D} + \Delta H\right)$$

$$= (E - A)^{-1}\left(\frac{q_1 + \overline{q}\left(E - \overline{A}\right)^{-1}A_1\right)\Delta x_1}{1 - q(E - A)^{-1}y^0k}y^0k + \Delta H\right) \tag{3.56}$$

under the condition that the employment in the industry $i_0 = 1$ increases.

The related import increase will be equal to

$$\Delta I = \Delta x - \overline{x},$$

where $\overline{x} = (\overline{x}_1, \ldots, \overline{x}_n)$, $\Delta x$ is defined according to (3.56).

The general change of export–import balance will be equal to

$$F_3 = (\overline{p}, \Delta H) - (\tilde{p}, \Delta I), \tag{3.57}$$

where $\overline{p} = (\overline{p}_1, \ldots, \overline{p}_n)$, $\tilde{p} = (\tilde{p}_1, \ldots, \tilde{p}_n)$ are the export and import prices of the production of the corresponding industries.

Let us now assume that import volumes are increasing in proportion to the changes of production volumes that is $\Delta I = \gamma \Delta x$, where $\gamma > 0$ is a some coefficient.

The estimation of the export–import balance change built on these assumptions will be:

$$F_4 = (\overline{p}, \Delta H) - (\tilde{p}, \gamma \Delta x), \tag{3.58}$$

where the value $\Delta x$ is defined from (3.56).

Analyzing the data about the industry structure of export and import of Ukraine during the last years, we can make a conclusion that the value $F_3$ will be the largest for industries which are supplied with own source of raw materials, have the sufficient potential for their development and produce the production of high liquidity, such as ferrous metallurgy. The value $F_4$ will be the largest for industries which have the largest real exceeding of export over import in the established conditions. For Ukraine they are industries of agro-industrial complex. The value $F_3$, however, is substantial also for the industries with unrealized potential, for example, for the light industry.

**C.** ***The estimations of extra expenses connected with job-creation.***

Along with positive effect that will be partially acquired in the future, one should consider the expenses of the present. Considering the constraints of the budget resources it makes sense to study apart the quantity of general expenses on creation of one job in the industry (let us define it $F_5$) and the quantity of the state budget expenses (including indirect ones, for example, tax privileges) made with the same goals (let us define the latter as $F_6$). In contrast to the previous criteria, not the least, but the largest values $F_5$ and $F_6$ should be the better solution.

**D.** ***Correspondence of traditional regional-industry specialization.***

This correspondence is one of informalized factors, which must be taken in consideration in making decisions. The construction of this criterion is based on the experts' estimations, which can be gotten from specialists on productive forces. While defining the estimations the specialist should take into account supplementary information concerning availability of material, labor forces of appropriate qualification, manufacturing powers in the region and other development factors of definite region. The following scale can be used in making estimations.

1. **The industry is leading**. It means that this industry was or is leading in the region's economics, all factors be necessary for its development including elements of infrastructure are observed, and there are no objective reasons for the changing the regional specialization.
2. **The industry partly corresponds to the regional specialization.** It means that the industry was and is present in the list of leading industries of the region; the region has enough manufacture and infrastructure factors for its development and there are no objective conditions for the changing of the regional specialization.
3. **The industry partly not corresponds to the regional specialization.** It means that the industry is present in the list of the most developed industries of the region, but there are no objective conditions for its further development (or this conditions aren't strong enough) or the industry wasn't developed before, but there are objective conditions for its considerable development.
4. **The industry can be developed only in limited measures.** It means that there are objective conditions only for limited, besides of the main regional specialization, development of this industry in the definite region.
5. **The development of the industry is impossible in the region.** There are objective factors, which make the industry's development impossible in the region.

The expert should refer each industry to a definite grope from mentioned before. As an exception, some intermediate estimation, for example, "something between partly correspondence to the regional specialization and the possibility to be developed in limited measures," can be possible. Independent estimations from a group of experts also can be used.

The $F_7$ criterion of correspondence of industry and regional specialization is ordinary, built on the strict scale, based on the experts' estimations.

**E.** ***Correspondence to the priorities of innovations.***

This criterion should convey the influence of development of the definite industry (by the establishing new work places) on the scientific and technical, innovational and technological economic development. It is estimated in an expert's way, but, in spite of the previous one, is not attached to the strict scale. The expert should arrange the industries from the point of view of their correspondence to the priorities of scientific, technical, and technological development of the region and point, which—absolute, grate or small—is the priority of one industry over the other. Let's denote these ordinary criteria as $F_8$.

Other indices, except of the mentioned, can refer to the group of economical criteria if necessary.

It should be noted that proposed criteria correspond to the requirements, stated in Ukrainian Law [71]. In Chap. 5 of this document [71] (see item 5 of Law 3076-III, http://zakon4.rada.gov.ua/laws/show/3076-14) it is pointed that "...the definition of the industry's priorities of establishing new jobs, ... connected with the government support of the industries, able to provide the growth of GNP extremely quickly ... define the level of Ukraine's economical safety, ... provide the material and technical foundation of manufacture, ... satisfy the top-priority needs of the population." It's easy to notice that $F_1$ criterion numerously formalizes the first demand, $F_3$ and $F_4$ criteria—the second, $F_7$—the third and $F_2$—the fourth.

**Social criteria.** These criteria should reflect the influence of development of the definite industry in the definite region on the general and regional social situation and its separate essential aspects. The following ones can be referred to these criteria:

**A.** ***The general quantity of supplementary labor places, made owing to the changes of economics dynamics, caused by the development of the definite industry.***

This index determines the quantity of new jobs established due to the changes in the amounts of the output production of individual industries, caused by the priority development one of them. As it was mentioned before, the general change of amounts of production in industries $\Delta x = (\Delta x_1, \ldots, \Delta x_n)$ can be estimated, using the Eq. (3.56). If $L_i$ is the labor productivity (real or prospective at the end of economical changes), then the general quantity of supplementary established places is equal to

$$F_9 = \sum_{i=1}^{n} \frac{\Delta x_i}{L_i}. \tag{3.59}$$

Using upper and down estimates of the labor productivity in the formula (3.59) instead of unknown exact value of this criterion we can receive the maximum and minimum values of the criterion, which is measured in such case in interval scale.

Then $F_9$ would be interval, and methods of fuzzy sets theory can be used for its estimation [60, 61]. It should be mentioned that in a case of its both numeral and interval estimation, the maximization of this criterion would be desirable.

**B.** ***Number of supplementary labor places, established in the region.***

This criterion should reflect regional interests on employment increase. It is defined similarly to the previous one, but in this criterion only that part of the total productivity growth, which corresponds to the definite region, and regional productivity labor are taken into account. This criterion is denoted as $F_{10}$, and its maximization would be desirable.

**C.** ***The industry's value as the city-forming factor in the region.***

Ukraine, similar to other post-communist countries, has a number of settlements (mainly small towns), in which employment is concentrated on the restricted quantity (which often is equal to one) of usually big industry factories. These settlements exist only owing to this factories, therefore, reducing of their production and related employment would cause local social catastrophe. The catastrophe deepens owing to the fact that mentioned factories provide the greater part of receipts to the local budgets and support an essential part of social infrastructure.

It is an acute problem for Ukraine. According to different information sources, about one third of Ukraine's small towns [for some regions (Chernivetska, Luganska, Chernigivska, Donetska) it is more than one half] can be referred to the above-mentioned city forming industrial centers. Availability of small quantity of factories in them is, usually, one of the main preconditions of monopsony on the labor market. Therefore, this information is the evidence of possible expansion of imperfect competition on the definite market segment.

It should be noted that factories of almost all industries: mining, electrical energy, heavy industry (ferrous and nonferrous metallurgy, chemistry and petrochemistry, mechanical engineering), food industry, production of goods of public consumption (timber and wood manufacturing, glass industry, etc.), are city forming. However, the value of these factories as the city forming varies depending on the region. For its estimation, it is necessary to use numerous formal and informal considerations, so this criterion should be determined by experts' way, as ordinary in case of lacking of the strict estimations scale. Let us denote it as $F_{11}$.

**D.** ***The value of the industry from the point of view of creation compensative employment.***

Technical rearmament of production is usually accompanied with considerable growth of structure unemployment. The development of industries, in which the demands to the qualifications are similar to the characteristics of unemployed workers, can soften the negative consequences of the technical rearmament of production. For example, when the considerable employment reduction in the coal industry took place in the 70th in FGR, of about 60 % of those, who become unemployed, started working in the factories producing constructive materials and at accomplishing jobs, 15 % (above all high qualified specialists—metalworkers, mechanics, etc.) started working in the factories on mechanical engineering, the other 25 % started working in other industries. Therefore, the two mentioned industries played the leading role in establishing compensative employment owing to their qualification demands. The great role in softening the social consequences of restructure in that time played the implementation of public and municipal programs on building roads and communications, particularly underground (*U-Bahn*) in the Ruhr's cities (see [3]).

This example demonstrates the importance of taking into consideration the factors of establishing compensative employment when determining regional industry priorities. However, the great numbers of informal factors influence on these criteria; consequently, it is useful to consider it as ordinary criteria, which is defined by experts' way. Let us denote them as $F_{12}$.

**E.** ***The influence of the industry's development on the general social situation.***

Social aspects of employment changes, which are not taken into account in the pervious indices, are taken into account using this criterion. It is also ordinary measured and defined by experts' way; it is denoted as $F_{13}$.

**Ecological Criteria** The necessity of analyzing mentioned criteria is conditioned first, by the influence of different industries on the environment; second, by the bad ecological situation in Europe and particularly in Ukraine; third, by the understanding of leading specialists the necessity of taking into account ecological factors together with traditional—economic and social, when elaborating the plans of society's development, particularly long-range planning. Therefore, ecological consequences should be taken into account in determining regional and industry's priorities of new jobs establishing.

For estimation of these consequences we will distinguish such concepts, as **ecological load**, the influence on environment, which the people's activity provides during the predictable passing of the productivity processes and their predictable consequences, and **ecological risk**—the influence on the environment, which may occur due the accidental attack of some negative events including the realization of unpredictable consequences of above-mentioned processes. For example, ecological load from the development of nuclear energy would be less, than from the heat one (because, the first one doesn't cause much throw of chemical agents in a case of accident-free functioning of equipment), but the ecological risk is appreciably greater (because the consequences of possible accidents on nuclear stations are greatly than the damages from the accidents on heat stations). From the mentioned examples we can make the conclusion that the load and risk should be considered as independent indices, moreover, we should distinguish current and long-term ecological consequences. Similar to other indices, concerned with the employment changes, it is necessary to take into account experts' estimates and ordinary scales for determining ecological criteria. Taking into account everything, it has proposed to take up the following criteria:

(a) *the influence of the industry's development on the change of ecological load*:

- direct influence, let's denote it as $F_{14}$;
- long-term influence (the influence on the long period of time), which we'll denote as $F_{15}$;

(b) *the influence of the industry's development on the change of ecological risk*:

- direct influence, let's denote it as $F_{16}$;
- long-term influence, which we'll denote as $F_{17}$.

In Ukraine's conditions such industries, as ferrous and nonferrous metallurgy, chemistry and petrochemistry, mining industry, heat energy supply, some enterprises of the building materials industry production, provide the greatest ecological load. The nuclear industry and waterpower energy production, chemical industry (especially, inorganic chemistry and organic synthesis), transport, some technologies of metallurgy and food industry provide the greatest ecological risk. In the two last the risk is caused by using toxic substances in the production aims and formation of wastes with indefinite long-term ecological effect.

Naturally, this list can be completed with new criteria, and present can be detailed if necessary.

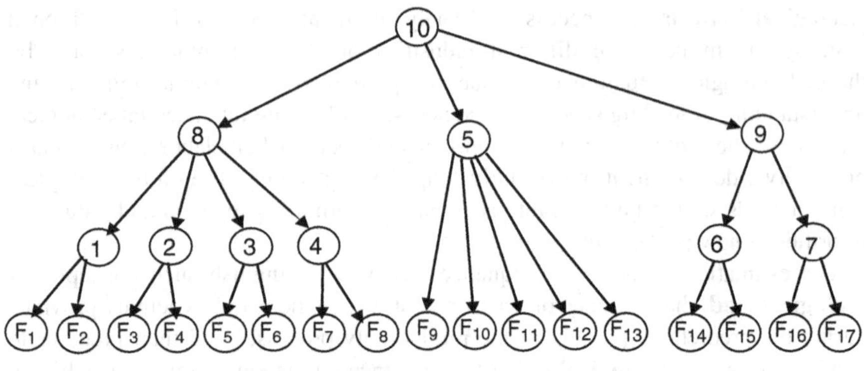


**Fig. 3.15** Criteria and hierarchical relations between them

Therefore, 17 criteria may be taken into account for determining regional industry priorities. The list of these criteria and hierarchical relationship within them are schematically represented in Fig. 3.15. We denote as

1. the set of multipliers criteria,
2. the set of export–import criteria,
3. budget criteria,
4. criteria of productive force development,
5. social criteria,
6. criteria of ecological load,
7. criteria of ecological risk.
8. the set of economic criteria,
9. the set of ecological criteria,
10. the total project estimate.

**Multicriterial Analysis of Regional Industry Priority**. Regional industry priorities are reflected by the criteria with different ways of definition, different demands (maximization or minimization) and different properties. There are different methods of multicriteria analysis for estimation of the single integral index, which would reflect the general priority, based on them. The AHP method, which was proposed by Saaty [60, 61], is one of them. Its advantages—simplicity, possibility to apply different methods of criteria definition, well-developed structure—determine the possibility of its application for defining regional industry priority. The method is intended for the solution of the following problem.

Let the finite multitude with $N$ alternatives (in our problem the alternative means the program on establishing of new jobs in the definite industry and region) be defined. Each alternative is characterized by $L$ criteria. They can be numerical, ordinary, interval, etc. The criteria combine in groups and sub-groups, which have the hierarchical relations in the form of tree, resembling the one represented in Fig. 3.15. It is expected to define the alternative's integral index value by which in future it would be possible to range them in order of acceptability (the best alternative would have the highest value).

The AHP method starts with depicting the values of all numerical criteria for each alternative on the interval [0;1]. First, it runs up to the positiveness of the estimated before values of those numeral criteria, which it is necessary to maximize. If necessary, the equal high enough number is added to all these values. For the numeral criteria, which we should minimize, we apply the transformation into the indices, which would be maximized and would take positive values, according to the formula

$$\overline{f}_k^{(i)} = c - f_k^{(i)},$$

where

$f_k^{(i)}$ is the value of $k$-th criterion for $i$-th alternative ($i = \overline{1, N}$, $k \in \{1, 2, \ldots, L\}$;

$\overline{f}_k^{(i)}$ – the value of the index, equivalent to this criterion, which should be maximized;

$c$ – some large enough positive number (it is chosen from the condition that $\overline{f}_k^{(i)}$ should be positive for each $i = \overline{1, N}$).

After the equivalent transformation into the positive-valued criteria they are normalized at the interval [0;1] according to the formula

$$U_k^{(i)} = \frac{\overline{f}_k^{(i)}}{\left(\sum_{i=1}^{n} \left(\overline{f}_k^{(i)}\right)^2\right)^{1/2}}, \quad i = \overline{1, N}, \quad k \in \{1, 2, \ldots, L\},$$

where

$U_k^{(i)}$–the normalized value of $k$-th criterion for $i$-th alternative;

$\overline{f}_k^{(i)}$–previously modified value of this criterion, which is positive and needed to be maximized.

The interval criteria elaboration is made the same as it is made for the numerical ones, except for the fact that in spite of the criteria their average values, fuzzy dispersions, quintiles with fixed confiding level are rated.

We act in the following way to receive the normalized values for ordinary-measured criteria. The preference matrix of dimension $N \times N$ is arranged for each of these criteria. For the criteria, got from the estimations on the tough scale, each alternative gets its rate, according to its group. At the same time the alternatives rang from the more priority group is higher than the groups rang, which goes directly after that. For example, for the examined criteria of the regional industry specialization correspondence the alternatives rang is equal to $2^5 = 32$, for the second–16 and so on, for the last fifth ("the development in the region is impossible")–1. In a case, when the expert proposes to correspond the alternative immediately to several criteria or when different experts correspond it to different groups, the alternatives rang is defined as the average from the ranges of these groups. After the definition of all alternatives' ranges, the elements $h_{ij}^{(k)}$ of the preference matrix $H^{(k)}$ are defined in the following way:

$$h_{ij}^{(k)} = \frac{r_i^{(k)}}{r_j^{(k)}}, \quad i, j = \overline{1, n}, \tag{3.60}$$

where $r_i^{(k)}$ is the range of alternative $i$ received according to $k$-th criterion; $r_j^{(k)}$ is the range of alternatives $j$ received according to the same criteria; $H^{(k)} = \left\{h_{ij}^{(k)}\right\}$ is the preference matrix for the ordinary-measured criteria $k$ build for the tough scale.

The preference matrix for other ordinary-measured criteria is defined in the following way.

First, according to the experts' estimates the alternative's ranking from the "best" (according to this criterion) to the "worst" is accomplishing. The "worst" alternative gets the rank, which is equal to 1. Then the expert approximately defines in how many times the following alternative is better than the previous one. The range of the following alternative is defined as the multiplication of the previous range on this number. This procedure would be continuing till the best alternative's range would be defined. After this the $h_{ij}^{(k)}$ elements of the preference matrix $H^{(k)}$ are determined the same as for the previous case, according to the formula (3.60).

From the $H^{(k)}$ matrix the normalized values $U_k^{(i)}$ for all current criteria are defined as the corresponding components of eigenvector of matrix's $H^{(k)}$, which (vectors) correspond to the largest eigenvalue of the matrix.

The wage coefficients $\gamma_{kl}$ for each $k$-th criterion in $l$-th sub-group, in which it is included according to the present criteria hierarchy are estimated after the normalization values for all criteria. It is possible to apply the methods of ordinary regression (see [29]) for defining these coefficients. Then the waged coefficients from each sub-group from the group of criteria are defined by the same way. The rank $g_i^l$ of each alternative $i$ from $l$-th group of criteria are defined for known value of waged coefficients according to the formula

$$g_i^l = \sum_{k \in K(l)} \gamma_{kl} U_k^{(i)}, \quad i = \overline{1, N}, \tag{3.61}$$

where $g_i^l$ is the $i$-alternative rate of $l$ group criteria.

In [44] the modified method of AHP, based on fuzzy sets theory, was proposed. This modification differs from the method by the criteria normalization on the [0;1] interval and the way of the alternative rates of the criteria groups defining.

The normalized criteria values are defined according to the formula

$$U_k^{(i)} = \frac{\overline{f}_k^{(i)} - f_k^{\min}}{f_k^{\max} - f_k^{\min}}, \quad i = \overline{1, N}, k \in \{1, 2, \ldots, \},$$

where

$$f_k^{\min} = \min_{i=\overline{1,N}} \overline{f}_k^{(i)}, \quad f_k^{\max} = \max_{i=\overline{1,N}} \overline{f}_k^{(i)}.$$

For the defining of criterion rates, we use the expression

$$g_i^l = \min_{k \in K(l)} \gamma_{kl} U_k^{(i)},$$

where the denotations are the same as in the formula (3.61).

It should be noted that the modified AHP method allows not only getting the integral index for projects of the new job creation but also analyzing the influence of the importance of the criteria and preferences, revealing strong and weak sides of each alternative.

Because of the differences in the values $U_k^{(i)}$ definition for the "traditional" and modified AHP method, the larger values $\gamma_{kl}$ in AHP should correspond to the more significant criteria, but in the modified AHP method they correspond to the less significant criteria.

As we have seen a two-argument function of individual labor supply is considered. Real wage and unemployment rates are the arguments of this function. Conditions of emergence of direct and inverse relations between these two economic indicators are investigated. This function is used to analyze processes in a competitive and monopsonic labor market. The expediency and conditions are substantiated for the use of the exogenous increase in labor remuneration by increasing minimal wage.

## 3.6 Labor Market Modeling Using a Labor Supply Function of Two-Arguments

Empirical investigations of labor markets in many countries demonstrate ambiguous relations between labor remunerations and unemployment rates. According to the classical approach to the analysis of the mentioned segment of a national market, unemployment is a consequence of an excess of labor supply over labor demand. Since labor supply increases with increasing labor remuneration and labor demand decreases in this case, unemployment must arise when the rate of labor remuneration exceeds its equilibrium value. With increasing the mentioned excess, unemployment must increase and, hence, a direct relation will exist between the rate of labor remuneration and unemployment rate. This relation is substantiated by empirical data for some countries and for different time intervals [7].

At the same time, statistical investigations demonstrate the opposite situation for many countries. There is an inverse relation between the rates of unemployment and labor remuneration [10], i.e., the case when an increase in wage rates matches with an increase in employment rate. In this case, the character of this relation is insignificantly different for countries with different levels of economic development [9] for which an ambiguous influence of the specificity of organization of their labor markets (including the presence or absence of different forms of imperfect competition) would be expected: The character of the mentioned relation can also vary according to-market conditions. In particular, in [47], empirical evidence is presented that a direct relation between unemployment and labor remuneration rates reflects the change in nominal wage rates up to the level necessary for compensating for a redundant labor supply. At the same time, in most cases, to an inverse relation correspond fluctuations around an equilibrium level of wage and unemployment rates [11]. Under these conditions, an increase in the minimum labor remuneration can have positive consequences. It should be noted that, at the present time, the inverse relation between labor remuneration and unemployment rates prevails in the majority of labor markets.

The existence of inverse relation is traditionally explained with the help of noncompetitive models of labor markets under assumptions on a greater market forces of employers and their capabilities to use unemployment as a means for forcing persons employed to agree with lower wage rates. Similar views underlie, in particular, the concept of the NAIRU (Non-Accelerating Inflation Rate of Unemployment) [18]. In [64], an inverse relation is interpreted as a result of "rewarding" workers by their employers for the qualitative fulfillment of their obligations. In the opinion of the authors, the necessity of such rewarding arises as a result of the relative scarcity of labor, i.e., as a result of a low unemployment rate. On the whole, an explanation of an inverse relation with the help of classical and neoclassic models is not always possible.

The absence of structural models that would adequately describe the mentioned relation puts the brakes on its use in theoretical investigations in economics and does not explain changes in labor remuneration policy.

The question of why this relation has not been observed in real situations when assumptions on the prevailing market force of employers was seemingly well substantiated remains open. The explanation of an inverse relation within the framework of "demand–supply–price" models that are traditional in investigating markets also remains conjectural. A drawback of theoretical analysis of consequences of practical use of an inverse relation between unemployment rates and labor remuneration rates puts the brakes on the use of tools providing an exogenous increase in labor remuneration in economic policy, in particular, owing to an efficient policy concerning the minimum wages.

In this section, an attempt is made to explain these phenomena as a result of considering a function of individual labor supply whose arguments are not only labor remuneration but also unemployment rate.

This section consists of four parts. Section 3.6.1 presents a microeconomic substantiation of properties of such a function on the basis of analyzing a model of the behavior of a person employed. This model takes into account not only the utility of money and free time for the mentioned person but also the influence of risk to lose his job on his actions. In Sect.3.6.2, with the help of the constructed labor supply function of two arguments, conditions are specified under which direct and inverse relations between labor remuneration and unemployment rates are possible. In Sect.3.6.3, the two-argument function is used for the investigation of competitive and monopsonic labor markets. In Sect.3.6.4, an estimate of consequences of regulation of labor remuneration and employment rates is presented.

### 3.6.1 *Microeconomic Substantiation of the Two-Argument Function of Labor Supply*

Let us consider a behavioral model of a person who determines the number of work hours that is acceptable for him (in what follows, we will denote this number by $x$) with a known labor remuneration rate $w$. We denote by $T$ the total number of hours available to the mentioned person during the period of planning his actions by him. Let $u(t)$ be a function commensurating the value of free time equal to $t$ and utility of money for the person employed. Next, we assume that this function is an increasing, concave, and differentiable. As has been noted earlier, in addition to incomes and utility of free time, a risk of losing his job is also taken into account in the model. Characteristics of such a risk are probability of retention of his job $p$ and the level of his well-being if he loses his job $u_0$. Hereafter, we assume that the quantity $p$ depends on the labor supply per labor remuneration unit (i.e., on the profitability the person employed for his employer) and on the unemployment rate $U$. Assume that the function $p = p\left(\frac{x}{w}, U\right)$ is increasing with respect to $\frac{x}{w}$ (i.e., the employer first dismisses workers who ensure the smallest profit) and decreasing with respect to $U$ (the greater unemployment rate, the greater risk to lose work) and, in this case, we have $p(0, U) = 0$ for any values of $U$. We also assume that the quantity $u_0$ is constant and smaller than $w_0 x + u(T - x)$ for any $0 \leq x \leq T$, where $w_0$ is a minimum established rate of labor remuneration. We also assume that the function $p = p\left(\frac{x}{w}, U\right)$ is differentiable with respect to its first variable.

A person employed determines his labor supply $x$ by maximizing, for known values of $w$ and $U$, the expected utility $G(x)$ defined as follows:

$$G(x) = \Big(wx + u(T - x)\Big)p\left(\frac{x}{w}, U\right) + u_0\left(1 - p\left(\frac{x}{w}, U\right)\right).$$

In this case, we have $0 \leq x \leq T$. Taking into account that $w \geq w_0$, under the above assumptions, the point $x = 0$ cannot be the point of maximum of $G(x)$. The case when this point is $x = T$ can be excluded from consideration under the assumption that the value of $u'(0)$ is sufficiently large. Thus, $G(x)$ can assume its maximal value only in an interior point of the interval $[0, T]$ at which its derivative is equal to zero. The relationship

$$G'(0) = \Big(w - u'(T-x)\Big) p\left(\frac{x}{w}, U\right) + \Big(wx + u(T-x)\Big) \frac{\partial p\left(\frac{x}{w}, U\right)}{\partial x} - u_0 \frac{\partial p\left(\frac{x}{w}, U\right)}{\partial x}$$

implies the equality

$$w - u'(T-x) = \frac{\partial p\left(\frac{x}{w}, U\right)}{\partial x} p^{-1}\left(\frac{x}{w}, U\right) \Big(u_0 - wx - u(T-x)\Big). \tag{3.62}$$

Let us pay attention to the fact that the quantity

$$\frac{\partial p\left(\frac{x}{w}, U\right)}{\partial x} p^{-1}\left(\frac{x}{w}, U\right) = E\left(\frac{x}{w}, U\right)$$

is the elasticity coefficient of the probability of job retention for a labor supply $x$. Under the assumptions made, this quantity will be positive and decreasing with respect to $x$ and $U$. The last multiplier in the right side of equality (3.62) assumes negative values for any $0 \le x \le T$ and $w \ge w_0$.

Thus, the maximum point $G(x)$ will be the point of intersection of the curves $y = w - u'(T-x)$ and $y = E\left(\frac{x}{w}, U\right)\Big(u_0 - wx_1 - u(T-x)\Big)$ (Fig. 3.16)

We note that the point of intersection of the curve $y = w - u'(T-x)$ with the axis $0x$ is the point $\overline{x}$ to which corresponds the decision of the person employed that is made without taking into account a risk of losing his work. This decision is traditionally considered during the microeconomic substantiation of properties of the individual labor supply function $x^*(U)$. Thus, the labor supply $x^*(U)$ determined with allowance for the risk of unemployment will always be larger than $x$. Taking into account the properties of $E\left(\frac{x}{w}, U\right)$, an increase in $U$ shifts the curve $y = E\left(\frac{x}{w}, U\right)\Big(u_0 - wx - u(T-x)\Big)$ downwards and, hence, for $U^1 > U$, we have $x^*(U^1) > x^*(U)$ (see Fig. 3.16), and labor supply is an increasing function with respect to $U$.

Since, under the assumptions made, the value of $\Big(u_0 - wx - u(T-x)\Big) E\left(\frac{x}{w}, U\right)$ decreases with respect to $w$ with decreasing $w - u'(T-x)$, it may be proved that, for sufficiently small $w$, labor supply is an increasing function of this quantity. Next, the former quantity decreases, whereas the latter increases. If this increase exceeds the decrease, then labor supply decreases. However, it is possible that a decrease in the former quantity will exceed that in the latter, and then the supply of labor continues to increase. Thus, the effect (typical for neoclassic models) of a decrease in labor supply owing to a high wage rate will take place only under additional

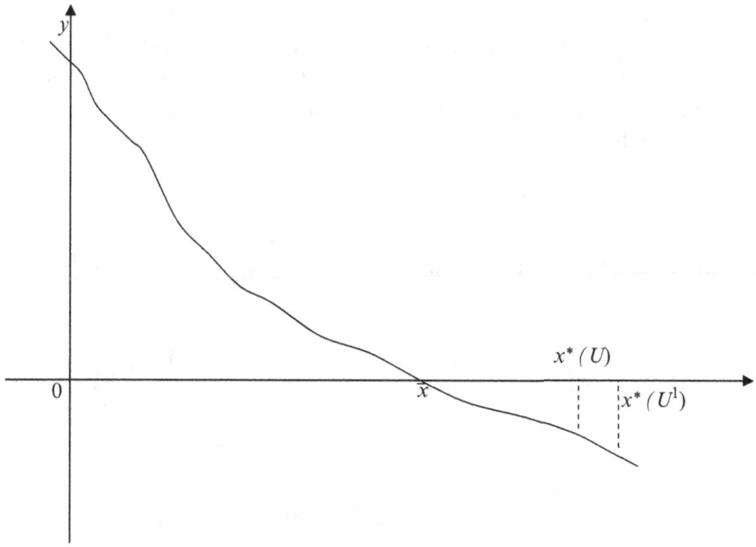


**Fig. 3.16** Two-argument function of individual labor supply

assumptions on a small sensitivity of the person employed to the possibility to lose his job.

The continuity of the curves that are considered in Fig. 3.16 implies the continuity of the dependence of an individual labor supply on $w$ and $U$.

Thus, under the assumptions considered above, the two-argument function of individual labor supply $L(w, U)$ is continuous and increasing with respect to $U$. For sufficiently small values of $w\left(w_0 \leq w \leq \tilde{w}\right)$, this function will also be increasing with respect to $w$. In this case, the quantity $w$ increases with increasing $U$. Under some conditions, it can be infinitely large. These properties of the function $L(w, U)$ are used for the further analysis of the relation between the labor remuneration $w$ and unemployment $U$.

### *3.6.2 Analysis of the Relation Between Labor Remuneration and Unemployment*

We consider a labor market whose labor supply is formed by persons with identical functions of individual labor supply $L(w, U)$. According to the results of the previous section, these functions are continuous and increasing with respect to $w$ and $U$ when $w < \tilde{w}$. It is this case that will be considered in what follows. Let $\overline{z}(w)$ be the number of persons offering their services under the condition of their payment in the amount of $w$. Next, we assume that $\bar{z}(w)$ is a differentiable function increasing with respect to $w$ and that the function $L(w, U)$ is also differentiable.

The situation in the market is determined by the labor remuneration $w$ and the number of persons employed $z \leq \bar{z}(w)$. The total amount of labor $F(w, z)$ used in this situation equals

$$F(w,z) = zL\Big(w, \bar{z}(w) - z\Big).$$

Let us investigate properties of the function $F(w, z)$. Its partial derivative with respect to the first variable is as follows:

$$\frac{\partial F(w,z)}{\partial w} = z\left(\frac{\partial L\Big(w,\bar{z}(w)-z\Big)}{\partial w} + \frac{\partial L\Big(w,\bar{z}(w)-z\Big)}{\partial u} \cdot \frac{\partial \bar{z}(w)}{\partial w}\right).$$

For $w \leq \tilde{w}$, it is positive.

The derivative with respect to the second variable is as follows:

$$\frac{\partial F(w,z)}{\partial z} = L\Big(w,\bar{z}(w)-z\Big) - z\frac{\partial L\Big(w,\bar{z}(w)-z\Big)}{\partial u}.$$

Taking into account that $z > 0$ and $L\Big(w, \bar{z}(w) - z\Big) > 0$, we draw the conclusion that the sign of the quantity $\frac{\partial F(w,z)}{\partial z}$ will be determined by the sign on the expression $z^{-1} - E_1(w, z)$, where $E_1(w,z) = \frac{\partial L\Big(w,\bar{z}(w)-z\Big)}{\partial u} L^{-1}\Big(w,\bar{z}(w)-z\Big)$ is the elasticity coefficient of the function of individual labor supply with respect to the unemployment rate.

If the inequality $E_1(w, z) \leq z^{-1}$ is satisfied, then we have $\frac{\partial F(w,z)}{\partial z} \geq 0$. In this case, the influence of the decrease in $w$ by the value of $F(w, z)$ can be compensated by an increase in $z$, i.e., a decrease in the unemployment rate $\overline{U} = \bar{z}(w) - z$ (in this case, $\bar{z}(w)$ also decreases). Under these conditions, a direct relation is observed between labor remuneration and unemployment rates in accordance with classical labor market models.

If the inequality $E_1(w, z) > z^{-1}$ is fulfilled, then we have $\frac{\partial F(w,z)}{\partial z} < 0$. The influence of decreasing $w$ by the value of $F(w, z)$ is compensated in this case by a decrease in $z$ and an increase in $U$. Thus, an inverse relation between labor remuneration and unemployment rates arises here, which is observed in constructing empirical relations.

The main cause that conditions a change in the character of this relation is a change in the elasticity of the two-argument function of labor supply with respect to the unemployment rate. To small values of the elasticity coefficient will correspond a direct relation between $w$ and $U$ and to large values will correspond an inverse relation. The elasticity of the unemployment rate with respect to the amount of labor remuneration that is frequently estimated during empirical investigations [9] will be determined by the elasticity of the function of labor supply with respect to the quantity $w$.

### 3.6.3 Labor Market Analysis

Using the above-mentioned two-argument function of labor supply, we will analyze processes occurring in differently organized labor markets. We begin with the investigation of a competitive labor market.

Let, in this market, the labor demand $L_D$ be determined by the relation $L_D = f(w)$, where $f(w)$ is some decreasing, convex, and differentiable function of the labor remuneration rate $w$. The labor supply $L_S$ also depends on the employment rate $z$ and is determined by the relationship $L_S = F(w,z)$ where $F(w,z)$ is the two-argument labor supply function that has been considered, in the previous section. The equilibrium condition in this market assumes the form

$$F(w,z) - f(w) = 0. \tag{3.63}$$

Equality (3.63) implies the existence of sets of equilibrium states $\hat{H}$ to which, on the plane $z0w$, corresponds the locus of points $(w,z)$ that satisfy the condition $f(w) = F(w,z)$. Note that, from the viewpoint of neoclassic models, this set includes pseudoequilibrium points with a nonzero unemployment rate. However, at all these points, the aggregate labor supply for all persons employed is equal to the aggregate labor demand, and the efficiency of using manpower resources (i.e., the ratio $L_D(w)$ for some of such points exceeds this indicator for points with zero unemployment. Thus, we can consider that all points of the set $\hat{H}$ provide economic and social stability, i.e., are analogues of equilibrium points in classical models. Therefore, we call the set $\hat{H}$ the set of equilibrium states in what follows.

Taking into account that $F(w,z)$ is monotonically increasing with respect to $w < \tilde{w}$ and that $f(w)$ is a decreasing function, it accordingly follows from the assumptions on sufficiently large and small values of these functions for sufficiently small and large $w$ that the mentioned set will not be empty and its graphic representation will be some line on the plane being considered.

Under these conditions, two types of changes in equilibrium states are possible. First, it is a shift of an equilibrium point along the line $\hat{H}$, i.e., the passage from some equilibrium state to another with fixed forms of the relations $L_D = f(w)$ and $L_S = F(w,z)$. The cause of such a passage can be the action of factors (including noneconomic ones) that do not directly exert influence on labor demand and on labor supply but lead to a change in the employment rate or labor remuneration. One of such factors can be a change in the minimum labor remuneration rate provided that the previous equilibrium rate of labor remuneration $w^0$ will be smaller than

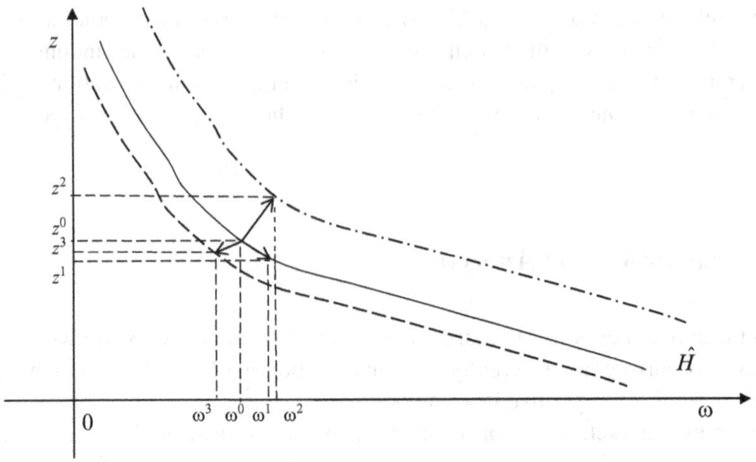


**Fig. 3.17** Diagram of changing equilibrium states under the condition $\dfrac{\partial F(w,z)}{\partial z} > 0$

its new minimum level. Second, a change in the current equilibrium state can be conditioned by varying the functions $f(w)$ and $F(w,z)$ under the action of factors that are not directly taken into account in equality (3.63). We consider these types of changes in more detail. We first assume that, for all $(w,z)$ from some neighborhood of the previous equilibrium point, the condition $\dfrac{\partial F(w,z)}{\partial z} > 0$ is fulfilled. Then an increase in $w$ is compensated by a decrease in $z$, and the line $\hat{H}$ will be the plot of some continuous decreasing function $z = z(w)$ (Fig. 3.17).

Let a point $(w^0, z^0)$ be the current equilibrium state. We first consider changes in employment and labor remuneration rates as a result of the passage to a new equilibrium state with the invariable location of the line $\hat{H}$. As is obvious from Fig. 3.16, one of parameters increases and the other decreases in this case. For example, for a new equilibrium point $(w^1, z^1)$, we have $w^1 > w^0$ and $z^1 < z^0$. In this case, a direct relation is observed between the labor remuneration rate $w$ and the unemployment rate $U = \bar{z}(w) - z$, which has been discussed in the previous section.

Let us analyze consequences of changes in locations of lines $\hat{H}$ (sets of equilibrium states). We assume that, under the action of some factors that have not been directly taken into account in equality (3.63), the labor demand rate $f(w)$ has increased for all $w$. This increase must be compensated by an increase in $z$ and, hence, equilibrium under a fixed $w$ will be attained for larger values of $z$. As a result, the line $\hat{H}$ will be shifted upward up to the position depicted by the dash-and-dot line in Fig. 3.17. The passage to the new equilibrium state $\left(w^2, z^2\right)$ in the market is mainly carried out so that this state will be insignificantly different from the previous one. As follows from Fig. 3.17, such a passage is accompanied, first of all, by increasing the employment rate and labor remuneration rate $\left(w^2 > w^0\right)$

and $(z^2 > z^0)$. If, in this case, the number of persons offering their services $\bar{z}(w)$ increases more slowly than the employment rate (i.e., the inequality $\bar{z}(w^2) - \bar{z}(w^0) < (z^2 - z^0)$ takes place, then the unemployment rate $U$ decreases. In this case, an inverse relation is observed between $w$ and $U$.

In the case of a change in the labor demand rate $f(w)$ for all $w$, the line $H$ is shifted down up to the position shown by the dashed line in Fig. 3.17. After the passage to the new equilibrium state $(w^3, z^3)$, the employment rate $(z^3 < z^0)$ and labor remuneration $(w^3 < w^0)$ decrease in most cases. If, in this case, we have $\bar{z}(w^3) - \bar{z}(w^0) < z^3 - z^0$, then the unemployment rate $U$ increases, which also leads to an inverse relation between $w$ and $U$.

Thus, a direct relation is observed between unemployment and labor remuneration rates in competitive labor market under the condition $\dfrac{\partial F(w,z)}{\partial z} > 0$ after the passage to a new equilibrium stale with a fixed position of the line $\hat{H}$. If the passage to a new equilibrium state is conditioned by a change in the line $\hat{H}$, then this relation can be both direct and inverse.

We now consider the case, when, for all $(w, z)$ from some neighborhood of $(w^0, z^0)$, the inequality $\dfrac{\partial F(w,z)}{\partial z} < 0$ is fulfilled. The line $\hat{H}$ will be the plot of some increasing function $z(w)$ (Fig. 3.18).

As obvious from Fig. 3.18, in the case of passage to a new equilibrium state with a fixed position of the curve $\hat{H}$, employment and labor remuneration rates change in one direction (in the case being considered, we have $(w^1 < w^0)$ and $(z^1 < z^0)$. If the conditions imposed on the function $\bar{z}(w)$ are similar to those considered in the investigation of the previous case, then an inverse relation arises between labor remuneration and unemployment rates.

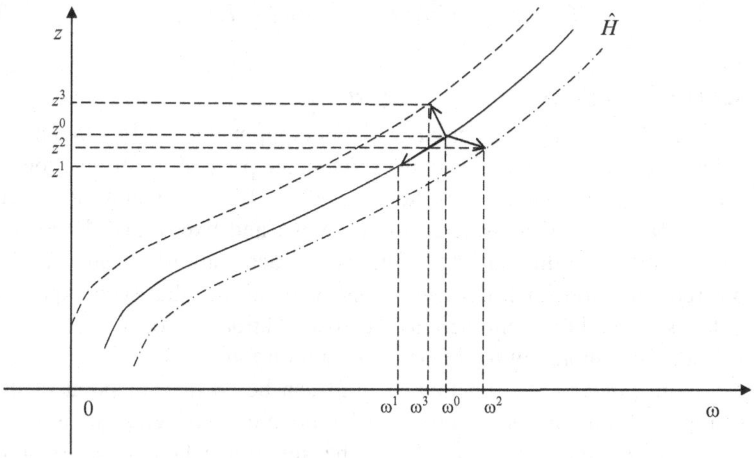


**Fig. 3.18** Diagram of changing equilibrium states under the condition $\dfrac{\partial F(w,z)}{\partial z} < 0$

With increasing labor demand under the action of factors that are not taken into account in equality (3.63), the line $\hat{H}$ is shifted downward since $F(w, z)$ increases with decreasing $z$. The new position of $\hat{H}$ is shown by the dash-and-dot line. After the passage to the new equilibrium point $\left(w^2, z^2\right)$, labor remuneration increases $\left(w^2 > w^0\right)$ and employment decreases, i.e., we have $z^2 < z^0$. As a result, the unemployment rate $U = \bar{z}(w) - z$ increases and a direct relation is observed between unemployment and labor remuneration rates. If labor demand decreases, $\hat{H}$ assumes the position denoted by the dotted line. For the new equilibrium state $\left(w^3, z^3\right)$, we have $z^3 > z^0$ and $w^3 < w^0$. In this case, a direct relation can also arise between $w$ and $U$.

Thus, when we have $\dfrac{\partial F(w, z)}{\partial z} < 0$, the passage to new equilibrium states with a fixed position of the line $\hat{H}$ is accompanied mainly by an inverse relation between labor remuneration and unemployment rates. If the passage to a new equilibrium state is conditioned by a change in the position of the line $\hat{H}$, when it is characterized by a direct relation between the mentioned rates.

If the sign of $\dfrac{\partial F(w, z)}{\partial z}$ changes for different $(w, z)$ from any neighborhood of the point $\left(w^0, z^0\right)$, then the change in unemployment and labor remuneration rates depends on the sign on this derivative at the point $\left(w^0, z^0\right)$ and in some regions around it and will be a combination of the two cases being analyzed.

We now consider the case of a monopsonic labor market in which one monopolist employer operates according to his interests and maximizes the profit obtained from the use of the acquired labor. Taking advantage of the model of such a market, which is considered in [36], we can draw the conclusion that, under these conditions, the employer determines the labor remuneration $w$ and employment rate $z$ so as to maximize the function

$$Q(w, z) = \min\Big(lF(w, z), \alpha v\Big) - Hwz \tag{3.64}$$

on the set $D = \Big\{(w, z) : w \geq 0, \quad 0 \leq z \leq \bar{z}(w)\Big\}$.

In function (3.64) is the productivity of labor computed according to the created added value, $\alpha$ is the portion of the added value in the price of the production that is manufactured by the employer, $V$ is the demand for this production (hereafter, the value of this parameter is considered to be known and fixed), and $H = 1 + h$, $h$ is the amount of additional expenditures for acquisition of labor force per unit labor remuneration fund. Following [55], we consider that such expenditures (indirect taxes on the labor remuneration fund, allocations to social needs, etc.) are proportional to the amount of the labor remuneration fund.

A maximum point of the function $Q(w, z)$ can be an interior point of the set $D$ at which this function is either nondifferentiable or its gradient is equal to zero or a point on the boundary of $D$. The situation when $w = 0$ or (and) $z = 0$ is satisfied at this point is of no interest for the further consideration in view of the zero value of production volumes. Therefore, in what follows, we exclude this situation and assume that, for sufficiently small $(w, z)$, the inequalities $\frac{\partial F(w,z)}{\partial w} > 0$ and $\frac{\partial F(w,z)}{\partial z} > 0$ must be fulfilled and that the partial derivatives mentioned above are sufficiently large. For a point $(w, z)$ at which the function $Q(w, z)$ is nondifferentiable, we have $lF(w, z) = aV$ or $w = F_w^{-1}\left(\frac{\alpha V}{l}, z\right)$, where $F_w^{-1}\left(\frac{\alpha V}{l}, z\right)$ is the inverse of the function $F(w, z)$ with respect to the variable $w$.

In the case when $\frac{\partial F(w,z)}{\partial z} > 0$ this function is increasing with respect to $V$ and decreasing with respect to $z$. The relation $w = F_w^{-1}\left(\frac{\alpha V}{l}, z\right)$, for a fixed $V$ is schematically represented in Fig. 3.19 and, with increasing $V$, the line $L$ that reflects this relation is shifted to the right and upwards to the position $L'$ shown by the dotted line. If the maximum point of the function belongs to $L$, then, for this point, the quantity $zF_w^{-1}\left(\frac{\alpha V}{l}, z\right)$ must assume the least value among values at all points of $L$. Assuming that the inverse to the function $F(w, z)$ is differentiable with respect to $z$ (this assumption is purely technical, its absence allows one to obtain similar results, but more complicated mathematical manipulations are required in this case), we obtain that, at the maximum point $(w^*, z^*)$ of the function $Q(w, z)$,

$$F_w^{-1}\left(\frac{\alpha V}{l}, z^*\right) + z^* \frac{\partial F_w^{-1}\left(\frac{\alpha V}{l}, z^*\right)}{\partial z} = 0 \quad \text{or} \quad z^* = -\left(E_3\left(V, z^*\right)\right)^{-1}$$

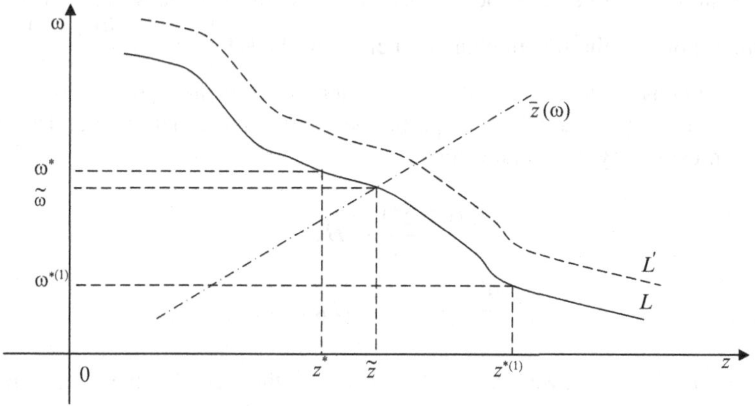


**Fig. 3.19** Diagram of the formation of employment and labor remuneration rates in a monopsonic market

must be fulfilled on the straight line $L$, where $E_3(V, z)$ denotes the value of elasticity of the function $F_w^{-1}\left(\frac{\alpha V}{l}, z\right)$ with respect to the variable $z$. We note that, under the assumptions made, the value of $E_3(V, z)$ is negative and the function $Q(w, z)$ increases with moving along $L$ to $(w^*, z^*)$.

In the case when we have $z^* > \bar{z}(w)$, i.e., $z^*$ is located below the line $z^* = \bar{z}(w^*)$ shown by the dash-and-dot line in Fig. 3.19 (such a situation arises for the point $\left(x^{*(1)}, w^{*(1)}\right)$, the maximum point of the function $Q(w, z)$ is the point $(\tilde{w}, \tilde{z})$ of intersection of the lines $L$ and $z = \bar{z}(w)$. At this point, the employment rate $\tilde{z}$ is equal to the number of persons offering their services $\bar{z}(\tilde{w})$, i.e., unemployment will be absent (this effect is peculiar to traditional models of a monopsonic labor market in which labor remuneration is considered as the main quantity controlled by the employer). But if the point $(w^*, z^*)$ is located above the line $z = \bar{z}(w)$, then the unemployment rate $U = \bar{z}(w^*) - z^*$ can be rather high. Thus, according to model (3.64), in a monopsonic labor market, considerable unemployment can take place if controlled variables for the employer are not only labor remuneration rate but also employment rate, and the function $F_w^{-1}\left(\frac{\alpha V}{l}, z\right)$ will undergo changes with moving along the line $L$. Let us also pay attention to the fact that the value of $-\left(E_3(V, z^*)\right)^{-1}$ can both increase and decrease with increasing $V$. Thus, direct and inverse relations are possible here between $w$ and $z$.

We now consider the case when $\frac{\partial F(w, z)}{\partial z} < 0$. Under these conditions, the function $F_w^{-1}\left(\frac{\alpha V}{l}, z\right)$ simultaneously increases with respect to $z$ and with respect to $V$, the value of $E_3(V, z)$ is positive, and the function $Q(w, z)$ increases with moving along the line $L$, which is accompanied by a decrease in $z$. Thus, the maximum point of the function cannot belong to the set $\left\{(w, z) : \frac{\partial F(w, z)}{\partial z} < 0\right\}$. It only remains to investigate the case when the function $Q(w, z)$ reaches its maximum at the point at which its partial derivatives are equal to zero. This point $(w^*, z^*)$ must satisfy the relationships

$$
\begin{aligned}
l\,\frac{\partial F(w^*, z^*)}{\partial w} - Hz^* &= 0, \\
l\,\frac{\partial F(w^*, z^*)}{\partial w} - Hw^* &= 0.
\end{aligned}
\tag{3.65}
$$

Taking into account that we have $l > 0$ and $H > 0$, the second of these relationships will not be fulfilled if $\frac{\partial F(w, z)}{\partial z} < 0$. Thus, the point $(w^*, z^*)$ also cannot belong to the set $\left\{(w, z) : \frac{\partial F(w, z)}{\partial z} < 0\right\}$.

In this case, the point $(w^*, z^*)$ must be located above the line $z = \bar{z}(w)$. But if it is located below this line, then the maximum of the function $Q(w, z)$ is reached on the boundary of the set $D$, provided that we have $z = \bar{z}(w)$. Necessary conditions for this maximum point assume the form

$$l\frac{\partial F(w^*, z^*)}{\partial w} - Hz^* = -\lambda^* \frac{\partial \bar{z}(w^*)}{\partial w},$$
$$l\frac{\partial F(w^*, z^*)}{\partial w} - Hw^* = \lambda^*, \tag{3.66}$$

where $\lambda^*$ is an optimal value of the Lagrange multiplier to which corresponds the constraint $z \leq \bar{z}(w)$. We note that, in this case, the inequality $\lambda^* > 0$ must be fulfilled [34] and, hence, the second equation from constraints (3.66) cannot be satisfied at the point at which we have $\frac{\partial F(w^*, z^*)}{\partial w} < 0$.

It follows from Eq. (3.66) that the point $(w^*, z^*)$ must satisfy the relationships

$$l\frac{\partial F(w^*, z^*)}{\partial w} - Hz^* = -\frac{\partial \bar{z}(w^*)}{\partial w}\left(l\frac{\partial F(w^*, z^*)}{\partial z} - Hw^*\right), \quad z = \bar{z}(w^*). \tag{3.67}$$

The point $(w^*, z^*)$ satisfying the systems of Eq. (3.65) or (3.67), will be the maximum point of the function $Q(w, z)$ if is located below the line $L$. Since this line is moved upward with increasing $V$, for rather large values of this parameter, the amounts of the monopsonic labor remuneration $w^*$ and monopsonic employment rate $z^*$ are not changed with the further increase in $V$. Here, an analogy with models of a monopsonic labor market that are considered in [36] is also traced. A change in other model parameters, for example, in $H$, can lead to both equally and oppositely directed changes in values of $w^*$ and $z^*$; in this case, both direct and inverse relations can arise between labor remuneration and unemployment rates.

Thus, the model of a monopsonic labor market in which the two-argument function is used demonstrates the possibility of occurrence of high unemployment that is used by the employer as a means of pressure on persons employed with a view to increasing individual labor supply without changing remuneration. Employment and labor remuneration rates will change in this model in the same or opposite directions owing to a change in the amount of demand for the production manufactured by the employer, additional expenditures for labor acquisition, and other model parameters. Here, direct and inverse relations can occur between unemployment and labor remuneration rates. An optimal strategy of the employer $(w^*, z^*)$ cannot belong to the set $\left\{(w, z) : \frac{\partial F(w, z)}{\partial z} < 0\right\}$.

Hence, various types of relations between labor remuneration and unemployment rates are possible for various forms of organization of a labor market. First of all, they are determined by the elasticity the two-argument function of labor supply and causes of changes in labor market conditions.

### 3.6.4 Analysis of Consequences of Administrative Regulation of Labor Remuneration and Employment

The results of investigation of a labor market using the two-argument function allow one to estimate the efficiency of actions in the field of employment and labor remuneration that are directed toward the mitigation of consequences of an economic crisis. This crisis exerts influence on all main segments of a national economy, but its consequences concerning the state of manpower and human capital of a country are particularly destructive. Fast curtailment of production leads to an essential decrease in labor demand and massive layoffs, and the deterioration in the financial state of enterprises forces employers to decrease wages and to make non-periodic payments. The mentioned processes stimulate the deterioration of professional skills of qualified workers, increase labor migration of best specialists, and increase social tension. Under such conditions, proposals are made to apply methods of government regulation of the labor market, in particular; to establish a minimum labor remuneration rate (or to increase a minimum payment established earlier with an extension of the sphere of action of this payment) and to restrict job cuts. Let us estimate the efficiency of these methods for the forms of organization of a labor market that are considered above.

We first consider a competitive market under the condition that the inequality $\frac{\partial F(w,z)}{\partial z} > 0$ holds true, i.e., in the case when an increase in labor remuneration $w$ and employment $z$ increases the aggregate labor supply. As has been noted above, under these conditions, a decrease in labor demand shifts the set of equilibrium states $\hat{H}$ to the left and downward (see Fig. 3.17). To the new equilibrium state correspond smaller values of $w$ and $z$. It follows from Fig. 3.17 that any administrative restriction on the increase in unemployment (e.g., fixing employment at the level $z^0$ that corresponds to the previous equilibrium state by the prohibition of job cuts) leads to a more essential decrease in labor remuneration in comparison with the case when such an prohibition is absent. In this case, to the new equilibrium point correspond a smaller amount of demand for labor and, hence, the reduction in production volumes increases. The shrinkage of labor remuneration and production volumes increases with increasing the inclination of the curve $\hat{H}$, i.e., with increasing the elasticity of an individual function of labor supply with respect to unemployment. As has been noted in Sect. 3.6.4, the mentioned elasticity is high in the case when losses of a person are considerable in the case of his dismissal and his stress reaction to the risk to lose his work. Thus, under these conditions, administrative restrictions on dismissals can strengthen the setback in production, decrease the well-being of employees, and increase negative social consequences.

We now investigate the case when, for a competitive labor market, we have $\frac{\partial F(w,z)}{\partial z} < 0$, i.e., when the labor remuneration $w$ and employment $z$ exert differently directed influences on the aggregate labor supply. Under these conditions, a decrease in labor demand shifts the set of equilibrium states (the curve $\hat{H}$) to the left and upward (Fig. 3.18). The passage to a new equilibrium state is accompanied by a decrease in $w$ and by an increase in $z$ and, hence, the employment restriction becomes pointless. However, the orientation of an economy toward the use of a less qualified low-paid work will strengthened, which will lead to negative social consequences. A restriction on the decrease in $w$ (e.g., by increasing of the minimum labor remuneration rate) will also restrict the increase in $z$ and slightly increase unemployment.

A similar situation is also observed in a monopsonic labor market. Depending on whether we have $\frac{\partial F(w,z)}{\partial z} > 0$ or $\frac{\partial F(w,z)}{\partial z} < 0$, any restriction on a decrease in employment under crisis conditions can lead to an additional decrease in labor remuneration or exert no influence on the situation in the labor market. Additional negative effects can arise in the case when the intersection point of the curve $L$ after its shift (as a result of a decrease in the demand $V$) and the vertical straight line $z = \tilde{z}$, where $\tilde{z}$, is the minimum employment rate, is located to the right of the intersection point of the lines $L$ and $z = \bar{z}$ (Fig. 3.19). In this case, the income function of the employer–monopolist $Q(w, z)$ decreases with respect to any $w$ and $z$ and, hence, it reaches its largest value at the point $(0, 0)$ to which corresponds a complete collapse of the employer's production. Any restrictions on the decrease in labor remuneration under conditions of economic recession inevitably increases unemployment, which follows from the analysis of a model with the two-argument function of labor supply.

The situation somewhat changes if the increase in the minimum labor remuneration is realized under the condition of achievement of economic stabilization and at the beginning of a rise in production.

We turn back to the consideration of a competitive labor market. In the case illustrated in Fig. 3.17, an increase in the minimum labor remuneration rate $w^0$ above the new equilibrium level $w^1$ simultaneously with moving the curve $\hat{H}$ to the right and upward changes the position of the new equilibrium point. In this case, the labor remuneration rate increases and the employment rate decreases in comparison with their rates in the case of absence of changes in the minimum payment. Since the new equilibrium employment rate $z$ exceeds the previous one, the aggregate income of persons employed increases, which, taking into account a relatively high inclination of the mentioned social group toward consumption, can accelerate the general economic growth. Apparently, it is such a sequence of events that took place in post-socialist leading countries mentioned at the beginning of this article. It should be noted that this policy can also lead to negative phenomena. In the case when the minimum labor remuneration rate will considerably exceed the new equilibrium level and the trajectory of shifting the equilibrium point will be almost parallel to the axis $Ox$, unemployment will considerably increase. Therefore, the increase in the minimum labor remuneration must be gradual and be accompanied by an estimation of its consequences with the help of the models considered above.

We now consider the case illustrated in Fig. 3.18. Here, under the condition of a change in labor demand, a direct relation between labor remuneration and unemployment takes place and, hence, an increase in the minimum labor remuneration inevitably decreases employment. However, if the market is at the state of equilibrium, then the result is a shift of the equilibrium point to the right and upward along the curve $\hat{H}$. In this case, owing to an inverse relation between $w$ and $U$, both employment and labor remuneration increase. The aggregate incomes of persons employed increases, which, as has been already noted, stimulates the general economic growth. This growth will be accompanied by an increase in labor demand, which, according to the considered model, will increase labor remuneration and decrease employment. Thus, negative social consequences are also possible here. They can be restricted if the increase in the minimum labor remuneration is gradually introduced without causing essential market disproportions.

Similar results are also obtained for a monopsonic labor market. Any increase in the minimum labor remuneration must be realized gradually and in the case when the situation in the labor market is such that there are grounds for supposing the presence of an inverse relation between labor remuneration and unemployment. Despite the presence of the mentioned relation, unemployment can increase not only over a bounded time interval but also in perspective.

Thus, the use of administrative levers cannot completely prevent negative social consequences of economic crises. Moreover, reducing social problems concerning one of aspects of the labor market (employment or labor remuneration), these levers worsen the situation in another aspect. Hence, their use must be accompanied by the estimation of all consequences taken into account with the help of model calculations. Under these conditions, the verification of the models proposed with the help of statistical data obtained for some countries and regions takes on particular significance. Note that a decrease in the elasticity of an individual two-argument function of labor supply with respect to unemployment restricts negative consequences of administrative regulation of labor remuneration and employment. Thus, anti-recessionary actions must also be directed toward the decrease in the sensitivity of persons to possible unemployment. This particularly holds true for highly skilled workers. This can be achieved with the help of actions such as the increase in amounts and durations of payments of unemployment benefits, increase in the control over the observance of rights of employees in questions of dismissal and hiring, revealing and prohibition of forms of hidden unemployment that are profitable for the employer, and the extension of the practice of public works.

Estimating the consequences of decreasing labor remuneration, it is also necessary to take into account its influence on highly skilled workers who form an essential part of the "middle" class. The mentioned groups are most vulnerable to the decrease in labor remuneration, and a consequence can be a large-scale labor immigration or training for a new profession. The renewal of the labor potential that can be lost in this case is a very difficult problem. These processes can be prevented by the development of alternative employment with the preservation of qualification, first of all, in the sphere of small and medium businesses. A number of measures concerning the regulation of venture businesses and engineering activities, free professions, stimulation of individual labor activity in the field of applied scientific and research-and-development activities, creation and implementation of new technologies must be considered by organs of legislative and executive authority as top-priority tasks for the preservation of the existing human capital under conditions of economic crisis.

### 3.6.5 Conclusions

The performed analysis of the leading increase in labor remuneration allows us to draw the following conclusions.

An increase in the minimum labor remuneration is an effective method of administrative influence on a labor market. However, such an increase must be bounded in view of a possible increase in unemployment. An analysis based on the introduced two-argument function of labor supply substantiates an inverse relation between unemployment and labor remuneration.

It is shown that the presence of a direct or an inverse relation between labor remuneration and unemployment rates is independent of the presence or absence of perfect competition in a labor market and is determined, first of all, by the presence or loss of equilibrium in the market in the current situation.

Proceeding from the results of modeling, the increase in the minimum labor remuneration can be used as an effective method of exogenous influence on manpower costs. In this case, it is desirable to use it under the condition of preliminary attainment of equilibrium in the labor market, and the increase itself would be stage-by-stage and uniform in time.

The increase in labor remuneration must be accompanied by the stimulation of attraction of expert labor (e.g., by imposing a tax on the use of sweated labor).

## References

1. Andronov, A.A., Leontovich, Ye.A., Gordon, I.I., Mayer, A.G.: Bifurcation Theory of Dynamic Systems on a Plane. Nauka, Moscow (1967, in Russian)
2. Arnold, V.I., Afraimovich, V.S., Il'yashenko, Yu.S., Shil'nikov, L.P.: Bifurcation theory. In: Modern Problems of Mathematics, vol. 5, pp. 5–218, VINITI, Moscow (1986)
3. Arno Kappler, Stefan Reichart. Facts about Germany. Frankurt/Main Societäts Verlag, 1999
4. Barth, E., Dale-Olsen, H.: Monopsonistic Discrimination and the Gender Wage Gap. NBER Working Paper, No. 7197 (June 1999). Available at: www.nber.org/papers/w7197
5. Belan, E.P., Mikhalevich, M.V., Sergienko, I.V.: Cyclic economic processes in systems with monopsonic labor markets. Cybern. Syst. Anal. **39**(4), 488–500 (2003)
6. Belan, E.P., Mikhalevich, M.V., Sergienko, I.V.: Models of two-sided monopolistic competition on a labor market. Cybern. Syst. Anal. **41**(2), 175–182 (2005)
7. Belton, M., Fleisher, Th., Kneisner, L.: Labor Economics: Theory, Evidence, and Policy. Prentice Hall, New York (1984)
8. Bignebat, C.: Labour Market concentration and migration patterns in Russia. Working Paper, No. 4. Marches Organization Institutions et Strategies d'Acteurs (2006)
9. Blanchflower, D.G.: Unemployment, well-being, and wage curves in Eastern and Central Europe. J. Jpn. Int. Econ. **15**, 364–402 (2001)
10. Blanchflower, D.G., Oswald, A.J.: The wage curve. Scand. J. Econ. **92**, 215–235 (1990)
11. Blanchflower, D.G., Oswald, A.J.: Estimating a wage curve for Britain. Econ. J. **104**, 1025–1043 (1994)
12. Bogolyubov, N.N., Mitropol'skii, Yu.A.: Asymptotic Methods in the Theory of Nonlinear Oscillations. Fizmatgiz, Moscow (1963, in Russian)

13. Bortis, H.: Institutional behaviour and economic theory. Cambridge University Press, Cambridge (1997)
14. Bradfield, M.: Long-run equilibrium under pure monopsony. Can. J. Econ. **23**(3), 700–704 (August 1990)
15. Brown, D.J., Earle, J.S.: Competition and firm performance: Lessons from Russia. Working Paper, No. 296. Stockholm School of Economics (March 2000)
16. Butenin, N.V., Neimark, Yu.I., Fufaev, N.A.: Introduction to the Theory of Nonlinear Oscillations. Nauka, Moscow (1987, in Russian)
17. Chalkley, M.: Monopsony wage determination and multiply unemployment equilibria in a non-linear search model. Rev. Econ. Stud. **58**(1), 181–193 (1991)
18. Chamberlin, G., Yueh, L.: Macroeconomics. Thomson Learning, New York (2006)
19. Chow, S., Mallet-Paret, J.: Integral averaging and bifurcation. J. Differ. Eq. **26**, 112–159 (1977)
20. Ciupagea, C., Turlea, G.: A study of the labor market in the industrial sector of the Romanian economy. ACE Project P95-2001. Res. Paper, Bucharest (1997)
21. Commander, S., Coricelli, P.: Unemployment, restructuring, and the labor market in Eastern Europe and Russia. The World Bank, Washington, DC (1995)
22. Delfgaauw, J., Dur, A.J.R.: From public monopsony to competitive market: More efficiency but higher prices. Tinbergen Institute Discussion Paper, TI 2002-118/1 (June 2003)
23. Dolan, E.G., Lindsey, D.E.: Macroeconomics, 6th ed. Dryden Press, Chicago (1991)
24. Dong, X.-Y., Putterman, L.: China's state-owned enterprises in the first reform decade: An analysis of a declining monopsony. The Davidson Institute, Working Paper, No. 93 (October 1997)
25. Eichberger, J.: Game Theory for Economists. Academic, New York (1993)
26. Filippov, A.F.: Differential Equations with Discontinuous Right-Hand Sides. Nauka, Moscow (1985, in Russian)
27. Garibaldi, P., Wasmer, E.: Equilibrium employment in a model of imperfect labor markets. Discussion Paper No. 950. Institute for the Study of Labor (IZA) (December 2003)
28. Green, F., Machin, S., Manning, A.: The employer size-wage effect: Can dynamic monopsony provide an explanation? Oxf. Econ. Pap. **48**, 433–455 (1996)
29. Gruber, J., Koshlai, L.B., Mikhalevich, M.V., Sergienko, I.V., Stoljar, A.N.: Models and methods of ordinary regression. Preprint No. 95-3. Institute of Cybernetics (1995, in Russian)
30. Haluk, I.E., Serdar, S.: A microeconomic analysis of slavery in comparison to free labor economies. Discussion Paper, No. 97-08. Bilkent University, Ankara (1997)
31. Hyman, D.H.: Economics. Irwin Inc., Boston (1992)
32. Ioannides, Y.M., Hsarides, C.A.: Monopsony and the lifetime relation between wages and productivity. J Labor Econ **3**(1 Part 1), 91–100 (January 1985)
33. Jones, D.W.: Monopsony and plant location in a Thunen land use model. J. Reg. Sci. **28**(3), 317–327 (1988)
34. Karmanov, V.G.: Mathematical Programming. Nauka, Moscow (1980, in Russian)
35. Keynes, J.M.: The General Theory of Employment, Interest and Money. The Selected Works. Macmillan Cambridge University Press, Cambridge (1936)
36. Koshlai, L.B., Mikhalevich, M.V., Sergienko, I.V.: Simulation of employment and growth processes in a transition economy. Cybern. Syst. Anal. **35**(3), 392–405 (1999)
37. Kuznetzov, Y.A.: Elements of Applied Bifurcation Theory. Springer, New York (1998)
38. Libanova, E.M.: The Labour Market. Center for student's literature, Kyiv (2003, in Ukrainian)
39. Manning, A.: Monopsony and the Efficiency of Labor Market Interventions. Centre for Economic Performance. London School of Economics and Political Science (August 2001). Available at: www.tinbergen.nl
40. Manning, A.: Monopsony in Motion: Imperfect Competition in Labor Market. Princeton University Press, Princeton (2003)
41. Markusen, J.R., Robson, A.J.: Simple general equilibrium and trade with a minopsonized sector. Can. J. Econ. **13**(4), 668–682 (November 1980)
42. Marsden, J.E., McCracken, M.: The Hopf Bifurcation and Its Applications. Springer, New York (1976)

43. Mikhalevich, M.: Peculiarities of Some Pricing Processes in the Transition Economy. WP-93-66. IIASA, Laxenburg (1993)
44. Mikhalevich, M.: Remarks on the Dyer-Saaty controversy. Cybern. Syst. Anal. **30**(1), 75–79 (1994)
45. Mikhalevich, M., Koshlai, L., Khmil, R.: Multisectoral models of labor supply for countries in transition. Research Memorandum, ACE Project No. 98/9. University of Leicester (1998)
46. Mikhalevich, M., Koshlai, L.: Modeling of multibranch competition in the labor market for countries in transition. In: Owsinski, J. (ed) MODEST 2002: Transition and Transformation: Problems and Models, pp. 49–59. The Interfaces Institute, Warsaw (2002)
47. Montuega-Gomez, V.M., Ramos-Parreno, J.M.: Reconciling the wage curve and the Phillips curve. J. Econ. Surv. **19**, 735–736 (2005)
48. Neimark, Yu.I.: The Method of Point Mappings in the Theory of Nonlinear Oscillations. Nauka, Moscow (1972, in Russian)
49. Nelson, R.R., Winter, S.G.: An Evolutionary Theory of Economic Changes. Belknap Press of Harvard University Press, Cambridge (1982)
50. Osaulenko V. A. (Editor) Yearbook of Ukraine for 2000. V.A.Golovko (resp. For issue) Kiev, "Technology", pp. 2001–598 (in Ukrainian)
51. Osaulenko V. A. (Editor) Yearbook of Ukraine for 1998. V.A.Golovko (resp. For issue) Kiev, "Technology", pp. 1999-576 (in Ukrainian)
52. Polterovich, V.M.: Paradoxes of the Russian labor market and the theory of collective firms. Econ. Math. Methods **39**(2), 210–217 (2003, in Russian)
53. Popkov, A.Yu.: Gradient methods for nonstationary unconstrained optimization problems. Autom. Remote Control **66**(6), 883–891 (2005)
54. Raiser, M.: Statistical review. Econ. Transit. **5**(2), 521–552 (1997)
55. Resnicoff, M.: European Union Minimum Monthly Salaries (2008). http://www.suitel01.com.10
56. Ridder, G., Berg, G.J.: Measuring labor market frictions: A cross-country comparison. Discussion Paper No. 814. Institute for the Study of Labor (IZA) (July 2003). Available at: www.iza.org
57. Robinson, J.: Economics of Imperfect Competition. University Press, Cambridge (1933)
58. Robinson, J.: Essays in the Theory of Economic Growth. Macmillan, London (1962)
59. Robinson, J.: Market structure, employment and skill mix in the hospital industry. South. Econ. J. **52**(2), 315–325 (1988)
60. Saaty, T.: The Analytic Hierarchy Process. McGraw-Hill, New York (1980)
61. Saaty, T: Theory and Applications of the Analytic Network Processes. University of Pittsburg, Pittsburg (2005)
62. Samuelson, P.: Economics. McGraw-Hill, New York (1967)
63. Sergienko, I.V., Mikhailevich, M.V., Stetsyuk, P.I., Koshlai, L.B.: Interindustry model of planned technological-structural changes. Cybern. Syst. Anal. **34**(3), 319–330 (1998)
64. Shapiro, C., Stiglitz, J.E.: Equilibrium unemployment as a worker discipline device. Am. Econ. Rev. **73**, 433–444 (1984)
65. Shieh, Y.-N.: An application of the Beckmann-Ingene argument to a monopsonist's spatial pricing problem. Pap. Reg. Sci. **78**, 319–321 (1999)
66. Shieh, Y.-N.: Employment effect of wage discrimination in the Weber-Moses triangle. Ann. Reg. Sci. **33**, 15–24 (1999)
67. Staiger, D., Spetz, J., Phibbs, C.: Is there monopsony in the labor market? Evidence from a natural experiment. NBER Working Paper No. 7258 (July 1999). Available at: http://www.nber.org/papers/w7258
68. Strobl, E., Walsh, F.: Getting it right: employment subsidy or minimum wage? Discussion Paper No. 662. Institute for the Study of Labor (IZA) (December 2002). Available at: www.iza.org
69. Strygin, V.V., Sobolev, V.A.: Separation of motions by the method of integral manifolds. Nauka, Moscow (1988, in Russian)

70. Tikhonov, A.N.: Systems of differential equations with a small parameter at higher derivatives. Mat. Sb. **31**(73, no.3), 575–586 (1952)
71. Ukrainian Law "On Approving of State Program of Employment". Verkhovna Rada of Ukraine, Law, Program (07.03.2002, No. 3076-III), http://zakon4.rada.gov.ua/laws/show/3076-14
72. Vasil'eva, A.B., Butuzov, V. F.: Asymptotic Expansions of Singularly Perturbed Equations. Nauka, Moscow (1973, in Russian)
73. Voicu, A.: Employment dynamics in the Romanian labor market: A Markov chain Monte Carlo approach. J. Comp. Econ. **33**(3), 604–639 (2005)

# Chapter 4
# Modeling of Foreign Economic Activity in a Transition Economy

## 4.1 Impact of Externalities on the Process of Economic Reforms

Practically all the former socialist countries started their market reforms under the conditions of isolation from the world markets, out of the international system of labor division, which existed then in the whole world. For some countries (such as Hungary and Poland) this isolation became substantially weak at the certain moments of time (for instance, at the end of the 1970s), but these conditions were preserved in a whole and the political factors were not the only reason for them. Serious economic reasons for autarchy also existed.

The emergence of independent contours of money circulation analyzed in Chap. 1 and the submission of monetary policy to the goal of providing planned production volumes (which as a rule were not balanced with a demand) made impossible even restricted convertibility of national currency, especially from "internal" monetary contours. These factors create more problems in foreign trade, than bulky and inflexible system of state monopoly that existed in most of the mentioned countries.

Implementation of imperfect industrial technologies with large energy and resource consumption per capita, a low quality of goods, produced by such technologies, made most of the types of final production uncompetitive at the world markets. The absence of competition both at an internal market (where the mechanisms of planned distributing prevailed) and abroad (where state enterprises even did not try to go out to) did not stimulate managers to increase a technical level of production and made technological lag even more grave.

Thus, the competitiveness of produced goods in the former socialist countries continued to go down and this was one of the main reasons for which the radical economic reforms were started.



I.V. Sergienko et al., *Optimization Models in a Transition Economy*,
Springer Optimization and Its Applications 101, DOI 10.1007/978-1-4899-7544-7_4

Attraction of foreign investments and especially of high technologies was impossible under the conditions of total state property and extremely low number of competitive products. The favorable situation, folded from the middle of the 1970s at the world oil market, stimulated reorientation of domestic investments in the former USSR to fuel and energy complex.

Technological policy in industries determining technical progress was more and more oriented to illegal (with violation of intellectual property rights) reproduction of wares and technologies already created in market countries. As a result, many post-communist countries started the reforms with the developed industry that was not able to produce final consumption goods competitive at the world markets. Such a situation was especially typical for the new independent states, the former Soviet republics.

Implementation of these countries into the world division of labor, their integration into the existing system of international economic relations became one of the main goals of market reforms.

It should be noted that external factors have their determining influence on the purposes of reforms as well as the possibility to make them.

Real domestic accumulations substantially went down and the internal demand was diminished by high inflation that accompanied the initial stage of the reforms (see Sects. 1.2 and 1.3). If an increased export volume does not compensate such a diminishing, it becomes the determinant of the economic recession. Only the external shocks (such as sharp increase of export and (or) decline of import) can provide transition of the system with the monopsony labor market considered in Chap. 3 from one equilibrium state to another, with a higher level of production.

Attraction of foreign investments and, especially, technological innovations is vitally important for the development of productive forces in post-communist countries. The account of advantages of international labor division, producing those type of goods that have comparative advantages, import of cheaper products allow to use existing resources more rationally and to improve the efficiency of the economy as a whole on this basis.

Taking into account the external impact on market reforms, it is necessary to pay the special attention to two processes determining the prospects of economic development.

The first of them is the acceleration of scientific and technical progress, accompanied by high-quality changes in the role of applied science as the important consistent of productive forces. The sphere of appearance and applications of new technologies broadens along with increase of intensity of technical innovations and reduction of their life cycle. Technological gallops become the results of discoveries not only in physics, chemistry, and informatics (as it was in the second half of the twentieth century), but also in biology, astronomy, and in many other sciences. Not any state can provide here its innovative development using only own forces. International labor division in the sphere of high technologies becomes deeper and grows into the major factor of general development.

Strengthening of interdependence of countries, internationalization of capital flows and business activity, and development of modern telecommunication and transport systems bore the second mentioned process: globalization of economy. The economic policy that does not take into account these realities of the existing world cannot be effective.

The situation in a transition economy especially depends on the external factors such as volumes of export and import, inflow and outflow of investments, and changes in the world market conjuncture. Such dependence is especially substantial for countries with open unstable transition economy without large oil and gas deposits to which Ukraine belongs. Hence, it is impossible to attain the substantial economic growth without considering external aspects.

It should be noted that the problems to be solved in transition countries in the process of foreign policy development are more difficult in many aspects than the problems of domestic policy. This is due, in particular, to much greater degree of risk and uncertainty. Many segments of the world market are so unstable that it seems to be quite impossible to receive even short-term forecasts for them. Internal data, as a rule, are more accessible and exact in comparison with external economic information. Many persons with heterogeneous and not fully exposed interests participate in external economic operations. It is necessary to take numerous factors into account, a considerable part of which has the numerical measuring when analyzing the foreign economic relations. Processes in the sphere of foreign economy are complicated and multidimensional, random impacts are inherent in many of them and difficulties are entailed for researchers. Like other spheres of a transition economy, the analysis of its external aspects supposes combination of different approaches, qualitative and quantitative methods. Mathematical modeling is only one of them. However, by virtue of the mentioned reasons, modeling tool suffers some changes. In particular, the methods of probability and game theory, scenario analysis and variant forecasting are applied here more intensively in comparison with modeling of internal economic processes. Nevertheless, it should be noted that sometimes a more simplified model gives better results than an advanced one. One of such examples is given in Sect. 4.5 of the present chapter.

Many models of foreign economic activity are now developed for all levels of aggregation, for different types of countries using different theories of international economy. In this connection, it is possible to mention the fundamental papers by Krugman [34, 35], Leontief [37], and Dolan and Lindsey [13]. The mathematical methods were applied by Sachs and Larrain [52] to analyze globalization processes and by Friedman [23] to investigate international finances. Synergetic effects, arising as a result of cooperation and labor division, were studied by Zhang [60]. Essential aspects of globalization impact on transition economies were studied by Stiglitz [56], Stigliz and Weiss [57], Kolodko [31], and Kornai [32]. Application of the existing models for a transition economy requires their modification and adaptation. Alongside with this, the original models of foreign trade and investment processes suggested by authors are introduced in the present Chapter. They are developed taking into account such peculiarities of a transition economy, as a high degree of monopolism, its instability and non-stationarity. It is impossible to collect all the models dealing with foreign aspects within the restricted volume of the work. Therefore, there are the models of currency exchange rates, financial operations, and also some aggregated models, for example, the gravity model and its modifications that remained beyond the scope.

Competition between domestic and imported goods inside a country is one of the major external factors influencing on realization of reforms. On one hand, such a competition restricts monopolism of internal producers, which, as it was pointed out in Chap. 2, is organically inherent in a transition economy. On the other hand, cheaper and high-quality imported commodities can force local producers using imperfect technologies for production, out from the market. Therefore, it is important to avoid extremes while forming the foreign trade policy, decisions must be self-weighted and grounded, and that can be achieved by using mathematical models. The approach to the analysis of such processes based on application of the oligopoly price competition model is considered in the present chapter.

In particular, Sect. 4.2 describes the model of foreign economic activity for firms occupying the monopolistic position in internal market. This model is a part of Decision Support System (DSS) "Ukrainian Budget" considered in Chap. 2.

Models for the development of an export potential in a transition country and revelation of export-oriented industries are presented in Sect. 4.6. Uncertainty of the world market conjuncture is stressed. Two approaches are used to solve the problem: variant modeling and stochastic optimization.

The stochastic optimization methods are also applied in Sect. 4.4 to model of investor behavior under risk and uncertainty. The influence of expected profitability of investment and exchange rates in recipient countries is analyzed. The models of export volumes forecasting for shallow processing goods (such as metals, oil refraction products, primary agricultural products) are considered in Sect. 4.5. The mentioned goods make the basis for export potential of many transition countries including Ukraine. These models demonstrate how the account of payable producers appears to be closely associated with the size of export under "soft" budget constraints. The recommendations are formulated for diminishing of a negative impact of this phenomenon in a state financial system.

The model of the rational structure of expenses called to promote competitiveness of produced goods is presented in Sect. 4.6. Such problems often arise in the marketing analysis. It can be formulated as an optimization problem specified by preference relations. The case, when these relations are not transitive, is considered there. Some approaches to formalization of the "best" decision concept are analyzed for such conditions.

Trade of common resources plays an important role in foreign economic activity of transition countries. Thus, Sect. 4.7 is dedicated to the models of pricing for common resources, such as water from transboundary rivers. These models are based on methods of nondifferentiable optimization.

## 4.2 Account of Foreign Aspects in Production and Financial Models

A set of mathematical models describing specific features of a transition economy, such as the presence of different forms of production organization, ownership and production, instability of prices and consumer demand, was developed by Mikhalevich and Mikhalevich [43], Mikhalevich and Podolev [44], and Pugachev and Pitelin [49]. These models are called by the authors as production and financial models. In addition to commodity flows, which are typically expressed in stable (constant or conventional) prices, these models also consider the main financial flows (payments made by producers and consumers for supplies, payments to the stage budget, bank deposits, etc.), which essentially depend on current prices.

This section presents a further development of the mentioned models intended for the analysis of structural interindustry commodity flows and the dynamics of aggregated financial variables, with special emphasis on predicting the ability to meet the main revenue items in a state budget. Despite of some differences in focus and degree of aggregation, all production and financial models are based on a single scheme, which is presented in a simplified form in Fig. 4.1.

To make the presentation clear to all readers, we start with a brief description of each block in the diagram.

The block PRODUCTION consists of the balance equations (mainly input–output Leontief's model) in constant prices. This block also uses balance standards and industry price indices relative to constant prices to calculate the index of production cost $\overline{P}$ of each industry.

The block PRODUCER AND CONSUMER FINANCES allows for expenses and revenues in current prices for all producers (industries) and consumers taking into account in the model. Consumers include different groups of the population classified by earning capacity, income level, social status, and other features, and also public-sector consumers (education, health care, law enforcement, etc).

The block DEMAND calculates the demand of consumers from their nominal incomes and current prices.

The block BUDGET calculates the basic normative components of the consolidated state budget. The production and financial models allow for the effect of these budget norms on production volumes and production costs of the industries (through taxes, on one hand, and direct and indirect subsidies and government investments, on the other), on consumer demand (e.g., through planned expenditure on public consumption), and on other important financial indicators.

The current prices are calculated in the block PRICES. The transition economy is characterized by co-existence of several pricing mechanisms, and this block includes the following options:

(a) cost-based pricing, when the price $P$ is the sum of production costs $\tilde{P}$ plus an acceptable profit;

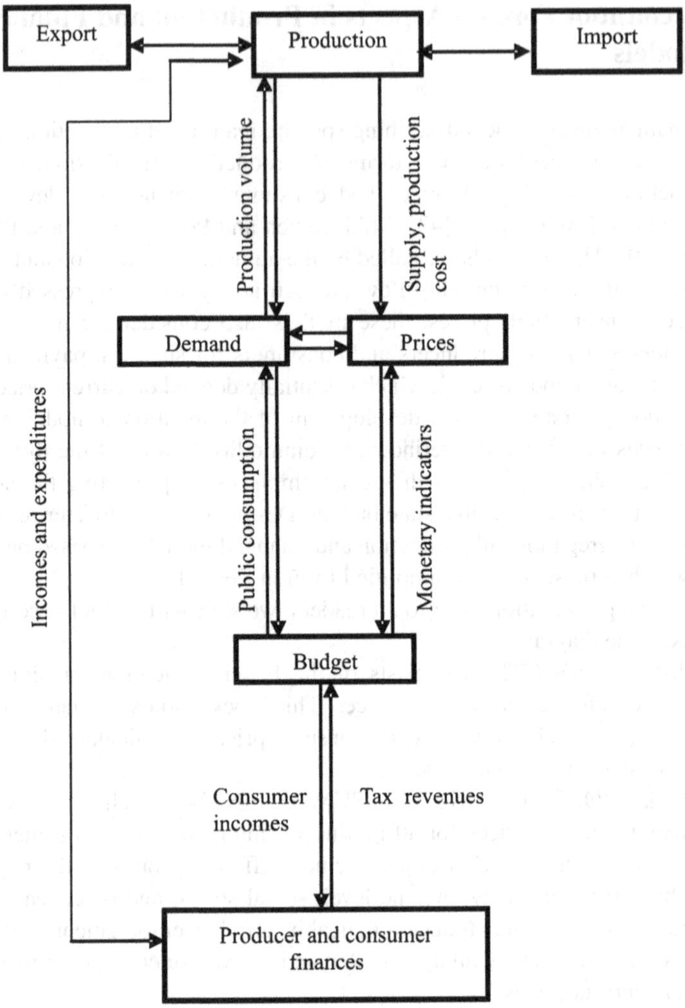


**Fig. 4.1** Simplified schematic diagram of the production–financial model

(b) pricing driven by supply and demand; this mechanism can be modeled, in particular, by Samuelson's equation (see [48]);

(c) "monetary" pricing, i.e., prices change due to variation of the quantity of money in circulation; models of this end were considered in [8, 53].

Pricing in a transition economy is also driven by monopolistic and oligopolistic mechanisms including "soft budget constraints." They are introduced in the cost-based pricing model through the assumption that producers attempt to maintain a stable profit margin in the price of their products, irrespective of the demand for these products. This model was considered in Chap. 1. This assumption is valid under high inflation and a liberal monetary policy, but it requires further analysis and refinement for the case when the combination of high prices and restrictive monetary measures makes it impossible to find a buyer. The block PRICES accordingly incorporates independent examination of monopolistic and oligopolistic pricing models.

Now let us consider more precisely blocks EXPORT and IMPORT.

Export and import in the models developed by Mikhalevich and Mikhalevich [43] and Mikhalevich and Podolev [44] are considered as given or as determined by the given level of consumption through the input–output balance equations. This is an acceptable assumption for critical import and government-controlled export. Liberalization of foreign trade, however, makes the volume and structure of import and export progressively more dependent on relative price in domestic and world markets, on foreign exchange policy, and on other financial factors. All these factors must be incorporated in pricing models. Competition with import may affect the pricing policy of domestic monopolists. Monopolistic mechanisms are replaced in this case with oligopolistic mechanisms, which are less oriented toward the demand of the most affluent groups of consumers. However, an imperfect technological base, high taxes, and some other objective and subjective reasons preclude many domestic producers from effectively competing with import. As a result, domestic producers are crowded out, and decline of production is accelerated. All these aspects are incorporated in further development of the model.

The model presented below differs in the following respects from the model described in [43, 44]:

1. in addition to the above-mentioned factors, the model allows for monopolistic (with and without export) and oligopolistic (in the presence of competing importers) pricing mechanisms;
2. the volume of export and import is determined not only by surplus or shortage of certain commodities, but also by world prices.

Continuing the approach of Pugachev and Pitelin [49], we consider the manufacturing industries as monopolists that set a single price for their products at a level not less than the industry-average cost. The economy consists of $n$ such industries, each producing one product.

All industries are conventionally divided into three groups:

1. industries with a high export potential, which are not deterred by competition from import in the domestic market; an example of such an industry in a transition economy is extraction of highly liquid mineral resources;
2. manufacturing industries facing competition from foreign importers in domestic markets;
3. industries that are natural monopolists in the domestic market and have no opportunities (or only limited opportunities) for exporting their products and services; these industries primarily include the infrastructure sectors (transport, communication, etc.).

Pricing in each group of industries is described by a separate model. Before discussing these models, we introduce some common notation used for all groups.

Let $i = \overline{1,n}$ be an industry index. Denote by $p_i$ the relative price of the product of industry $i$ (in relation to the constant price) and by $\tilde{p}_i$ the relative cost of this product (in relation to the same constant price); in fact, $\tilde{p}_i$ are the components of the previously mentioned vector $\tilde{P}$. Denote by $\varphi_i(p_i)$ the demand for the product of industry $i$ as a function of its price without competition. This function is continuous, monotone decreasing, and nonnegative, taking arbitrarily small values for sufficiently large $p_i$.

For industries of the first and second groups, $\overline{p}_i$ is the relative price of the product of industry $i$ in the world market (again expressed in relation to the constant price):

$$\overline{p}_i = \frac{\overline{p}_i^b p_b}{p_i^0}, \quad i = \overline{1,n},$$

where $\overline{p}_i^b$ is the average price of the same product in the world market in a freely convertible currency, $p_b$ is the exchange rate of the freely convertible currency to domestic currency, $p_i^0$ is the constant domestic price of the product of industry $i$ expressed in domestic currency. We assume that imported products are sold in the domestic market at the price $(1 + \Delta_i)\overline{p}_i$, where $\Delta_i$ is determined by the applicable taxes, customs duties, transport and other costs, acceptable profit of intermediaries, and other mark-ups. We denote by $\hat{p}_i$ the additional costs associated with the export of the product of industry $i$ related, like $p_i$ and $\overline{p}_i$, to the constant price for this product.

The value of domestic sales of the product of industry $i$ is denoted by $z_i$, the value of export is denoted by $\hat{z}_i$, and the value of import of equivalent products is denoted by $\overline{z}_i$. All these values are expressed in constant prices.

Let us now consider the pricing models for each group of industries.

The first-group monopolists fully control the domestic price of their products and the sales in domestic markets (where the limit is set by purchasing capacity of consumers) and abroad (the export market is assumed to have unlimited capacity). Choosing these three parameters, they strive to maximize their earnings:

$$F(p_i, z_i, \hat{z}_i) = (p_i - \tilde{p}_i) \min\left(\varphi_i(p_i), z_i\right) + \left(\overline{p}_i - \tilde{p}_i - \hat{p}_i\right)\hat{z}_i \to \max \tag{4.1}$$

subject to the constraints

$$z_i + \hat{z}_i \le x_i, \quad z_i \ge 0, \quad \hat{z}_i \ge 0, \quad p_i \ge 0, \tag{4.2}$$

where $x_i$ is the maximum possible output determined by technological factors (for instance, the available productive capacities and the raw materials). The component $p_i^*$ of the optimal solution $\left(p_i^*, z_i^*, \hat{z}_i^*\right)$ is the monopolistic price set for the products of these industries.

Note that problem (4.1), (4.2) can be simplified on the basis of the following propositions concerning the properties of $\left(p_i^*, z_i^*, \hat{z}_i^*\right)$.

**Statement 4.1.** If $z_i^* > 0$ and $\overline{p}_i > \tilde{p}_i + \hat{p}_i$, then $\varphi_i\left(p_i^*\right) = z_i^*$.

*Proof.* Assume that $\varphi_i\left(p_i^*\right) > z_i^*$. Denote by $p_i^{(1)}$ the solution of the equation $\varphi_i\left(p_i^*\right) = z_i^*$. By the properties of the function $\varphi_i(p_i)$ this solution always exists for $z_i^* > 0$; since $\varphi_i(p_i)$ is monotonically decreasing, we have the inequality $p_i^{(1)} > p_i^*$, and thus

$$F\left(p_i^{(1)}, z_i^*, \hat{z}_i^*\right) = \left(p_i^{(1)} - \tilde{p}_i\right)z_i^* + \left(\overline{p}_i - \tilde{p}_i - \hat{p}_i\right)\hat{z}_i^*$$
$$> \left(p_i^* - \tilde{p}_i\right)z_i^* + \left(\overline{p}_i - \tilde{p}_i - \hat{p}_i\right)\hat{z}_i^* = F\left(p_i^*, z_i^*, \hat{z}_i^*\right),$$

which contradicts the optimality of $\left(p_i^*, z_i^*, \hat{z}_i^*\right)$.

Now assume that $\varphi_i\left(p_i^*\right) > z_i^*$. Then

$$z^{(1)} = \varphi_i\left(p_i^*\right), \quad \hat{z}^{(1)} = \hat{z}_i^* + \left(z_i^* - z_i^{(1)}\right).$$

Note that

$$F\left(p_i^*, z_i^{(1)}, \hat{z}_i^{(1)}\right) = \left(p_i^* - \tilde{p}_i\right)\varphi_i\left(p_i^*\right) + \left(\overline{p}_i - \tilde{p}_i - \hat{p}_i\right)\hat{z}^{(1)}$$
$$= \left(p_i^* - \tilde{p}_i\right)\varphi_i\left(p_i^*\right) + \left(\overline{p}_i - \tilde{p}_i - \hat{p}_i\right)\hat{z}_i^* + \left(\overline{p}_i - \tilde{p}_i - \hat{p}_i\right)\left(z_i^* - \varphi_i\left(p_i^*\right)\right)$$
$$= F\left(p_i^*, z_i^*, \hat{z}_i^*\right) + \left(\overline{p}_i - \tilde{p}_i - \hat{p}_i\right)\left(z_i^* - \varphi_i\left(p_i^*\right)\right).$$

Since $\left(\overline{p}_i - \tilde{p}_i - \hat{p}_i\right)\left(z_i^* - \varphi_i\left(p_i^*\right)\right) > 0$, the last equality also contradicts the optimality of $\left(p_i^*, z_i^*, \hat{z}_i^*\right)$. We are thus left with the equality $\varphi_i\left(p_i^*\right) = z_i^*$ as the only possibility.

**Statement 4.2.** *If* $\overline{p}_i > \tilde{p}_i + \hat{p}_i$, then $z_i^* + \hat{z}_i^* = x_i$.

*Proof.* Assume that $z_i^* + \hat{z}_i^* < x_i$. Set $\hat{z}_i^{(1)} = \hat{z}_i^* + (x_i - z_i^* - \hat{z}_i^*)$.

Then

$$F\left(p_i^*, z_i^*, \hat{z}_i^{(1)}\right) = \left(p_i^* - \tilde{p}_i\right)\min\left(\varphi_i\left(p_i^*\right), z_i^*\right) + \left(\overline{p}_i - \tilde{p}_i - \hat{p}_i\right)\hat{z}_i^{(1)}$$
$$> \left(p_i^* - \tilde{p}_i\right)\min\left(\varphi_i\left(p_i^*\right), z_i^*\right) + \left(\overline{p}_i - \tilde{p}_i - \hat{p}_i\right)\hat{z}_i^*$$
$$= F\left(p_i^*, z_i^*, \hat{z}_i^*\right),$$

which contradicts the optimality of $\left(p_i^*, z_i^*, \hat{z}_i^*\right)$.

**Statement 4.3.** *If* $\overline{p}_i < \tilde{p}_i + \hat{p}_i$, then $\hat{z}_i^* = 0$.

*Proof.* Assume that $\hat{z}_i^* > 0$. Let $\hat{z}_i^{(1)} = 0$. Then

$$\begin{aligned} F\left(p_i^*, z_i^*, \hat{z}_i^{(1)}\right) &= \left(p_i^* - \tilde{p}_i\right) \min\left(\varphi_i\left(p_i^*\right), z_i^*\right) \\ &> \left(p_i^* - \tilde{p}_i\right) \min\left(\varphi_i\left(p_i^*\right), z_i^*\right) + \left(\overline{p}_i - \tilde{p}_i - \hat{p}_i\right)\hat{z}_i^* \\ &= F\left(p_i^*, z_i^*, \hat{z}_i^*\right), \end{aligned}$$

which contradicts the optimality of $p_i^*, z_i^*, \hat{z}_i^*$.

Note that for $z_i = 0$ the function $F\left(p_i, z_i, \hat{z}_i\right)$ is independent of $p_i$, and for $\overline{p}_i = \tilde{p}_i + \hat{p}_i$ it is independent of $\hat{z}_i$. Therefore $\left(p_i(0), 0, \hat{z}_i^*\right)$ is among the optimal solutions of problem (4.1), (4.2) in the first case, and $\left(p_i^*, z_i^*, 0\right)$ is among its optimal solutions in the second case. Here $p_i(0)$ is the solution of the equation $\varphi_i(p_i) = 0$.

In view of the above, if $\overline{p}_i > \tilde{p}_i + \hat{p}_i$, then instead of the problem (4.1), (4.2) we can determine $p_i^*$ from the problem

$$H_i^{(1)}(p_i) = \left(p_i - \tilde{p}_i\right)\varphi_i(p_i) + \left(\overline{p}_i - \tilde{p}_i - \hat{p}_i\right)\left(x_i - \varphi_i(p_i)\right) \to \max \tag{4.3}$$

subject to the constraints

$$\varphi_i(p_i) \le x_i, \quad p_i \ge 0. \tag{4.4}$$

If $\overline{p}_i \le \tilde{p}_i + \hat{p}_i$, then $p_i^*$ can be obtained as the solution of the problem

$$H_i^{(2)}(p_i) = \left(p_i - \tilde{p}_i\right)\varphi_i(p_i) \to \max \tag{4.5}$$

subject to (4.4).

Problems (4.3)–(4.5) seek the maximum point of a continuous function of a single variable on the real half-line. They can be solved by standard numerical methods (see [30]).

Let us now consider the pricing model for the second-group industries. Here the producer's objective is again profit maximization, but the demand function now depends on both $p_i$ and $\overline{p}_i$. The following relationships for implicit specification of this function have been considered in [49]:

$$\frac{\overline{z}_i}{z_i} = k\left(\frac{p_i}{\overline{p}_i(1+\Delta_i)}\right)^{\alpha} \tag{4.6}$$

and

$$\overline{p}_i(1+\Delta_i)\overline{z}_i + p_i z_i = \Phi_i, \tag{4.7}$$

where $k$, $\alpha$ are some nonnegative constants that reflect the market response to a change in the price ratio of imported and domestically manufactured products: $\Phi_i$ is the purchasing-power demand for the product of industry $i$. The profit function in this case is given by the formula

$$\overline{F}_i(p_i) = \left(p_i - \tilde{p}_i\right) z_i\left(p_i, \Phi_i\right),$$

where the dependence $z_i(p_i, \Phi_i)$ is implicitly defined by relationships (4.6), (4.7). This dependence can be written in explicit form by expressing $z_i$ from Eq. (4.7), substituting the resulting expression in Eq. (4.6), and solving this equation for $\overline{z}_i$:

$$z_i(p_i, \Phi_i) = \frac{\Phi_i}{k(1+\Delta_i)\overline{p}_i\left(\frac{p_i}{\overline{p}_i(1+\Delta_i)}\right)^{\alpha} + p_i}. \tag{4.8}$$

Using (4.8), we write the oligopolistic pricing problem in the form

$$\overline{F}_i(p_i) = \frac{\Phi_i(p_i - \tilde{p}_i)}{k(1+\Delta_i)\overline{p}_i\left(\frac{p_i}{\overline{p}_i(1+\Delta_i)}\right)^{\alpha} + p_i} \to \max,$$

$$p_i \geq 0.$$

Note that in model calculations the function $\overline{F}_i(p_i)$ can be conveniently replaced with the function

$$G_i(p_i) = \frac{\overline{F}_i(p_i)}{\Phi_i} = \frac{(p_i - \tilde{p}_i)}{k(1+\Delta_i)\overline{p}_i\left(\frac{p_i}{\overline{p}_i(1+\Delta_i)}\right)^{\alpha} + p_i}, \tag{4.9}$$

which does not depend on the problematic parameter $\Phi_i$. The maximum seeking problem for the function $G_i(p_i)$ also can be solved by numerical methods.

If both $\overline{p}_i$ and $\overline{z}_i$ are given (for instance, the government imposes import quotas), then $p_i$ and $z_i$ are determined by solving the system of Eqs. (4.6) and (4.7). Note that Eq. (4.7) implicitly defines the demand function

$$z_i = \frac{\Phi_i - \overline{p}_i(1+\Delta_i)\overline{z}_i}{p_i}. \tag{4.10}$$

In other words, we assume the demand to be equal to the purchasing-power demand of the consumers. In reality, the demand may decrease with increasing prices faster than the purchasing-power demand because of the substitution effect of high-price commodities with lower cost commodities or the refusal of consumers to purchase commodities that are not essential. It is therefore relevant to consider the demand function

$$z_i = \max\left(\frac{\Phi_i - \overline{p}_i(1+\Delta_i)\overline{z}_i}{p_i^{\beta}} - C, 0\right), \tag{4.11}$$

which is a generalization of Eq. (4.10).

Just as equalities (4.8), (4.9) have been obtained from (4.6), (4.7), we can use (4.6) and (4.11) to derive the formulas

$$z_i(p_i, \Phi_i) = \max\left(\frac{\Phi_i - C p_i^{\beta}}{p_i^{\beta} + \overline{p}_i(1+\Delta_i)k\left(\frac{p_i}{\overline{p}_i(1+\Delta_i)}\right)^{\alpha}}, 0\right)$$

and

$$\overline{F}_i(p_i) = (p_i - \tilde{p}_i)\max\left(\frac{\Phi_i - C p_i^{\beta}}{p_i^{\beta} + \overline{p}_i(1+\Delta_i)k\left(\frac{p_i}{\overline{p}_i(1+\Delta_i)}\right)^{\alpha}}, 0\right),$$

which generalize the previous oligopolistic pricing model to the case of a more general demand function.

Let us now consider the third group of industries. The price of the product produced by these industries is determined so as to maximize the profit from domestic sales:

$$\tilde{F}_i(p_i, z_i) = \left(p_i - \tilde{p}_i\right)\min\left(\varphi_i(p_i), z_i\right) \to \max$$

subject to the constraints $0 \le z_i \le x_i$, $p_i \ge 0$.

Note that this problem is equivalent to (4.4), (4.5), i.e., the pricing policy of these monopolists is like the pricing policy of export oriented industries when export become unprofitable.

These pricing models have been implemented in DSS "Ukrainian Budget," modeling the effect of the state budget proposal on the main macroeconomic and interindustry characteristics and develops recommendations concerning efficient macroeconomic policy. The core of the system is a model of the consolidated state budget, considered in Chap. 2, and a set of different pricing models. Using hierarchical menus and computer graphics, the user (an expert in macroeconomic analysis) can select for each industry the pricing mechanisms that are most essential in his view.

The computation results (including various alternatives) can be presented in tabular and graphic forms. DSS produces a forecast of price dynamics, domestic sales volumes, and export and import volumes. An important function of the system is checking the effectiveness of various strategies of direct and indirect subsidies to producers and developing recommendations for reasonable taxation of production and export–import transactions. The modeling results predict the fulfilment of the main revenue and expenditure components of the budget, and also explain the deviations of the predicted values from the initial assumptions of the draft budget.

DSS has been used to carry out a series of computations based on the main economic indicators of Ukraine for 2007–2010 and for the 1990s.

Model runs using the data for 1992–1999 were the part of comprehensive analysis of high inflation and postinflationary processes of the 1990s in Ukraine. Internal aspects of these processes were investigated in Chaps. 1 and 2. Now let us pay attention mainly to the impact of external factors.

According to our computations, the inflationary processes in the period were developing by a scenario which could be called structural-budgetary inflation. The main inflationary component in this scenario is the increase of production costs due to both external and internal factors, leading to an inflation of consumer costs. Cost-based pricing mechanisms predominated in most industries under these conditions. Analyzing the observed price changes, we distinguish two groups of industries in Chap. 1. The first group includes industries with high production costs (coal and food industries), and the second one includes industries with a high industrial demand for their products (other sectors of the fuel and energy complex, machinery, agriculture). In these industries, the prices increased according to the scenario of the structural inflationary crisis (see Sect. 1.3 of Chap. 1) and were determined by the structure of the individual production costs. The rates of price increases for these groups of industries were higher than for the economy as a whole. In other industries, price increases were mainly attributable to monetary factors (budget deficit, increase of the nominal money supply), higher costs of critical import and devaluation of the domestic currency due to the monetary factors.

Introduction of monopolistic pricing mechanisms had only a slight effect on the predicted prices in most industries. For some industries the monopolistic price was somewhat lower than the cost-based price calculated for a stable level of profitability. This could be attributed to the fact that under conditions of rapidly contracting purchasing-power demand for consumer goods the price calculated on the basis of this profitability does not support a sufficient volume of sales and therefore produces a lower profit. The monopolistic price is 30–40 % higher than the cost only for industries with low production costs and traditionally modest profitability (such as transport and communication).

Competition with import and oligopolistic pricing in high-cost industries, such as the coal industry and the agro-industrial complex, does not slow down price increases in the 1990s. We see from (4.9) that oligopolistic prices are independent of purchasing-power demand. At the same time, the oligopolistic prices may be lower than cost even when the cost is essentially higher than the import price $\overline{p}_i\ (1 + \Delta_i)$.

Therefore, the response of these industries to higher production costs and lower demand is to reduce their output while raising their prices.

In addition to predicting prices, DSS "Ukrainian Budget" had been used to check the effectiveness of different strategies of government grants and subsidies to producers. The results demonstrated low efficiency of the "price freeze" policy and the policy of cost reimbursement from the state budget to the high-cost industries. These policies triggered a structural inflationary crisis. The inflationary impact of large money emissions that are necessary to support subsidies and grants to all industries in the presence of limited budget revenues is comparable to the consequences of price increases for the products of these industries under "free" pricing (without subsidies). Additional negative effect of subsidies was connected with necessity to increase budget revenues to cover them. Growth of taxes and import tariffs increases both import price and cost of production for industries, which are involved into oligopolistic import competition. As a result, these industries reduce the volume of production and increase prices. A more reasonable practice is to subsidize the industries that face a high industrial demand for their products. This breaks the "chain" of price increases in industries enmeshed in the structural inflationary crisis. Price increases in high-cost industries are thus isolated and are gradually suppressed by decreasing demand for their products or displacement of their products by cheaper import (under oligopolistic pricing).

These calculations have produced some recommendations concerning tax policy.

The predicted budget revenues from VAT, sales tax, and especially profit tax should be based on a forecast of sales, and not a forecast of production volumes. The sales forecast should be based on an estimate of the cash demand on domestic and export markets. Because of decreasing purchasing power and poor competition with import, a substantial portion of the output may remain unsold, which will necessarily affect the dynamics of budget revenues. Under high cost inflation, it is apparently necessary to revise not only the basic tax rates, but also the structure of the entire tax system. Calculations demonstrate the stabilizing role of budget revenues that are independent of current business results, such as land tax, property tax, and tax on machinery and equipment. The proportion of these taxes in production costs decreases with the increase of production volume, which encourages producers to reduce prices as the output increases. Moreover, these taxes are indexed to the GDP deflator, and have only a slight effect on monopolistic price increases. Another advantage of these taxes is the simple assessment and collection procedure.

Calculations carried out with DSS "Ukrainian Budget" on the basis of the 18-industry balance for 1993–1994 [9, 10] made it possible to estimate the effect of different inflationary factors in a period of galloping inflation. Price increases of critical import proved the most important factor in this respect. Thus, a 20-fold increase of prices of imported energy resources (in dollar equivalent terms) in 1993 produced no less than an eightfold overall price increase. This accounts for about 40–45 % of the observed price increases at that time (assuming multiplicative effect of the inflationary factors).

The most significant among internal factors was the increase of production costs according to the structural inflationary crisis scenario described in [43]. It accounted for a fivefold price increase, i.e., nearly 30 % of the overall price increases (assuming multiplicative effect of the inflationary factors).

Calculations made in [43] also show that the quantity of money actually exceeded its optimal value at certain times. The effect of monetary mechanisms can explain an overall price increase by a factor of 1.8, which accounts for about 10 % of the observed price increase. Monopolistic price increases account for 5–7 % of the overall price change. The remaining inflationary factors, both objective and subjective, accounted for 7–15 % of the observed price increase.

In 1994, external factors affecting inflationary processes were substantially weaker, because the prices of critical import approached the world level (and in some cases exceeded it). Further growth of nominal costs due to import was driven largely by devaluation of domestic currency. The restrictive monetary policy stopped this process in late 1994. Internal factors affecting cost inflation persisted, although their impact also became much weaker. Excessive production costs and high demand for industrial products were observed in 7 and 8 industries (out of 18) in 1994 compared respectively to 12 and 14 in 1993. Direct unit costs, which were the main factor influencing cost inflation, decreased in some industries. At the same time, the continuation of the structural inflationary crisis produced a steady growth of production costs by 4–6 % per month. One third of total price increases are attributable to the growth of the nominal money supply above its optimal values and to monopolistic price formation. Calculation results based on our model show that import liberalization under conditions of depressed purchasing-power demand for consumer goods makes it possible to reduce the prices by 20–25 % below monopolistic prices. Since high production costs prevent domestic producers from reducing prices, domestic products are replaced with import.

The numerical experiments carried out using data for 2007–2010 were aimed at the scenario forecasting of the future development of export-oriented industries of Ukrainian economy and their influence on the value and a branch structure of attracted foreign investments.

Minimal amount of investments necessary for overcoming of the decline and production stabilization was needed in size of 4–6 billion US dollars per year, most effect would be received by a stream of investments in a few ten billion US dollars per year. Stream of the external investments conditioned by maximizing of investors' income, must be concentrated in the relatively narrow sector of economy. The degree of specialization was great. The model calculation demonstrated the presence of two scenarios of export specialization and orientation of foreign investment flows, which can appear with approximately equal possibility.

As the realization of these scenarios required the development of substantially different industrial infrastructure, possibility of their combination was small, because of the large volume of additional capital was needed for this purpose, which can reduce the efficiency of investments. Model calculations demonstrated the following scenario of further development of Ukrainian economy.

First scenario was called "metallurgical." According to it, the basic volume of primary investments is directed to ferrous metallurgy and chemical productions (which now are the main Ukrainian export-oriented industries) using local raw materials. The attractiveness of these industries is conditioned by cheap labor force and undeveloped ecological legislation. The primary stabilizing and even increase of production volumes is therefore possible in the nearest 1–2 years. Development of the indicated industries requires additional investments in other branches, foremost in heavy and transport machinery, energy production, sea ports, and other transport infrastructure. All the above-mentioned branches are characterized by a high level of monopolism, thus the pricing mechanisms, described by models (4.1) and (4.3), will dominate in Ukrainian economy. According to this scenario, export-oriented enterprises will be concentrated in Eastern and Southern regions of Ukraine. Folded production complex there will be oriented mainly to the export, but such development of economy needs the sharp increase of import of energy resources and consumers goods, as the internal production of the last will grow short. Approximately three years later the sources of development of priority industries on the basis of existent technologies were outspent and their deep reconstruction has been required with the purpose of fall-off of power-hungriness of production, increase of degree of processing of raw material and specific gravity of the products fitted for consumption without its further treatment.

The new increase of volume of external investments will be required for these purposes. The deficit of payment balance will be saved in a visible prospect, although it can be substantially diminished. The possibility of using the standard developed abroad technologies and high level of the technogenic loadings on an environment are important characteristic of "metallurgical" scenario.

The second possible scenario was called "agro-industrial." According to this scenario, the basic volume of investments will be directed to agriculture, processing (in particular, food) industry and a sphere of services. Development by such a way needs the deep technological changes called to reduce the level of material expenditures in agriculture and the level of energy consumption in processing (food) industry. Therefore, industrial decline, according to this scenario, will continue during some time, necessity for the achievement of the indicated goals, and the volume of primary investments must be greater than for a "metallurgical" scenario. The complex of priority industries will be oriented both to the export and to the internal consumption that some measures for stimulation of consumer demand will be necessary (e.g., increasing of social payments and redistribution of monopoly superprofits among population). Competitive market pricing will dominate in the export-oriented industries, which will be approximately uniformly distributed over all Ukrainian regions. Samuelson equations, introduced in Chap. 1, were used in calculations as an appropriate model of the above-mentioned pricing. At the same time, this scenario will be related with less volumes of import, which more quickly stabilize the payment of balance. Development of "agro-industrial" complex must be accompanied by growth in such branches as labor intensive (in particular, agricultural) machinery and transport. Realization of this scenario needs development and modification of new technologies taking into account a local specific and advantages and will be accompanied by the decreasing of technogenic pressure at the environment.

Further development of Ukrainian economy depends on many factors, not taken into account in the considered model, such as the interests of separate political and social groups, changes of the state of affairs at the world markets, from general political situation, etc. Large importance, undoubtedly, will have the measures to create the favorable situation in investment, in particular, changes in property relations. In any case, forming of investment policy, it is necessary to make the choice of priority scenario and to determine those factors and measures, which can assist its realization.

Foreign aspects are also taken into account in econometric macromodel proposed by Movchan and Giucci [47]. The analysis of reasons for changing economic dynamic in Ukraine in the 2000s from decline to growth is the main goal of this model.

After almost a decade of decline, Ukraine's economy started to grow in 2000 provoking questions about the reasons. Two broad categories of explanations appeared: the growth is external demand driven vs. reform driven. The former assumes short-term external demand factors like favorable developments within the ferrous metals market (see [27]). Adherents of this idea also argue that, as analogous to the Russian economic experience [24], the real devaluation of the hryvnia increased the competitiveness of Ukrainian products, both by contributing to the surge in export and by enhancing import-substitution processes in the economy [59]. On the other hand, many economists associate the upturn of Ukraine's economy with the series of reforms initiated by Viktor Yuschenko's government appointed in 1999 [3].

If one divides the transition countries into those that are adherents of the "big bang" policy of reform, and those that prefer gradualism, Ukraine definitely belongs to the latter. However, in 1999 the appointment of Viktor Yuschenko as Prime Minister precipitated a reform process in Ukraine. The reforms initiated by Yuschenko's government targeted several important areas including:

(a) establishment of clear divisions of responsibilities within the executive branch;
(b) promotion of fiscal payments discipline, including the prohibition of non-monetary transactions between the state and economic agents;
(c) reforms in the energy sector reducing rent-seeking activities;
(d) deregulation;
(e) large-scale privatization;
(f) ownership reforms in agriculture including establishment of private land ownership and disbandment of collective farms [3].

The most important changes were the enhancement of payments discipline in the fiscal and energy sectors. In particular, this led to the elimination of excessive offsets and other forms of non-monetary payments, and to an overall reduction of rent-seeking activities within the economy. Prior to 2000 non-monetary means of payments to the state, as well as overdue payables and receivables were the usual practices, contributing to further rent-seeking behavior by all economic agents [61]. The same situation was observed in the transactions between enterprises in general, indicating that one of the most important institutional features of a market economy, namely effective contract enforcement and, consequently, security of property rights and the rule of law, was not in place, at least not in the official part of the economy. Instead, shadow contracts were very common and a shadow system of contract enforcement diverted funds from the official economy. The major achievement of the Yuschenko's government was to increase the level of contact enforcement in the economy promoting a shift from the shadow to the official economy [14].

The considered model was constructed in order to measure the impact of the mentioned factors. The coefficients of the autoregressive distributed lag (ADL) equation for the growth rates in the GDP was estimated in [47] against three blocks of explanatory variables, and reform variables.

In order to maximize the degrees of freedom in the model, monthly periodicity of the real GDP was chosen. However, there were difficulties associated with this data set. First, the monthly data of the real GDP, provided by the State Committee of Statistics (SCS), were only preliminary for the last years of the twentieth century. The SCS revised only the annual and quarterly information. Second, monthly data were published only as growth rates in cumulative terms, thus, there were no monthly data indexed to a base year. Furthermore, there was no reliable officially published GDP deflator that can be applied to nominal GDP figures to generate a real time series. Therefore, several well-known experts faced the necessity to develop their own indexes of real GDP dynamics for Ukraine (see [1]). To partially resolve these difficulties, they estimated the monthly figures of real GDP of the basis of their quarterly dynamics. First, an index of the real GDP was constructed on the basis of year-on-year changes in the quarterly GDP, assuming that the true series in quarter-on-quarter terms are known for the year 1996. Second, missed monthly data was linearly interpolated. These constructed monthly dynamics are quite close to official data taking into account that spikes in monthly data are usually smoothed out in the process of annual data revision. Thus, authors of the above-mentioned papers proceeded with the analysis based on these constructed series.

For stability proxies, they used inflation measured as a change in the consumer price index (INFLATION), the consolidated fiscal deficit (DEFICIT), and the dummy for the election campaign (ELECTION) [1].

Following Fischer [18] that postulates stability to be a precondition of growth, a negative correlation between the dependent variable and inflation and / or deficit was expected. The dummy variable for election campaigns was constructed as follows: ELECTION $=$ 1 for the 6 months preceding an election, and ELECTION $=$ 0 otherwise. The time period covered by the paper of Movchan and Giucci [47] includes three election campaigns: two parliamentary election campaigns in 1998 and 2002, and one presidential election campaign in 1999. A negative correlation between this dummy variable and economic growth was expected, since election campaigns tend to increase risks, and hamper investment activity diverting funds to "political investments."

The list of external demand proxies includes world steel prices (STEEL), the real effective exchange rate (REER), and the volume index of ferrous metal export (METAL). The focus on metal is explained by the dominance of metal export over all other goods in the structure of Ukraine's export. The world steel prices were approximated by the CRU steel price index [11]. The estimation of the real effective exchange rate was based on a methodology proposed for Ukraine by Mankovska [38]. Finally, the volume index for metal export of category 72 (ferrous metals) in the harmonized trade nomenclature was obtained from SCS publications.

The list of reform proxies designed to capture the impact of the major reforms initiated in 1999/2000 includes: overdue receivable state arrears (BUDGET), overdue payable inter-enterprise arrears (OVERDUE), and agricultural output (AGRIC). The first two variables deals with major problems tackled by the above-mentioned reforms, namely poor contract enforcement in the form of pervasive non-payments in the economy. However, they differ as to the source of the problem. Overdue receivables from the state are an indication of the degree of fiscal discipline. Overdue payables of enterprises represent the other side of the problem, namely the low payments discipline of non-state economic agents. This behavior is closely correlated with rent-seeking activities since it gives birth to hopes of subsequent write-offs or of mutual settlements at artificial prices. The agricultural output is taken as a proxy for agricultural reforms in the country.

To test the hypotheses concerning the factors of economic growth, the following basic regression structure was applied:

$$A(L)Y_t = m + B(L)\overline{I}_t + \varepsilon_t, \tag{4.12}$$

where $Y_t$ is the economic growth variable, $\overline{I}_t$ is a vector for the explanatory variables, and $L$ as a lag operator so that

$$A(L) = 1 - \alpha_1 L - \alpha_2 L^2 - \ldots - \alpha_p L^p;$$

$$B(L) = \beta_0 + \beta_1 L + \beta_2 L^2 + \ldots + \beta_q L^q.$$

Here an ADL was used for specification with the real GDP growth as a dependent variable, three blocks of explanatory variables, and the lags of the real GDP, assuming that lags of the dependent variable perform a control function, i.e., capture the effect of all other factors that can potentially influence the growth dynamics.

Using the reparametrization procedure from Johnson and DiNardo [28], Eq. (4.12) is transformed into the form:

$$\Delta Y_t = m + B(L)\overline{I}_{t-1} + \delta_0 \Delta\overline{I}_t + \delta_1 \Delta\overline{I}_{t-1} + \ldots + \delta_q \Delta\overline{I}_{t-q} -$$
$$-A(L)Y_{t-1} - \gamma_1 \Delta Y_{t-1} - \ldots - \gamma_p \Delta Y_{t-p} + \varepsilon_t.$$

This transformation allows several important features that simplify calculations. First, the coefficients $B(L)$ of the lagged explanatory variables after reparametrization are the sum of the coefficients of the lagged explanatory variables before reparametrization. Moreover, as it was shown in [28], after leaving out redundant lags, these coefficients remains meaningful. Second, the $p$-values attached to the lagged levels after reparametrization are the same as the $p$-values for Wald testing, when these sums are zero in the non-reparametrized equation. In addition, the standard errors of regression, information criteria, and log-likelihood values are identical for the equations before and after reparametrization according to [28].

The final number of lags in each case is chosen on the basis of the Akaike and Schwarz information criteria, as well as the Jarque–Bera test on normal distribution, the Breusch–Godfrey test on serial correlation in residuals. The overall model is also tested for possible specification errors using the Ramsey RESET test.

Finally, Movchan and Giucci [47] adjusted the obtained coefficients of the explanatory variables on the coefficient of lagged dependent variable by the following equation, obtaining static relationship between variables:

$$\overline{Y} = \frac{m}{A(L)} + \frac{B(L)}{A(L)}\overline{I}. \tag{4.13}$$

Thus, following [1], a structural stance on the determinants of the GDP developments was not considered, but test whether changes in the variables (primarily, external conditions and the institutional environment) significantly affect economic growth controlling for the auto-regressive history of the dependent variable has been done.

After verifying the stationarity of variables in the first differences, Movchan and Giucci [47] proceeded with estimating the reparametrized ADL model of the real GDP growth in Ukraine. The choice of model was done controlling for information criteria, residuals normality, absence of serial correlation, and specification error. Following Eq. (4.13), the static relationship between growth of the real GDP and other variables can be expressed by the following equation (see Table 4.1 for the ADL results):

**Real GDP Growth** = 0.0045 + 0.0725 **Metal** + 0.2664 **Steel** ** – 0.0062 **Reer**– 0.1158 **Overdue** – 0.1276 **Budget** * ++ 0.0174 **Agric** ** – 0.0017 **Inflation** – 0.0022 **Deficit*** – 0.0174 **Election** **.

Here symbol * indicates a 1 % significance level for a corresponding coefficient of regression; symbol ** indicates a 5 % significance level.

All the variables excluding the real effective exchange rate have the expected sign; the equations' residuals are normally distributed (Jarque–Bera test's $p$-value is 0.6) and serially uncorrelated (Breusch–Godfrey test's $p$-value is 0.3), while the $p$-value of the Ramsey RESET test is equal to 0.43, hence there is no evidence of significant misspecification of the equation. The analysis for stability done through Chow breakpoint test did not provide evidence for structural break. Due to short data series, Chow tests allow checking for breakpoints only in the year 2000.

As can be seen, the most significant factors that determine the real GDP growth in Ukraine, at least during the period being studied, were the rise of world steel prices in the external demand block, the decline in overdue fiscal receivables and

**Table 4.1** ADL regression results

| | Coefficient | Std. error |
|---|---|---|
| CONST | 0.0007 | 0.0009 |
| **INFLATION(−1)** | −0.0002 | 0.0005 |
| D_INFLATION | −0.0009 | 0.0007 |
| D_INFLATION(−2) | −0.0016*** | 0.0008 |
| D_INFLATION(−4) | −0.0022* | 0.0008 |
| D_INFLATION(−6) | −0.0013** | 0.0006 |
| **DEFICIT(−1)** | −0.0004** | 0.0002 |
| D_DEFICIT(−3) | −0.0004** | 0.0001 |
| **ELECTION** | −0.0032** | 0.0015 |
| **DL_METAL(−1)** | 0.0122 | 0.0164 |
| DDL_METAL(−1) | −0.0153 | 0.0128 |
| DDL_METAL(−2) | −0.0089 | 0.0065 |
| **DL_STEEL(−1)** | 0.0490** | 0.0267 |
| DDL_STEEL | 0.0048 | 0.0350 |
| **DL_REER(−1)** | −0.0011 | 0.0008 |
| **DL_OVERDUE(−1)** | −0.0210 | 0.0187 |
| DDL_OVERDUE(−1) | −0.0402* | 0.0145 |
| **DL_BUDGET(−1)** | −0.0227* | 0.0080 |
| **DL_AGRIC(−1)** | 0.0031*** | 0.0016 |
| DDL_AGRIC(−12) | 0.0015*** | 0.0009 |
| **DL_GDP(−1)** | −0.1798*** | 0.0937 |
| DDL_GDP(−3) | −0.3323* | 0.1179 |
| DDL_GDP(−6) | −0.2235 | 0.1405 |
| DDL_GDP(−9) | −0.3097** | 0.1395 |
| DDL_GDP(−12) | 0.5051* | 0.1413 |
| Adjusted $R2$ | 0.7051 | |
| Jarque–Bera test ($p$-value) | 0.9049(0.6360) | |
| Breusch–Godfrey test ($p$-value) | 2.5692(0.2768) | |
| Ramsey test ($p$-value) | 0.6192(0.4313) | |
| No. of observations | 59 | |

* 1 % significance level; ** 5 % significance level; *** 10 % significance level

agricultural reforms in the reform block, and the reduction of the fiscal deficit and the election dummy in the stability block.

No evidence was found that real effective depreciation stimulated growth as had happened in the case of Russian economy. Moreover, the real effective exchange rate index appeared to have significant negative correlations with real growth. That contradicts anecdotic evidences that emphasized the importance for real depreciation for import-substitution process, e.g., in food industry. The negative sign might be explained by the fact that the macroeconomic destabilization caused by the real devaluation of the national currency overshot the positive effect of the higher external demand for domestically produced products and higher internal demand for domestically produced products.

The decline in overdue fiscal receivable arrears appeared to be more significant for economic growth in the country than the decline in overdue inter-enterprises payables. That could be explained by the fact that the problem of fiscal arrears can be solved easier by state reforms, than inter-enterprises arrear that include a lot of unofficial agreements and are not under the direct jurisdiction of the government. The decline in inter-enterprise arrears is more of an evolutionary process after the right institutions have been set up, while reducing fiscal arrears provides a clear and immediate signal that the state is eager to combat improper contract enforcement in the country.

The impact of inflation is negative (expected sign), but insignificant. That could be explained by the fact that price stabilization as such was achieved well before the overall macroeconomic stabilization became effective. The results of the analysis also indicate that the political cycle has a serious impact on the economic performance of the country, hampering overall growth. Most like explanation is that significant funds are diverted from the economy into unproductive financing of "political campaigns." Elections in Ukraine also carry an often unpredictably high uncertainty about future legislation and policies.

Thus, to ensure stable economic growth in the country, Ukraine has to follow the reform pattern chosen in 1999 against a background of a stable macroeconomic situation, and it must try to persuade its politicians of the value of maintaining a market orientation.

## *Conclusions*

The development of an anti-inflationary policy in a transition economy must take into account both external and internal factors which affected different inflationary processes including demand and cost inflation. Therefore, pure monetary measures in the framework of this policy should be supplemented by deep structural–technological changes, in particular as part of government-directed technological programs. The main priority of these programs is to reduce production costs in industries enmeshed in cost inflationary processes. Technological cooperation with market countries where more than 90 % of new technologies are elaborated is the necessary condition for the mentioned changes.

Import liberalization and stabilization of domestic currency are conducive to overall reduction of inflation rates, but this policy leads to rapid decline of domestic production in high-cost industries. Therefore, as an interim measure before the introduction of new technologies, it is necessary to find compromise solutions that will reduce inflation and at the same time protect domestic producers.

Streamlining of the stale budget is one of the main stabilization mechanisms in a transition economy. However, elimination of budget deficit is not the only goal: it is also necessary to take into account the effect of taxes and customer tariffs on prices and product sales, positive and negative consequences of differential subsidies to some industries, and changes in the structure of export and import due to price dynamics.

The tax system in a transition economy should be different from that under stable conditions. The tax system should rely on taxes assessed independently of business results (taxes on land, property, etc.). The tax burden should be 20–30 % of the profit of an average business entity. The sum of other payments to the budget should not exceed 30–40 % of these taxes.

## 4.3 Export and Import Structure Optimization

Nowadays, some directions in simulating foreign-trade turnover structure for countries with a transition economy have been developed, namely, models accounting for absolute and relative advantages of trade partners [7], models of oligopoly price competition by Pugachev and Pitelin [49] supplemented in [45] with the mechanisms of monopolistic price setting inside the exporting country, econometric models [25]. Profit obtained in this case is considered in practically all the mentioned models as the major factor influencing the volume of foreign trade operations. The assumption of the decisive role of direct profit in foreign trade is justified in most cases; however, it is not well founded for purchase of goods of critical import (which represent a major portion of the foreign trade turnover of Ukraine).

Critical import is a component of the foreign trade turnover with a specific formation of its volume and structure. Critical import comprises goods and services whose consumption is extremely necessary for the functioning of the national economy but whose production volume is inadequate inside the state or which are not produced at all due to absence of necessary resources (deposits of raw material, production capacities, know-how, etc.). Critical import comprises basically goods of intermediate consumption—power resources, structural materials, units, and components. This determines changes in the volume of critical import due to changes in the currency, tariff-customs, and pricing policy of the state. Demand for goods of critical import depends weakly on variation of their prices; however, a rise in their price under conditions of a transition economy is transmitted via the production chain to other goods for whose production they are used. Such process may become an important factor of costs inflation typical for the early stages of a transition economy. On the other hand, significant changes in volumes of production of consumption goods, even under conditions of stable prices, may produce a considerable effect on demand for goods of critical import. There are also some distinctions in objectives and methods of state regulation concerning critical import. In addition to diversification of sources of arrival of such goods, which should make impossible manifestations of monopolism, an important objective of regulation is to search for sources of compensation of costs of their acquisition.

Thus, an increase in volume of critical import should be accompanied in a long-term perspective by an adequate increase in volume of export and in relocation of the final product, in particular, in the form of a decrease of the share of domestic consumption of the products by the export-oriented branches. All this can be taken into account by applying methods of interindustry balances.

Let us consider an economic system comprising $N$ branches of material production. Denote by $a_{ij}$ the direct specific production costs of the $i$-th branch of production of products of the $j$-th branch ($i, j = \overline{1, N}$), by $x_i$ the gross product of the $i$-th branch (in fixed internal prices), and by $\overline{x}_i$ the actual production capacities of this branch. By production capacities, we will mean the greatest possible (for the resources and other production factors available) volume of production for the given branch. We will denote by $I_i$ the volume of import (basically the critical one) of the products of the $i$-th branch, and by $E_i$ the volume of export of such products. Denote by $\alpha_i$ the share of products of the $i$-th branch in the final consumption inside the country, and by $Z$ the level of this consumption relative to the existing one being assumed a unit. We will denote by $\overline{E}_i$ the capacity of foreign markets, i.e., the maximum amount of products of the $i$-th branch which can be exported. In determining this quantity, it is necessary to take into account the foreign customer's demand for the given products, tariff-customs conditions, export and import quotas, and export supply. The latter will depend on the price at which the products can be sold in foreign markets and on its prime cost for the exporting country expressed in terms of a hard currency or the importer's currency at the exchange rate prevailing. To predict changes in average-prime cost of the branch and other parameters of the export proposal function, the models considered in [45] can be used.

The problem consists in determining such volumes of internal production $x_i$, export $E_i$, and import $I_i$ of products of each of $N$ branches, for which production, consumption, export, import, and actual production capacities are balanced. The consumption level inside the country will be no less than its admissible value $\overline{Z}$, and returns from export and costs of import will be maximally balanced.

This problem can be formulated as an optimization model

$$x_i + I_i = \sum_{j=1}^{N} a_{ij} x_j + E_i + \alpha_i Z, \quad i = \overline{1, N} \tag{4.14}$$

(the condition of balance of production, import, export, industrial, and private consumption for each of the branches),

$$x_i \le \overline{x}_i, \quad i = \overline{1, N} \tag{4.15}$$

(constraint on actual production capacities),

$$E_i \le \overline{E}_i, \quad i = \overline{1, N} \tag{4.16}$$

(constraint on the capacity of foreign markets; the value of import is not upper bounded in the model since the total supply of goods in the world market exceeds, as a rule, the demand of any separate country),

$$Z \geq \overline{Z}, \quad x_i \geq 0, \quad E_i \geq 0, \quad I_i \geq 0, \quad i = \overline{1, N} \tag{4.17}$$

(constraint on the admissible consumption level and conditions of non-negativity of variables),

$$F_1 = \sum_{i=1}^{N} \overline{p}_i E_i - \sum_{i=1}^{N} \overline{\overline{p}}_i I_i \to \max \tag{4.18}$$

(the condition of maximum excess of returns from export over the costs of import).

In (4.18), $\overline{p}_i$ is the average price at which products of the $i$-th branch made inside the state will be sold in the foreign market, and $\overline{\overline{p}}_i$ is the average price of a purchase of imported products of the $i$-th branch.

Problem (4.14)–(4.18) is a linear programming problem, which can be solved by well-known methods (see, for instance, [6]).

Of interest is the multicriteria version of this problem, in which, in addition to objective function (4.18), attaining of the greatest possible level of the internal consumption is considered

$$F_2 = Z \to \max .$$

Such a problem can be solved by the convolution method when instead of criteria $F_1$ and $F_2$ their weighted sum is applied,

$$G = \beta F_1 + (1-\beta) F_2 = \beta \left( \sum_{i=1}^{N} \overline{p}_i E_i - \sum_{i=1}^{N} \overline{\overline{p}}_i I_i \right) + (1-\beta) Z \to \max . \tag{4.19}$$

In formula (4.19), $\beta$ is some constant parameter with the value from the interval (0; 1). For its determination, the methods stated in [33] can be used. The base information for these methods is expert estimates of the greater or smaller acceptability of variants (alternatives) with different values of the criteria $F_1$ and $F_2$.

Regional differentiation is also possible in the model of export and import. In this case, upper constraints may be superimposed not only on the value of export but also on the import from particular countries (regions); prices (coefficients of the objective function (4.18)) also may depend on the region where (or wherefrom) export (or import) is carried out.

Some ambiguity due to impossibility of exact prediction of prices for exported products, import, and structure of domestic consumption arise in using the model proposed and its modifications for long-term prediction. Therefore, prediction of changes in actual production capacities and active control of such parameters become important. Under these conditions, the values of $\overline{E}_i$, $\alpha_i$, and $\overline{p}_i$, $\overline{\overline{p}}_i$

($i = \overline{1, N}$) can be considered random with known distribution. Determination of necessary transformations in the existing production capacities will be considered as a two-stage stochastic problem (see [15]). The first stage of its solution is a direct determination of the required changes with regard for the associated costs, and the second is solving problems of the form (4.14)–(4.18).

Denote by $\Delta\overline{x}_i^+$ and $\Delta\overline{x}_i^-$ the predicted increase and, respectively, decrease in production capacities of the $i$-th branch. Let these changes are associated with the consumption of resources of $K$ types (such resources may be investment, labor, natural ones, etc.). Denote by $g_{ki}^+ (\Delta\overline{x}_i^+)$ the dependence between an increase in production capacity of the $i$-th branch and expenditures of the $k$-th resource ($k = \overline{1, K}$) necessary to provide it, and by $g_{ki}^-(\Delta\overline{x}_i^-)$ the dependence between the decrease in production capacity of the $i$-th branch and the expenditures of the $k$-th resource (if $g_{ki}^-(\Delta\overline{x}_i^-) > 0$) necessary to provide it or the amount of the $k$-th resource obtained additionally due to such decrease (if $g_{ki}^-(\Delta\overline{x}_i^-) < 0$). Let $R_k$ be the available amount of such resource.

Denote by $f_i^+(\Delta\overline{x}_i^+)$ the total costs (expressed in terms of money units) due to an increase in production capacity of the $i$-th branch by the value of $\Delta\overline{x}_i^+$, and by $f_i^-(\Delta\overline{x}_i^-)$ the total costs (if this value is positive or the profit is negative), due to a decrease in production capacity of the $i$-th branch.

Then the problem of the first stage can be defined as follows.

It is required to find such values of $\Delta\overline{x}_i^+$ and $\Delta\overline{x}_i^-$ ($i = \overline{1, N}$) that minimize the value of total costs

$$\sum_{i=1}^{N} f_i^+ (\Delta\overline{x}_i^+) + \sum_{i=1}^{N} f_i^- (\Delta\overline{x}_i^-) + MF(\Delta x, \theta) \to \min \tag{4.20}$$

subject to the constraints

$$\sum_{i=1}^{N} g_{ki}^+ (\Delta\overline{x}_i^+) + \sum_{i=1}^{N} g_{ki}^- (\Delta\overline{x}_i^-) \le R_k, \quad k = \overline{1, K}, \tag{4.21}$$

(the constraints on resources available),

$$\Delta\overline{x}_i^+ \ge 0, \quad \Delta\overline{x}_i^- \ge 0, \quad i = \overline{1, N} \tag{4.22}$$

(the condition of non-negativity of the variables).

Here $F(\Delta x, \theta)$ is the optimal value of the objective function of the problem of linear programming similar to (4.14)–(4.18)

$$F = \sum_{i=1}^{N} \overline{\overline{p}}_i(\theta) I_i - \sum_{i=1}^{N} \overline{p}_i(\theta) E_i \to \min \tag{4.23}$$

provided that

$$x_i + I_i = \sum_{j=1}^{N} a_{ij} x_j + E_i + \alpha_i(\theta) Z, \quad i = \overline{1, N}, \tag{4.24}$$

$$x_i \le \overline{x}_i + \Delta\overline{x}_i^+ - \Delta\overline{x}_i^-, \quad i = \overline{1, N}, \tag{4.25}$$

$$E_i \le \overline{E}_i(\theta), \quad Z \ge \overline{Z}, \quad x_i \ge 0, \quad E_i \ge 0, \quad I_i \ge 0, \quad i = \overline{1, N}. \tag{4.26}$$

The notation $\overline{\overline{p}}_i(\theta)$, $\overline{p}_i(\theta)$, $\alpha_i(\theta)$, $\overline{E}_i(\theta)$ means that these parameters are considered random with known distribution. The problem (4.23)–(4.26) is solved for each of the possible stales of the nature $\theta$ (i.e., separately for each of possible values of random variables). The latter will undoubtedly cause some computing difficulties in solving this problem; however, efficient numerical algorithms for solving this class of problems have been developed by Ermol'ev and Wets [15], Kall and Wallance [29], and experience in their successful application has been accumulated.

Two main approaches have been applied to solve the stochastic problem (4.20)–(4.26). The stochastic quasigradient method is the first one. According to it, the sequence

$$\{\Delta x^S\} = \left\{\left(\Delta x_1^{+S}, \ldots, \Delta x_N^{+S}, \Delta x_1^{-S}, \ldots, \Delta x_N^{-S}\right)\right\}$$

is obtained by the following rule.

Let $\Delta x^0$ be an arbitrary vector. Now we describe the procedure to construct $\Delta x^{S+1}$ from given $\Delta x^S$, where $S$ is the number of iteration ($S = 0, 1, 2, \ldots$).

1. Determine the set:

$$K^* = \left\{k : \sum_{i=1}^{N} q_{ki}^+(\Delta\overline{x}_i^{+S}) + \sum_{i=1}^{N} q_{ki}^-(\Delta\overline{x}_i^{-S}) > R_k\right\}.$$

   If $K^*$ is not empty, go to 4, otherwise go to 2.
2. Generate

$$\overline{\overline{p}}_i(\theta^S), \quad \overline{p}_i(\theta^S), \quad \alpha_i(\theta^S), \quad \overline{E}_i(\theta^S),$$

   the $(S + 1)$-th independent realization of random model parameters. We solve the dual problem to (4.23)–(4.26) for these values of the mentioned parameters. Let $U(\theta^S) = \left(U_1(\theta^S), \ldots, U_N(\theta^S)\right)$ be the optimal values of dual variables corresponding to constraints (4.25).
3. Calculate

$$\xi_i^{+S} = \nabla f_i^+\left(\Delta x_i^{+S}\right) + U_i(\theta^S), \quad i = 1, N,$$

$$\xi_i^{-S} = \nabla f_i^-\left(\Delta x_i^{-S}\right) - U_i(\theta^S), \quad i = 1, N,$$

   where $\nabla f_i^+(\cdot)$, $\nabla f_i^-(\cdot)$ are components of the generalized gradients of the functions $f_i^+(\cdot)$ and $f_i^-(\cdot)$. Go to 5.

4. Calculate

$$\xi_i^{+S} = \nabla g_{k*i}^{+}(\Delta \overline{x}_i^{+S}),$$
$$\xi_i^{-S} = \nabla g_{k*i}^{-}(\Delta \overline{x}_i^{-S}), \quad i = \overline{1, N},$$

where $\nabla g_{ki}^{+}(\cdot)$, $\nabla g_{ki}^{-}(\cdot)$ are components of the generalized gradients of corresponding functions,

$$k^* = \arg\max_{k \in K^*} \left( \sum_{i=1}^{N} g_{ki}^{+}(\Delta \overline{x}_i^{+S}) + \sum_{i=1}^{N} g_{ki}^{-}(\Delta \overline{x}_i^{-S}) - R_k \right).$$

5. Calculate

$$\Delta \overline{x}_i^{+S+1} = \max\left(0, \Delta \overline{x}_i^{+S} - \rho_S \xi_i^{+S}\right), \quad i = \overline{1, N},$$
$$\Delta \overline{x}_i^{-S+1} = \max\left(0, \Delta \overline{x}_i^{-S} - \rho_S \xi_i^{-S}\right), \quad i = \overline{1, N},$$

where a step multiplier $\rho_S$ satisfies the following conditions:

$$\rho_S \geq 0, \quad \sum_{S=0}^{\infty} \rho_S = \infty, \quad \sum_{S=0}^{\infty} \rho_S^2 < \infty.$$

Set $S = S + 1$ and go to 1.

As it was proved by Mikhalevich [39], every convergent subsequence of $\{\Delta x^S\}$ converges with probability 1 to the optimal solution of considered stochastic problem.

The proposed algorithm is rather simple, but the speed of its convergence is not high, especially in the case of medium and large dimension ($N \geq 10$).

The second approach has been applied to solve the problem (4.20)–(4.26) in the case, when a number of possible realizations of random parameters is finite. According to Kall and Wallance [29] it is typical for scenario-based stochastic models.

Let us denote $L$–the number of possible scenarios, i.e., the realizations of parameters $\overline{\overline{p}}_i(\theta)$, $\overline{p}_i(\theta)$, $\alpha_i(\theta)$, $\overline{E}_i(\theta)$. The probability of realization $l$ is denoted by $\pi_l$ and corresponding values are denoted by $\overline{\overline{p}}_i^{\,l}$, $\overline{p}_i^{\,l}$, $\alpha_i^l$, $\overline{E}_i^{\,l}$, $l = \overline{1, L}$. Thus, these variables are depended.

Then two-stage stochastic model can be reduced to the following problem of mathematical programming:

$$\sum_{i=1}^{N} f_i(\Delta \overline{x}_i^{+}) + \sum_{i=1}^{N} f_i^{-}(\Delta \overline{x}^{-}) + \sum_{i=1}^{N} \pi_l \left( \sum_{i=1}^{N} \overline{\overline{p}}_i^{\,l} I_i^l - \sum_{i=1}^{N} \overline{p}_i^{\,l} E_i^l \right) \rightarrow \min$$

subject to the constraints:

$$\sum_{i=1}^{N} g_{ki}^{+}(\Delta\overline{x}_i^{+}) + \sum_{i=1}^{N} g_{ki}^{-}(\Delta x_i^{-}) \leq R_k, \quad k = \overline{1,K},$$

$$x_i^l + I_i^l = \sum_{j=1}^{N} a_{ij} x_j^l + E_i^l + \alpha_i^l Z^l, \quad l = \overline{1,L}, \quad i = \overline{1,n},$$

$$x_i^l \leq \overline{x}_i + \Delta\overline{x}_i^{+} - \Delta\overline{x}_i^{-}, \quad i = \overline{1,n}, \quad l = \overline{1,L},$$

$$E_i^l \geq 0, \quad I_i^l \geq 0, \quad \Delta\overline{x}_i^{+} \geq 0, \quad E_i^l \leq \overline{E}_i^l,$$

$$Z^l \geq Z, \quad x_i^l \geq 0, \quad \Delta\overline{x}_i^{-} \geq 0, \quad i = \overline{1,n}, \quad l = \overline{1,L}.$$

The dimension of this problem is sufficiently large, thus effective decomposition methods can be applied in this case. Such methods are based on the algorithms of nonsmooth optimization, developed by Shor [55].

The calculations carried out with a help of these approaches using actual information allow us to formulate recommendations concerning the development of the export structure of Ukraine. The export should be based, along with metallurgy and the agrarian complex, on such branches as chemical industry, machinery, transport and communications. The share of each of these branches in cumulative export varies from 10 to 25 %.

Comparison of the recommended branch structure with the existed at the end of the twentieth century has shown that the share of ferrous metallurgy was overstated and for the chemical industry and transport was underestimated. The values of capacities have to be changed.

Model (4.14)–(4.18) considered above was employed for alternative calculations using real data for Ukraine. Their purpose was to estimate the influence of unstable parameters of the model on recommendations obtained with its help and to compare these recommendations with the actual structure of export and import in the middle of the 1990s. In addition to data on interindustry balance [9,10], information support of the models was provided by using of the estimates of foreign markets capacity obtained from the Ministry of Foreign Economic Relations and Trade of Ukraine and the estimates of available production capacities obtained from the Ministry of Economics of Ukraine. It should be pointed out that most likely these estimates are somewhat understated. For example, for the majority of branches, the estimates of production capacities exceeded actual volumes of production in 1995–1996 no more than 20–30 %. A similar situation also holds for estimates of foreign market capacity for products of particular branches. Thus, we can consider the data of this model as reflecting only those production reserves which undoubtedly can be used for attraction of extremely limited investment resources, and recommendations obtained with the help of this model concern, first of all, a more efficient use of the industrial potential involved.

With regard for these factors, the model being considered was used for the following alternative calculations.

**Variant 1 (Basic).** Maintenance of prices in the middle of the 1990s for the exported and imported products and correspondence of the capacity of foreign markets to the obtained estimates are assumed.

**Variant 2.** A 60 % reduction of the foreign market capacity for machine building products is assumed with conservation of all the remaining values of the model parameters characteristic of Variant 1. This variant stems from calculations on the previous variant and conclusions about significant underestimation of the export potential of this branch.

**Variant 3.** As compared with the base version, a 3.2-fold increase in the world-market prices for oil-and-gas industry products, 2.3-fold for ferrous metallurgy products, and 1.7-fold for chemical industry products are assumed as well as 20 % reduction of prices for food industry products. This variant, first, reflects changes in prices for power resources and foodstuffs which took place in the late of 1990s and in the beginning of the twenty-first century, and, second, in the experts' opinion, this is just the combination of prices that, to the greatest degree, promotes growth of export of products of the chemical–metallurgical complex.

**Variant 4.** An increase is assumed, as compared with the basic version, in the prices for products of oil-and-gas industry by a factor of 3.2, ferrous metallurgy 1.6, chemical industry 1.9, as well as 20 % reduction of prices for products of the food industry. As compared with the previous variant, the market capacity of ferrous metals is reduced for this variant by 45 %, and that of products of the chemical industry by 65 %. This variant, taking into account the loss of the most important foreign markets, can be considered as a pessimistic scenario of conditions of foreign trade activities. It should be noted that the situation appeared after the global credit crisis in 2008 is characterized by approximately the same values of the above-mentioned variables.

Integrated classification of the balance does not allow us to take into account some structural effects taking place inside branches represented in it and manifesting themselves as counterflows of export and import of the branch products. Because of this, the structures of the designed and actual foreign trade turnovers were compared at the level of export–import balance of particular branches. The share of each branch was calculated separately for each group of branches where export exceeds import and vice versa, depending on to what group this branch belongs. For Variants 3 and 4, export–import balance of branches was calculated separately in prices of the basic variant (stable prices) and in prices intrinsic to the variant being considered (current prices).

The calculated (according to the variants) and the real (for 1994–1995) structures of foreign trade turnover are compared in Table 4.2.

**Table 4.2** Share of export–import balance of branches (%) in the active (+) and passive (−) parts of export–import balance

| | By the results of simulation | | | | | | | |
|---|---|---|---|---|---|---|---|---|
| | | | Prices Var. 3 | | Prices Var. 4 | | Actually | |
| Branch | Var. 1 | Var. 2 | Stable | Current | Stable | Current | 1994 | 1995 |
| Electric power industry | 1.84 | 2.40 | −30.79 | −16.32 | −34.38 | −17.13 | 0.34 | −0.86 |
| Oil-and-gas industry | −63.44 | −64.20 | −40.30 | −68.36 | −45.76 | −72.97 | −78.63 | −74.69 |
| Coal mining | −13.68 | −13.90 | −19.20 | −10.10 | −0.30 | −0.15 | −2.27 | −7.62 |
| Other fuel industries | 0 | 0 | 0 | 6 | 0 | 0 | 0 | 0 |
| Ferrous metallurgy | 25.08 | 28.00 | 25.23 | 42.22 | 26.45 | 36.02 | 59.05 | 59.55 |
| Nonferrous metallurgy | 0.09 | 0.50 | 0.70 | 0.51 | 0.75 | 0.64 | −0.85 | 0.58 |
| Chemistry and petrochemistry | 9.23 | 10.20 | 11.42 | 14.12 | 5.77 | 9.33 | 3.17 | 1.93 |
| Engineering industry | 18.91 | 11.20 | 19.16 | 13.94 | 20.50 | 17.45 | −6.85 | −4.89 |
| Forest and woodworking industry | 1.20 | 1.30 | 1.16 | 0.85 | 1.24 | 1.06 | −2.84 | −3.93 |
| Constructional materials | −5.17 | −4.80 | −14.69 | −7.79 | −4.19 | −2.09 | 3.19 | 0.02 |
| Light industry | −9.67 | −9.80 | −7.93 | −4.20 | −8.86 | −4.41 | −6.14 | −7.75 |
| Food industry | 17.13 | 18.23 | 16.61 | 9.67 | 17.77 | 12.10 | 13.88 | 16.67 |
| Other industries | 1.00 | 1.07 | 0.97 | 0.70 | 1.04 | 0.89 | −0.55 | −0.25 |
| Construction | 1.18 | 1.30 | 1.15 | 0.84 | 1.23 | 1.05 | 1.14 | 0.39 |
| Agriculture (farming) | 2.84 | 3.00 | 2.75 | 2.00 | 2.94 | 2.50 | −1.38 | 0.40 |
| Transport and communica-tions | 20.60 | 21.90 | 19.98 | 14.53 | 21.38 | 18.19 | 19.21 | 19.81 |
| Trade and catering | −8.03 | −7.20 | −6.09 | −3.23 | −6.51 | −3.24 | 0.03 | 0.01 |
| Other branches | 0.88 | 0.90 | 0.85 | 0.62 | 0.91 | 0.77 | −0.04 | 0.63 |

Note that despite of significant differences in the conditions of particular variants, the structure of optimal solutions has much in common. An active part of the export–import balance is generated by branches of the chemical–metallurgical complex (35–40 % of its value, including 25–30 % for ferrous metallurgy), machine building, food industry, and transport and communications (18–22 % for each of these branches). A decrease in capacity of the foreign market for one of these branches (e.g., of machine building for Variant 2) results in some increase in the share of export of the remaining branches of the considered group, without significant

**Table 4.3** Profitability of export–import operations, $ million

| Index | By the results of simulation[a] | | | | Actually | |
|---|---|---|---|---|---|---|
| | Var. 1 | Var. 2 | Var. 3 | Var. 4 | 1994 | 1995 |
| Export | 11968.5 | 9104.3 | 19149.8 | 11953.2 | 10298.0 | 11966.5 |
| Import | 8999.6 | 6539.6 | 20547.5 | 15342.9 | 11042.9 | 13134.5 |
| Export–import balance | 2968.9 | 2564.7 | −1397.7 | −3389.7 | −744.0 | −1168.0 |

[a] The profitability indices were estimated using the average market exchange rate of the national currency vs US dollars in 1995

changes in their correlation. In Variants 3 and 4, the increase in contribution to the total export of such branches as ferrous metallurgy and chemical industry can be explained solely by the change in prices for their products. The relation in physical volumes of export (or export at fixed prices) of the above-mentioned and other export-oriented branches remains approximately the same as for Variants 1 and 2.

Note that the structure of the foreign trade turnover obtained in model calculations differs essentially from the actual structure in 1994–1995. Even in Variant 3 characterized by the best export conditions for metallurgical products, the share of ferrous metallurgy in the active part of the export–import balance represents no more than 43 %. At the same time, it actually exceeded 59 %. Import of machine building products in 1994–1995 permanently exceeded its export. The contribution of chemical and food industries to the active part of the export–import balance in 1994–1995 also was slightly less than the calculated one. Products of the fuel and energy complex (75–80 % of total value of the passive part of the balance) were the basis of import, both actually and according to the model calculation data. First of all, these are purchases of organic fuel; however, for Variants 3 and 4, with the increase in its price and rather low fixed prices for electric power, internal production in the electric-power industry was completely replaced by import. This shows its dependence on external expansion.

Diversification of structure of export with an increase in the share of less power-intensive branches leads to a decrease in import with the volume of export remaining practically the same (Table 4.3).

A decrease in capacity of foreign markets for some kinds of products results in a reduction of the total volume of export but in this case, import decreases adequately, first of all, due to its critical components. For example, Variant 2 is characterized by a large share of the food industry and agriculture, which are less power-intensive as compared with machine building and metallurgy. As a result, if cumulative export of branches decreases by more than 2 billion US dollars, then export–import balance of only this variant will decrease only by 0.4 billion US dollars as compared with Variant 1 and will remain positive. Import in Variants 3 and 4 increases due to the increase in prices rather than the increase in the volumes of imported products.

It should be pointed out that the structure of foreign investments in the economy of Ukraine in the 1990s also did not correspond to the actual structure of export nor to priorities and tendencies revealed as a result of model calculations (Table 4.4). This shows the relative independence and isolation of product and investment flows.

**Table 4.4** Branch structure (in percentage of the total volume) of direct foreign investments in Ukraine in 1994–1999

| Branch | Year | | | | | |
|---|---|---|---|---|---|---|
| | 1994 | 1995 | 1996 | 1997 | 1998 | 1999 |
| Fuel and energy complex | No data | 1.20 | 0.9 | 1.5 | 2.8 | 5.8 |
| Ferrous and nonferrous metallurgy | 4.8 | 5.0 | 2.4 | 2.0[a] | 4.1 | 5.1 |
| Chemical and petrochemical industry | 5.7 | 4.2 | 3.0 | 6.9 | 4.4 | 3.7 |
| Machinery (engineer.) | 23.3 | 12.9 | 10.2 | 8.2 | 12.7 | 10.9 |
| Food industry | 14.2 | 14.5 | 12.3 | 20.6 | 21.0 | 20.4 |
| Agriculture (farming) | 1.5 | 2.5 | 1.3 | 2.2 | 2.1 | 2.0 |
| Transport and communications | 5.2 | 4.1 | 3.3 | 2.9 | 5.3 | 5.3 |
| Other branches | 45.3 | 55.6 | 66.6 | 55.7 | 47.6 | 46.8 |

[a] Only ferrous metallurgy

**Table 4.5** Comparison of branch structure (%) of the active (+) and passive (−) parts of export–import balance

| Branch | Year | | |
|---|---|---|---|
| | 1994 | 1995 | 2005 |
| Fuel and energy production complex | −80.66 | −83.17 | −53.1 |
| Ferrous and non-ferrous metallurgy | 58.2 | 60.13 | 67.0 |
| Chemical and petrochemical ind. | 3.17 | 1.93 | −0.3 |
| Machinery (engineering) | −6.85 | −4.89 | −31.5 |
| Food industry and agriculture | 12.5 | 17.07 | 12.6 |
| Transport and communications | 19.21 | 19.81 | 15.6 |

Therefore, mechanisms of formation of investment priorities under the conditions of a transition economy should be considered separately. In the model presented below, the investment risk is considered as one of the factors determining volumes and structure of import of capital.

The structure of Ukrainian export and import has not essentially changed during the last 15 years (Table 4.5). Even more, occurred changes were in opposite direction to the modeling proposals: both the share of ferrous metallurgy in export and negative surplus of machinery have increased. Import of chemical production exceeds its export due to imperfect structure of trade. Ukraine exports mineral fertilizers, coke and production of oil refraction and import high-technological chemical production. The sharp decline of Ukrainian export (more than 60 %) in the beginning of global financial crisis demonstrated the vital necessity of deep structural changes in the sphere of Ukrainian foreign trade.

### *Conclusions*

Ukrainian export should be based, along with metallurgy and the agrarian and industrial complex, on such branches as chemical industry, machinery, transport and communications. The share of each of these branches in cumulative export varies from 10 to 25 %, and it may be increased up to 40–45 % only under conditions extremely favorable for export.

The existed structure of Ukrainian export and import is far from optimal for actual market conjuncture as well as production capacities in disposal. Starting in the beginning of the 1990s, the value of export of ferrous metallurgy production is greater than optimal and the value of machinery, high technological chemical production and food export is less than optimal. The share of transport services in export is close to optimal. This situation did not cardinally changed in the first decade of the twenty-first century and was one of the reasons why the consequences of global financial crisis were so destructive for Ukraine.

## 4.4 Stochastic Investor's Behavior Models

Raising of foreign investments is one of the most important directions of state foreign economic policy. Efficient solution of emerging problems has significant effect on competitiveness of domestic products in world markets, on technical modernization of production, replenishment of credit and financial resources of a banking system and circulating assets of a company, and, in the long run, on the position of the state in the world labor division. Of particular importance is the rise in investment attractiveness under conditions of a transition economy, which is characterized by financial instability, chronic deficiency of resources in economic entities, and fast changes of forms of ownership and owners of production plants. The volume of the attracted investments will be determined under these conditions by the speed of overcoming crisis and the degree of development of negative processes accompanying it. Along with this, the cited features of a transition economy together with instability of the legislation system, political and other non-economic factors, and changes in market conditions create indeterminacy in the level of profitability of investments and sometimes in the possibility of obtaining profit itself. This increases investment risk and adversely affects receipts of investments.

Investment risk can be compensated by raising the investment profitability and taking measures to strengthen the investor's confidence in obtaining profit. For the analysis of their efficiency, the behavior of the investor and his motivations to invest should be studied. This problem has many planes and its solution requires economic, psychological, sociological, political, and other studies, including methods of systems analysis and mathematical simulation.

Let us consider the possible model of formation of priorities in investment of available capital under the risk conditions.

Let the main purpose of the investor be obtaining maximum profit in hard currency per capita invested. Under conditions of uncertainty concerning results of investment, he should act by the trial-and-error method, by investing his capital in those countries and investment directions where he has obtained the largest profit and by decreasing investment where the profit was less. The purpose of the further proposed model is to reflect the dynamics of such change in priorities.

Denote by $n$ the number of countries where the investor realizes his investment (specifying in investment directions and particular investment projects is also possible). Let $\alpha_i^t$ be the coefficients of priority of the $i$-th country for the investor at the instants of time $t$, according to which he distributes the investments. At each instant of time, these coefficients should satisfy the conditions

$$\sum_{i=1}^{n} \alpha_i^t \leq 1, \quad \alpha_i^t \geq 0, \quad i = \overline{1,n}. \tag{4.27}$$

Fulfillment of the condition $\sum_{i=1}^{n} \alpha_i^t < 1$ means that at the instant of time $t$, the part of the free actual capital in the amount of $1 - \sum_{i=1}^{n} \alpha_i^t$ is not being used for investment. Let time $t$ varies discretely taking the values $t = 0, 1, 2, \ldots$. Denote by $I_i$ the profitability of investment in the $i$-th country provided that profit is obtained in its national currency, and by $P_i$ the rate of exchange of the national currency of the $i$-th country for hard currency. These quantities depend on uncertainty factors, which we will denote by $\theta$. Thus, we will believe that at the instant of time $t$ we have $I_i = I_i\left(\alpha_i^t, \theta^t\right)$ and $P_i = P_i\left(\alpha_i^t, \theta^t\right)$, where $\theta^t$ is the value of the factor $\theta$ at the current instant of time. In the general case, $\theta$ can be considered a random variable, and $\theta^t$ is the observations of its realizations. Taking into account the fact that, all other factors being the same, a greater investment priority of a country strengthens its national currency and contributes to increase in production efficiency, but the effect of changes in priorities on the mentioned processes is limited for their small changes, we can assume that $I_i(\alpha_i, \theta)$ and $P_i(\alpha_i, \theta)$ will be continuous nondecreasing functions of $\alpha_i$ for fixed $\theta$.

We assume that priority for the investor grows if a specific profit in hard currency, obtained from investment in this country, is greater than the specific profit $\hat{I}$ obtained in alternative way (without investing to the foreign countries), and decreases if the profit due to investment is less. The specific range of variation of priority will be determined by random coefficients $k_i(\theta)$ which can reflect some subjective tastes of the investor, and by the coefficient $\rho_t$ which reflects the general tendency in changes of his priorities in time. If new priority coefficients do not satisfy condition (4.27), then it is logical to assume that the investor will search for those values which satisfy these conditions and are as close to the obtained values as possible.

In view of the assumptions made and taking into account the fact that the specific profit from investment in the $i$-th country at the instant of time $i$ is equal to $P_i\left(\alpha_i^t, \theta^t\right)\left(1 + I_i\left(\alpha_i^t, \theta^t\right)\right)$, the change in the values of priority coefficients is determined by the equation

$$\alpha_i^{t+1} = \pi_D\left(\alpha_i^t + \rho_t\left(k_i(\theta^t) P_i\left(\alpha_i^t, \theta^t\right)\left(1 + I_i(\alpha_i^t)\right) - \hat{I}\right)\right), \qquad (4.28)$$

$$i = \overline{1, n}, \quad t = 0, 1, 2, \ldots,$$

where $\pi_D(\cdot)$ is the operation of projection of $n$-dimensional vector onto the set

$$D = \left\{\alpha : \sum_{i=1}^{n} \alpha_i \leq 1, \quad \alpha_i \geq 0, i = \overline{1, n}\right\},$$

and $\rho_t$ are determinate quantities.

Note that the sequence $\left\{\alpha^t\right\}$, where $\alpha^t = \left(\alpha_1^t, \ldots, \alpha_n^t\right)$ is random.

Let us study its convergence for some assumption as to variations of the coefficient $\rho_t$.

To do this, let us consider the optimization problem

$$F(\alpha) = \sum_{i=1}^{n} M\left(k_i\left(G_i\left(\alpha_i, \theta\right) + H_i\left(\alpha_i, \theta\right)\right) - \hat{I}\alpha_i\right) \to \max \qquad (4.29)$$

on the condition that $\sum_{i=1}^{n} \alpha_i \leq 1$, $\alpha_i \geq 0$, $i = \overline{1, n}$, where $M(\cdot)$ is the operation of finding the mathematical expectation of a random variable,

$$G_i\left(\alpha_i, \theta\right) = \int_0^{\alpha_i} P_i\left(\tau, \theta\right) d\tau, \quad H_i\left(\alpha_i, \theta\right) = \int_0^{\alpha_i} P_i\left(\tau, \theta\right) I_i\left(\tau, \theta\right) d\tau.$$

Based on the assumptions made earlier, the corresponding integrals exist and functions are differentiable by $\alpha_i$ for every $\theta$.

The following statement is true.

**Theorem 4.1.** *Let* $0 < C \leq P_i\left(\alpha_i, \theta\right) \leq \overline{C} < \infty$, *and* $0 \leq C \leq I_i\left(\alpha_i, \theta\right) \leq \overline{C} < \infty$, $i = \overline{1, n}$, *for any* $\alpha_i$ *and* $\theta$, *and also the following relations be fulfilled:*

$$\rho_t \geq 0, \quad \sum_{t=0}^{\infty} \rho_t = \infty, \sum_{t=0}^{\infty} \rho_t^2 < \infty, \quad t = 0, 1, \ldots,.$$

*Then all the limiting points of the sequence $\left\{\alpha^t\right\}$ will belong with probability 1 to the set*

$$X^* = \left\{\alpha : \left\|\alpha - \pi_D\left(\alpha - \nabla F(\alpha)\right)\right\| = 0\right\},$$

*where $\nabla F(\alpha)$ is the gradient of the function $F(\alpha)$.*

*Proof.* Note that

$$\frac{\partial G_i\left(\alpha_i,\theta\right)}{\partial \alpha_i} = P_i\left(\alpha_i,\theta\right), \quad \frac{\partial H_i\left(\alpha_i,\theta\right)}{\partial \alpha_i} = P_i\left(\alpha_i,\theta\right) I_i\left(\alpha_i,\theta\right).$$

Hence, the vector

$$\xi\left(\alpha,\theta\right) = \left(\xi_1\left(\alpha_1,\theta\right),\ldots,\xi_n\left(\alpha_n,\theta\right)\right),$$

where

$$\xi_i\left(\alpha_i,\theta\right) = k_i P_i\left(\alpha_i,\theta\right)\left(1 + I_i\left(\alpha_i,\theta\right)\right) - \hat{I},$$

satisfies the condition $M\left(\xi(\alpha,\theta)\ /\ \alpha\right) = \nabla F(\alpha)$, i.e., it will be the stochastic quasigradient of the function $F(\alpha)$. In view of this, relation (4.28) determines the procedure of search for the solution of problem (4.29) by the method of stochastic gradients with projection. Convergence of the sequence $\left\{\alpha^t\right\}$ follows directly from results by Ermol'ev and Wets [15] under assumptions concerning $\rho_t$, which are contained in conditions of the present theorem, and with assumption about boundednesses, with the probability 1, all components of the vector $\xi(\alpha,\theta)$. According to these results, all the limiting points of the given sequence should belong with probability 1 to the set $X^*$, whence validity of Theorem 4.1 follows. In the same study, equivalence of the conditions determining the set $X^*$ to the conditions of the Kuhn–Tucker theorem in the differential form was proved, whence validity of the theorem directly follows.

Using the latter conditions, let us study the properties of the set $X^*$ in the case, when $k_i\,(i = \overline{1,n})$ are not random.

**Theorem 4.2.** *Let $\alpha^* = \left(\alpha_1^*,\ldots,\alpha_n^*\right)$ be an arbitrary element from $X^*$. Denote by $J\left(\alpha^*\right)$ the set of its nonzero components. Then for any $i,\ j \in J\left(\alpha^*\right)$, the following equality should be fulfilled:*

$$k_i\left(M\left(P_i\left(\alpha_i^*,\theta\right)\right) + M\left(P_i\left(\alpha_i^*,\theta\right) I_i\left(\alpha_i^*,\theta\right)\right)\right)$$
$$= k_j\left(M\left(P_j\left(\alpha_j^*,\theta\right)\right) + M\left(P_j\left(\alpha_j^*,\theta\right) I_j\left(\alpha_j^*,\theta\right)\right)\right); \tag{4.30}$$

*but if* $i \notin J(\alpha^*)$ *and* $j \in J(\alpha^*)$, *then*

$$\begin{aligned} & k_i\Big(M\Big(P_i(0,\theta)\Big) + M\Big(P_i(0,\theta)\, I_i(0,\theta)\Big)\Big) \\ \geq\ & k_j\Big(M\Big(P_j\left(\alpha_j^*,\theta\right)\Big) + M\Big(P_j\left(\alpha_j^*,\theta\right) I_j\left(\alpha_j^*,\theta\right)\Big)\Big). \end{aligned} \tag{4.31}$$

*Proof.* The conditions of the Kuhn–Tucker theorem in differential form (see [46]) for problem (4.29) will have the form

$$M\left(k_i\left(\frac{\partial G_i\left(\alpha_i^*,\theta\right)}{\partial \alpha_i}\right) + \left(\frac{\partial H_i\left(\alpha_i^*,\theta\right)}{\partial \alpha_i^*}\right)\right) - \hat{I} \leq \nu^*, \tag{4.32}$$

$$\alpha_i^*\left(M\left(k_i\left(\frac{\partial G_i\left(\alpha_i^*,\theta\right)}{\partial \alpha_i}\right) + \left(\frac{\partial H_i\left(\alpha_i^*,\theta\right)}{\partial \alpha_i^*}\right)\right) - \hat{I} - \nu^*\right) = 0, \tag{4.33}$$

$$\nu^*\left(\sum_{i=1}^{n} \alpha_i^* - 1\right) = 0, \tag{4.34}$$

where $\nu^*$ is the best value of the dual variable corresponding to the constraint of problem (4.29).

Let us note that according to the assumptions made earlier concerning properties of the functions $P_i(\alpha_i,\theta)$ and $I_i(\alpha_i,\theta)$, the function $F(\alpha)$ is convex, nondecreasing, and does not take identical values at all points of the domain of its definition. According to the statements from [30], all the points of maxima (local and global) of such function on a convex set belong to its boundary. Hence, either the condition $\alpha_i^* = 0,\ i = \overline{1,n}$ or $\sum_{i=1}^{n} \alpha_i^* = 1$ should be fulfilled. For the former case, according to (4.34), $\nu^* = 0$ and from (4.32), it follows that for all $i = \overline{1,n}$ the following inequality is fulfilled:

$$k_i\Big(M\Big(P_i(0,\theta)\Big) + M\Big(P_i(0,\theta)\, I_i(0,\theta)\Big)\Big) \leq \hat{I}. \tag{4.35}$$

For the latter case $\nu^* \geq 0$ and, according to (4.33), the condition

$$k_i\Big(M\Big(P_i\left(\alpha_i^*,\theta\right)\Big) + M\Big(P_i\left(\alpha_i^*,\theta\right) I_i\left(\alpha_i^*,\theta\right)\Big)\Big) = \hat{I} + \nu^*$$

should be fulfilled for all $i \in J(\alpha^*)$. Note that the right-hand side of this equality does not depend on $i$. The validity of Eq. (4.30) for arbitrary $i,\ j \in J(\alpha^*)$ follows from here.

But if $i, j \overline{\in} J(\alpha^*)$, then $\alpha_i^* = 0$ and the validity of inequality (4.31) follows directly from (4.32).

The theorem is proved.

Note that according to (4.32) and (4.33), the following relation should be fulfilled:

$$\nu^* = \arg\max_{i=\overline{1,n}} \left( k_i \left( M\left(P_i\left(\alpha_i^*, \theta\right)\right) + M\left(P_i\left(\alpha_i^*, \theta\right) I_i\left(\alpha_i^*, \theta\right)\right)\right) - \hat{I} \right). \tag{4.36}$$

Now let us give an economic interpretation to the obtained equalities (4.30) and (4.36) and inequalities (4.31) and (4.35). Priority of an object of investment under risk conditions will be determined according to (4.30) by the average value of the exchange rate of the national currency of the country attracting investments for hard currency and by the coefficient of covariance between the exchange rate and profitability of investment. The average value of profitability of investment has no direct effect on the investment attractiveness. The level of revenue $\hat{I}$ from alternate use of capital has an effect only on the decision of a potential investor as to the possibility of investment itself but not on his investment priorities. As follows from (4.35), an increase in this level may result in an investor's refusal to invest somewhere although he made such investments earlier.

The sum of the average rate of exchange of the national currency for hard currency and the coefficient of covariance of this rate and profitability, multiplied by the coefficient $k_i$, can be considered as the index of attractiveness of the country for investments (the so-called index of investment attractiveness). In this case, nonzero investment priority will be given only to countries where this index is greatest. The values of such priorities can be found from the system of Eq. (4.30). In order to achieve a positive priority, namely, to pass from fulfilment of inequality (4.31) to equality (4.30), it is necessary either to increase the average value of the exchange rate (i.e., to strengthen the national currency) or to increase the profitability of investments, coordinating the latter index with the exchange rate. It is important in this case that a decrease in the exchange rate should be compensated by an increase in profitability of investments and vice versa. Thus, a deliberate exchange-rate policy is probably the most important factor of attracting foreign investments under risk conditions.

The results obtained allow us, in particular, to explain the change in opinion of investors concerning the investment climate in Ukraine in 1992–1999. In the beginning of this period, strong inflation took place: only during 1993 the karbovanets (national currency at that time) depreciated at least 30 times with respect to hard currency (see [50]). Under such conditions, even a high internal profitability of production, artificially supported by a galloping inflation, did not allow to provide a high value of the index of investment attractiveness, and the investment rating of Ukraine was very low. Financial stabilization in 1994 began from stabilization of exchange rate of the national currency for hard currency. Profitability of production began to decrease due to the decrease in effective domestic demand

and the increase in actual credit rates, but this decrease was compensated first by a decrease in external inflation, and investment rating of Ukraine began to grow (see [51]). However, the further decrease in production efficiency (down to 3–5 % of profitability in 1997) with the almost constant rate of exchange of the hryvnia (new national currency) for hard currency caused a new decrease in the index of investment attractiveness. This process is strengthened under the economic situation in Ukraine during global financial crisis when the interaction of both negative tendencies is observed: the external inflation (depreciation of the national currency) and the decrease in profitability of investments in a particular sector of the economy.

Thus, for improvement of the investment climate, along with actions that would remove subjective factors of low investment attractiveness, it is necessary to stimulate the growth of actual production efficiency and to coordinate the rate of exchange of the national currency for hard currency with this index.

The analysis of investment processes does not give an answer to the question why ferrous metallurgy is the main export-oriented industry in Ukraine. To obtain the answer, it is necessary to study a number of economic mechanisms specific for the given branch and determine the dynamics of its export. In the next section, some of them will be studied with the use of statistical data analysis and econometric simulation.

Below we consider the next model of mutual dependence between inflation and investment processes proposed by Kuznetsov [36] where the impact of risk and uncertainty was taken into account. Unpredicted price grows is one of the sources of investment risk; interest rates resulting from expected level of inflation impacted both at investment flows and inflation rates. Model uses the following assumptions (the author's denotations in the model are saved).

There is a large number $N$ of potential identical institutional entrepreneurs. Each potential entrepreneur decides how much he is willing to invest and then compares his investment plans with his endowment $A$. If it is larger than his investment plan $X$ he lends the remaining funds with the interest rate $i$. If his endowment $A$ is smaller than $X$ he starts to search for the necessary funds. If he locates them he borrows with the interest rate $i$. However, if he fails to procure the whole amount $X$ he abstains from investment altogether. Making investment the entrepreneur raises the capital stock $y$ of the sector of institutional entrepreneurs.

The aggregate investment into the sector of institutional entrepreneurs is determined in two steps. First the optimal plan $X$ of an individual entrepreneur is found. Second, the share of potential entrepreneurs who are able to raise funds to match their initial plans is specified.

Assume that entrepreneur is risk-neutral and has the following utility function

$$f_1 = \mu_1 - 0.5\sigma_1^2,$$

where $\mu_1$ and $\sigma_1^2$ are the expected value and the variance of $V$, the wealth of entrepreneur-investor at the end of the period. Making calculation of his desirable investment entrepreneur assumes that he can lend and borrow at the market rate $i$ and can invest any amount with the uncertain rate $\overline{r}$; $r$ is its expected value, $\sigma_r^2$ is the variance of $\overline{r}$.

If entrepreneur invests the amount $X$ at the uncertain rate $\overline{r}$, his final wealth will be

$$V_1 = A_1(1 + i) + X(\overline{r} - i)$$

with

$$\mu_1 = A_1(1 + i) + X(\overline{r} - i)$$

and

$$r^2 V_1 = X^2 \sigma_r^2.$$

The following equality follows from a necessary condition for optimum

$$X = \frac{\overline{r} - i}{r^2}.$$

The expected value and variance of investment can be specified taking into account realities of the noisy economy including the assumption that net expected return (profit) of investment is the function of the "noisiness" $(1 - y)$:

$$r - i = 1 - y. \tag{4.37}$$

In other words, if sector of institutional entrepreneurs is fully developed, the economy becomes perfectly competitive, any rents are dissipated away and profits are driven to zero. If the economy is continued to be noisy inputs and outputs are not free and there is a potential to retain the rents.

Variance of investment is assumed to be a function of inflation $p$ only, where $\overline{p}$ is the logarithm of the price level. It is realistic to assume that as the logarithm of inflation $\overline{p}$ approaches the hyperinflation level $C$, the variance of investment approaches infinity. If inflation is low (close to zero) variance is modest and determined by the technical characteristics of the project in question which are outside the model. The simplest function that satisfies these conditions is:

$$\sigma_1^2 = 1/(C - \overline{p}). \tag{4.38}$$

The following optimal investment plan of the individual investor is obtained combining (4.37) and (4.38)

$$X = (C - \overline{p})(1 - y). \tag{4.39}$$

The ratio of potential entrepreneurs who succeed in obtaining finance for their investment plans is determined on the second step. This ratio will be specified on the assumption that it increases with $y$ (size of the sector of institutional entrepreneurs) and $x$ (credits to the economy):

$$(ax + y)/N, \tag{4.40}$$

where $a$ is the ability of financial sector to distinguish between performing and non-performing assets, $0 < a < 1$ (this parameter will enter also the equation of inflation), $N$ is a number of potential entrepreneurs willing to invest. The equation for $y$ is specified combining (4.39) and (4.40):

$$y' = (1 - y)(C - \overline{p})(ax + y). \tag{4.41}$$

Another model assumption is that there are $M$ potential "destructive" entrepreneurs who may want to purchase the assets of the sector of institutional entrepreneurs and turn them into a tool of unproductive business along the lines of Diaz-Alejandro [12]. Each unproductive rent-seeker has the preference function

$$f_2 = \mu_2 - 0.5\sigma_2^2/b,$$

where $b$ is risk-loviness (invert of risk aversion) or "animal spirit" of Diaz-Alejandro entrepreneurs. Mapping the realities of the noisy economy in the model, the assumption is made that the return to this type of entrepreneurship is $(1 - a)x$ so that "bad" entrepreneurs survive on the amount of inflationary money that are destined to subside non-adjusting money-loosing enterprises. It is also assumed that variation of this investment is a function of the development of the sector of institutional entrepreneurs or the level of noisiness of the economy $r_2^2 = 1/(1 - y)$ so that when economy approaches perfect competition, the variation of the rent of return on this investment reaches infinity. Finally, the share of investors willing to finance exit from the sector of institutional entrepreneurs is $\overline{p}/M$: the more rampant is inflation, the more widespread the failures of productive entrepreneurs, the more attractive are speculative and rent-seeking activities. The equation for $y$ then becomes:

$$y' = -\overline{p}(1 - y)b(1 - a)x. \tag{4.42}$$

Combining (4.41) and (4.42) the final equation for $y$ can be obtained:

$$y' = (1 - y)\Big((C - \overline{p})(ax + y) - p(1 - a)bx\Big). \tag{4.43}$$

The inflation equation closes the system. It has the following form:

$$p' = (1 - a)x + c - y + B, \tag{4.44}$$

where $B$ is the level of government social expenditures (no taxation is assumed), $c-y$ is the extent to which economy is noisy, $(c < 1)$, or the strength of the implicit propagation mechanism of the inflationary impulse, $(1-a)x + B(1-a)x$ is the amount of dissaving.

It should be pointed out that the system (4.43), (4.44) is collapsible into one equation for $y$. Given $y$ one can also derive the equation for the rate of inflation $\pi = d\overline{p}/dt$:

$$\pi' = (1-a)x' - y'.$$

Let us assume that level of credit $x$ is a parameter. Then

$$\pi(t) = \pi(0) - y. \tag{4.45}$$

From (4.43) and (4.44) it is easy to see that $y$ may have two stable equilibria. Then from (4.45) there will be two stable inflation levels.

Several interpretations of the low output/high inflation trap are possible. One of them distinguishes between "easy" and "capital deepening" stages of entrepreneurial evolution. The first stage requires relatively small and non-specialized investment to set up consulting firms, banks, and other elements of financial infrastructure. However, continuing inflation and related uncertainty discourages product-specific investment with large sunk costs including investment in the human capital. When the relatively easily available rents are dissipated away entrepreneurship drive comes to a halt. But since the system remains noisy, the flow of inflationary credit has continued to avoid complete collapse of the economy. Equilibrium high inflation and related uncertainty precludes the transformation of entrepreneurship to the more difficult stage when product-specific investment are required to entry into new productive lines, weak competition and poor market infrastructure maintains situational monopolies keeping the economy persistently noisy.

Several controlled parameters include this model. Therefore consideration of optimal control problem constructed on the basis of this model is necessary for possible applications.

The first question to ask about any macroeconomic optimal control problem is who supposed to exercise the optimal control. There are at least two ways to answer this question. The standard one is to assume that the government performs control functions. Governments are presumably set macroeconomic objectives and choose instruments to accomplish them—money supply $x$ in our case. Following this approach we can assume that the government maximizes some measure of institutional development $y$. For example, it may maximize an intertemporal utility with some discount factor with $y$ as instantaneous utility.

The contrasting evolutionary way focuses on forces of natural selection. This way is based on the assumption that whatever types of entrepreneurs coexist in a transition economy in the long-run it is Schumpeterian (productive) entrepreneurs who have the highest survival rate. Thus, the system behaves as if it maximizes some terminal criteria $y(T)$, where $T$ is a measure of a distant future. Control variables is then determined in the process of economic selection too. Government does not choose anything according to assumptions were made: it just follows forces of evolutionary selection. If the first way is excessively optimistic about the ability of the government to intervene efficiently, the second one puts too much faith in the ability of natural selection to choose the right type of entrepreneurs. Nevertheless, the evolutionary way assumes that the government corrects potential failures of selection mechanism. Assuming, for the sake of notational simplicity that there are no inflationary impulses beside credit $x$, i.e., $c = B = 0$, the following control problem is obtained:

$$y(T) \to \max,$$

$$y' = (1-y)\Big\{(ax+y)\big(C-(1-a)x+y\big) - b(1-a)x\big(-y+(1-a)x\big)\Big\},$$

$$y(0) = y_0.$$

Maximizing Hamiltonian with respect to $x$ we get the following necessary condition for $x$:

$$x = \frac{aC + y(2a + (1-a)b - 1)}{2(1-a)(a+b)}. \tag{4.46}$$

It is quite obvious that as long as not ail credit $x$ ends up subsidizing non-adjusting enterprises or goes to the non-productive rent-seekers, i.e., as long as $a > 0$, and hyperinflation border $C$ is large, the optimal credit creation is greater than 0 irrespective how noisy is the economy, i.e., how large is the sector of institutional entrepreneurs. Optimal credit may decrease or increase with the increase of $y$ depending on the sign of

$$K = 2a + (1-a)b - l,$$

where $K < 0$ if, for example, both $a$ and $b$ are small. This means that capital market is prone to failure and unproductive rent-seekers are lethargic (too risk averse). Then productive entrepreneurial sector is sufficiently strong to invest out of retained earnings. External credit becomes irrelevant: as $y$ increases, $x$ should decrease. Taking the institutional development perspective, the hypothesizes is formulated that allocation of funds to new ventures occurs through internal capital market operating within large diversified business groups which finances investment largely from retained earnings. This pattern is called as the evolution the company group model [36].

If $K > 0$ so that financial system is initially sufficiently sophisticated, then amount of credit increases with institutional development. The hypothesizes is stated that economic units plunging into new ventures are smaller compare to the company group model and they tend to finance investment largely by borrowing.

Inserting (4.46) into (4.43), (4.44) and denoting

$$k = \frac{2a + (1-a)b - 1}{2(a+b)}$$

and noting that given the former assumption $0 < a < l$ $k$ is always less than unity we get the following equation of the dynamics of the rate of inflation:

$$\pi' = (k-1)y', \quad \pi(t) = \pi(0) - (1-k)y.$$

Since optimal $x$ is linear function of $y$ the differential equation for $y$ will be structurally equivalent to (4.33) and the optimization model still will have two stable inflation rate/output equilibria. Thus, the pursuit of the optimal credit policies does not change the multiple equilibria property of the system.

As an extension of the model it is assumed that $a$–the ability of the noisy economy to target credit to productive entrepreneurs is endogenous and $a = y$. For simplicity, it is also assumed that unproductive rent-seekers are risk neutral: $b = 1$. Then the following expression for $x$ is obtained:

$$x = \frac{y(C+y)}{(1-y)}.$$

Optimal credit increases with the increase of $y$ and reaches infinity as the economy becomes perfectly competitive.

In this version of the model high-level stable equilibrium $y = 1$ disappears and higher-level equilibrium becomes unstable.

## *Conclusions*

The model suggests two sets of policy implications relevant for the noisy economy: one concerning institutional development and the other stabilization. If initial quality of the financial intermediation is low, and rentseekers are not very aggressive, then economy is likely to evolve along company group model in which internal capital market partly replaces weak outside capital market.

The model also suggests that macroeconomic stabilization (curbing the high inflation) should proceed in two steps.

The first step consists of the reduction of credit creation to arrive on the optimal credit supply curve.

The second one is stabilization-cum-industrial policy measures, which would imply increase in credit to stimulate growth of the institutional entrepreneurs sector.

## 4.5 Models of Shallow Remaking Goods Export

Ferrous metallurgy is one of the most important components of the export potential of Ukraine. Its share in the total export of goods exceeded 50 % in the middle of the 1990s. Having its own deposits of high-quality raw material, cheap labor, and liberal environment protection legislation, Ukraine can enter the international division of labor just with this production. The history of economics gives examples of skilful use of export potential in this sphere to create prerequisites for fast general economic growth, as was the case, in particular, in Sweden in the 1920–1930s, in Japan in the 1940–1950s, and in South Korea in the 1950–1970s. At the same time, this resource- and power-intensive branch of the economy of Ukraine was characterized in the second half of the 1990s and in the beginning of the twenty-first century by the instability in time of volumes of production, export of individual kinds of products, and reinforcement of such general-crisis phenomena as the growth of trade liabilities, decrease in profitability of production, reduction in circulating assets of enterprises, etc. All this requires a deep economic analysis of the processes taking place in the branch, in order to evaluate more exactly its potentialities. Simulation of volume of export is a component of such analysis; its objective is to determine the factors having an effect on these processes and to establish the peculiarities of their variations.

Predictive models were developed for individual groups of products, such as plates, semifinished items made of iron and steel and ferroalloys. The peculiar features of the products considered lead, first of all, to application of the methods of price competition for their introduction into foreign markets; therefore, price factors should be taken into account in the models first of all. Thus, it is expedient to use models of oligopoly pricing.

Some of such models with their modifications are considered below. In the beginning of this section, the nonlinear econometric model is presented, which determines the volumes of export of a definite kind of products to a definite country depending on price factors, standards of tariff-and-customs and taxation policy of the exporter and importer, and aggregated indices characterizing the economic situation in the importing country. Further modifications of this model will be presented, for instance, its linearized version. A simplified linear model is proposed as an alternative of nonlinear one.

Application of the proposed models for studying the processes of export from Ukraine of particular kinds of products of the mentioned branch is also considered.

Let us consider a situation where some products exported from one country to another compete with products made in the latter country. The model of oligopoly pricing (see [49]), where it is assumed that the volumes of realization of the competing goods are in inverse relation to their prices, often describes such competition. Let us consider this model.

Let us introduce the following notation: $p$ is the average price of the exported goods in the domestic market of the country to which it is exported, $\overline{p}$ is the average price of realization of similar goods made inside this country (or exported from other countries), $z$ is the volume of export of the goods to the country in question, and $\overline{z}$ is the volume of realization of products made inside the country in its domestic market. It is assumed that the volumes of realization of the exported and competing products are related to the prices as follows:

$$\frac{p}{\overline{p}} = k\left(\frac{\overline{z}}{z}\right)^{\alpha}, \tag{4.47}$$

where $\alpha > 0$ and $k > 0$ are some coefficients.

It should be pointed out that retrieval of information concerning the volumes of realization of the competing products sometimes faces some problems. Therefore, this index can be evaluated indirectly, through the general effective demand $\Phi$ for these products. Assuming that the demand is completely satisfied by the totality of the exported and competing products, we have

$$pz + \overline{pz} = \Phi,$$

whence $\overline{z} = \dfrac{\Phi - pz}{\overline{p}}$.

Substituting the expression for $\overline{z}$ into (4.47) and solving the obtained equation for $z$, we obtain

$$z = \frac{\Phi}{p\left(1 + \left(\dfrac{p}{k\overline{p}}\right)^{1/\alpha}\right)}. \tag{4.48}$$

Since it is rather difficult to evaluate directly the value of $\Phi$, we can use for its determination the empirical dependence between the unknown demand and the most important macroeconomic indices influencing it. In particular, it is proposed to use the following relation:

$$\Phi = \frac{1}{\delta\overline{p}}\left(a_1 V + a_2 I + a_3 W + a_4\right), \tag{4.49}$$

where $\delta\overline{p}$ is the annual average rate of inflation in the country to which the goods are exported, $V$ is the value of actual gross national product in this country, $I$ is the actual net investment in it, $W$ are actual average wages in this country, and $a_1$–$a_4$ are coefficients.

Relations (4.48) and (4.49) form the model being considered. Using it, we can make the following predictive calculations.

**Analysis of the Results of Changes in the Tariff-and-Customs Standards.** It is expedient to consider two scenarios in predicting these results:

1. the change in standards is accompanied by a commensurate change in the export prices;
2. no change in prices will take place as the standards change but the profitability of export changes.

Let us consider the former scenario.

Denote by $p^{\delta}$ the price which would provide the same level of profitability of export as the existing one, for zero rates of customs tariffs, by $h$ the cumulative rate of ad valorem duties for the mentioned goods (the import duty in the importing country and export duty in the exporting country), and by $q$ the cumulative rate of special duties. Then the price $p$ at which the goods may be exported with the same level of profitability is determined by the quantity

$$p = p^{\delta} + hp^{\delta} + q.$$

A change in customs standards of the importing and exporting countries will affect the value of the parameters $h$ and $q$ and, therefore, the value of $p$.

Substituting the new value of $p$ into (4.48), we obtain a prediction of the volumes of export.

For the other alternative, the value of $p$, and therefore, $z$ will not vary, and the change in profitability of export $\Delta x$ can be evaluated as follows:

$$\Delta x = -\left(\Delta hp^{\delta} + \Delta q\right), \tag{4.50}$$

where $\Delta h$ is the change of the cumulative rate of ad valorem duties, and $\Delta q$ is the change of the cumulative rate of special duties. The negative sign in formula (4.50) means that an increase in rates of duties for a fixed price will reduce profitability of export–import operations and, vice versa, their decrease will increase the profitability.

The price $p^{\delta}$ can be determined by the formula

$$p^{\delta} = \frac{\tilde{p} - \tilde{q}}{1 + \tilde{h}},$$

where $\tilde{p}$ is the current price of the imported products, $\tilde{h}$ is the cumulative rate of the effective ad valorem duties, and $\tilde{q}$ is the cumulative rate of effective special duties of the exporter and importer.

**Analysis of Changes in Macroeconomic Situation in the Importing Country** In the case of significant changes in macroeconomic factors influencing demand, it is necessary to substitute their new (or predicted) values into (4.49), and then, according to (4.48), to determine new volumes of export for the new values of demand $\Phi$ and for conditions of constant prices and tariff-and-customs standards.

**Analysis of Influence of Changes of Prices in the Domestic Market** If the prices for products competing with the exported ones in the domestic market are changed, then it is necessary to substitute into (4.48) the new values of this price $\overline{p}$ and to obtain the prediction of changes in volume of export for constant demand and constant tariff-and-customs standards.

In the actual practice of analysis of a foreign economic situation, the necessity may arise to estimate the results of changes in all of the above-mentioned factors.

Then the considered calculations should be carried out sequentially, which will allow us to determine both the general changes in volume of export and the contribution of each factor to these changes.

For calculations according to the model (4.48), (4.49), the values of its parameters $k, \alpha, a_1, a_2, a_3, a_4$ should be identified. The following statistical information is used for this purpose:

- data for several years about volume of export of definite goods to a definite country (we denote them by $z_t$, $t = \overline{1,T}$, where $[1;T]$ is the period of observations);
- data for the same years about average prices at which these goods are realized, and average prices for similar goods in the home market of the importer (we denote them by $p_t$ and $\overline{p}_t$, $t = \overline{1,T}$);
- macroeconomic indices of the importing country in the same years:

  (a) the value of its actual gross national product $V_t$,
  (b) the net investments $I_t$,
  (c) the actual average wages $W_t$,
  (d) the annual rates of inflation $\delta\overline{p}^t$.

The parameters of the model should be identified based on these data using a nonlinear analogue of the least-squares method. According to this method, the parameters will be determined as a solution of the optimization problem

$$F(k,\alpha,a_1,a_2,a_3,a_4) = \sum_{t=1}^{T}\left(z_t - \frac{\delta\overline{p}^t\,(a_1V_t + a_2I_t + a_3W_t + a_4)}{p_t\left(1+\left(\frac{p_t}{k\overline{p}_t}\right)^{1/\alpha}\right)}\right)^2 \to \min \qquad (4.51)$$

under the constraints $k \geq 0$, $\alpha \geq 0$. Taking into account the discontinuity of the function being minimized for $k = \alpha = 0$, it is advisable to replace the latter with the inequalities $k \geq \varepsilon$ and $\alpha \geq \varepsilon$, where $\varepsilon$ is some small number.

Problem (4.51) is the problem of search for the extremum of a nonconvex differentiable function. Rather efficient numerical methods for solving such problems are presented in [54]; however, the absence of its analytical solution does not allow us to study the properties of the obtained parameter estimates, as was made for problems of linear regression. Therefore, of some interest is consideration of the linearized version of model (4.48), (4.49), where, due to the loss of simulation accuracy, a greater statistical validity of identification procedures of model parameters could be achieved.

Let us consider now the linearized version of the model of oligopoly pricing.

Based on experience in the application of oligopoly models (see [49]), the values of the parameters $k$ and $\alpha$ of this model differ not much from 1. In countries with a rather stable economy, the value of effective demand $\Phi$ also is not subjected to significant changes. Taking this into account, we can replace the right-hand side of Eq. (4.48) with the linear function of $k$, $\alpha$ and $\Phi$, obtained by expansion of the nonlinear function

$$G(k,\alpha,\Phi) = \frac{\Phi}{p\left(1+\left(\frac{p}{k\overline{p}}\right)^{1/\alpha}\right)} \tag{4.52}$$

in the vicinity of the point $(1,1,\Phi^0)$, where $\Phi^0$ is some constant value of the consumer demand $\Phi$. Before constructing such a linearized model, let us make some transformations of the parameters. Denote $\alpha_1 = 1/\alpha$ and $k_1 = k^{-1/\alpha}$. Then equality (4.52) takes the form

$$G_1(k_1,\alpha_1,\Phi) = \frac{\Phi}{p\left(1+k_1\left(\frac{p}{\overline{p}}\right)^{\alpha_1}\right)}.$$

Hereafter, precisely the function $G_1(k_1,\alpha_1,\Phi)$ is expanded in the vicinity of the point $(1,1,\Phi^0)$.

We have

$$\frac{\partial G_1(1,1,\Phi^0)}{\partial k_1} = -\frac{\Phi^0\left(\frac{p}{\overline{p}}\right)}{p\left(1+\left(\frac{p}{\overline{p}}\right)\right)^2}, \quad \frac{\partial G_1(1,1,\Phi^0)}{\partial \alpha_1} = -\frac{\Phi^0\left(\frac{p}{\overline{p}}\right)\ln\left(\frac{p}{\overline{p}}\right)}{p\left(1+\left(\frac{p}{\overline{p}}\right)\right)^2},$$

$$\frac{\partial G_1(1,1,\Phi^0)}{\partial \Phi} = -\frac{1}{p\left(1+\left(\frac{p}{\overline{p}}\right)\right)^2}.$$

Whence

$$\begin{aligned}
G_1(k_1,\alpha_1,\Phi) &\approx G_1(1,1,\Phi^0) + \frac{\partial G_1(1,1,\Phi^0)}{\partial k_1}(k_1-1) \\
&\quad + \frac{\partial G_1(1,1,\Phi^0)}{\partial \alpha_1}(\alpha_1-1) + \frac{\partial G_1(1,1,\Phi^0)}{\partial \Phi}(\Phi-\Phi^0) \\
&= \frac{\Phi^0}{p\left(1+\left(\frac{p}{\overline{p}}\right)\right)} - \frac{\Phi^0\left(\frac{p}{\overline{p}}\right)}{p\left(1+\left(\frac{p}{\overline{p}}\right)\right)^2}(k_1-1) \\
&\quad - \frac{\Phi^0\left(\frac{p}{\overline{p}}\right)\ln\left(\frac{p}{\overline{p}}\right)}{p\left(1+\left(\frac{p}{\overline{p}}\right)\right)^2}(\alpha_1-1) + \frac{\Phi-\Phi^0}{p\left(1+\left(\frac{p}{\overline{p}}\right)\right)} \\
&= A_1\Phi - A_2k_1 - A_3\alpha_1 + B_1,
\end{aligned}$$

where

$$A_1 = \frac{1}{p\left(1+\left(\frac{p}{\overline{p}}\right)\right)}, \quad A_2 = \frac{\Phi^0}{\overline{p}\left(1+\left(\frac{p}{\overline{p}}\right)\right)^2},$$

$$A_3 = \frac{\Phi^0 \ln\left(\frac{p}{\overline{p}}\right)}{\overline{p}\left(1+\left(\frac{p}{\overline{p}}\right)\right)^2}, \quad B_1 = \frac{\Phi^0\left(1+\ln\frac{p}{\overline{p}}\right)}{\overline{p}\left(1+\left(\frac{p}{\overline{p}}\right)\right)^2}.$$

Substituting expression (4.49) into the last equality, we obtain the final form of the equation

$$Z = \frac{A_1}{\delta \overline{p}}\left(a_1 V + a_2 I + a_3 W + a_4\right) - A_2 k_1 - A_3 \alpha_1 + B_1 \tag{4.53}$$

which is linear in the parameters $a_1$, $a_2$, $a_3$, $a_4$, $k_1$, and $\alpha_1$ and whose values can be identified by the methods of linear regression. The same statistical information will be necessary in this case as for identification of parameters of the nonlinear model (4.48), (4.49). Application of the mentioned methods allows us not only to determine the values of the model parameters but also to estimate their confidence intervals, as well as solve other problems of statistical analysis.

The proposed models were used for prediction of volumes of export from Ukraine to individual regions by the following integrated commodity items:

1. Ferroalloys (code 7202 of the commodity line according to the State Classifier of Goods and Services). The following regions to which import was carried out were considered for this commodity line:
   - Central Europe (the corresponding commodity flow is designated below as 7202.a).
   - The Baltic-Scandinavian region (commodity flow 7202.b).
2. Intermediate products of iron and steel (code 7207 of the commodity line). The main importers for this group are the countries of Southeast Asia (commodity flow 7207.a).
3. Rolled iron and steel plates (code 7208 of the commodity line). The following regions of import were taken into account in this study:
   - Near East and the Mediterranean (commodity flow 7208.a);
   - Central and East Europe (commodity flow 7208.b);
   - Asian-Pacific region (commodity flow 7208.c).

As an example of export of power-consuming products other than ferrous metals, export of raw aluminium to Hungary was considered (commodity flow 7601.a).

The regions were determined according to such criteria as homogeneous conditions for export of corresponding products to the countries included in the region, close prices for these products in home markets of the countries being considered, similar foreign economic policy of these countries, informal considerations, etc.

Data obtained from the Ministry of Foreign Economic Relations and Trade, Ministry of Economics, Ministry of Industrial Policy, and the National Bank of Ukraine were used as the sources of information for further statistical studies, as well as the sources from Consolidated Interindustry Balance [9, 10], Foreign Trade [22] and other sources. The period of observations was 1994–1999 with a quarterly partition of data. Taking into account currency reform carried out in Ukraine in 1996, which has changed the form of the national currency, and also the forms and methods of currency and price regulation, the price factors were considered in the form of relative indices (indices of the average price of the current quarter related to the price in the first quarter of 1994).

Construction of the models was preceded by correlation analysis in order to estimate the degree of influence on the volume of export of the factors taken into account. This analysis was carried out through calculation of coefficients of correlation between the volumes of export of the respective products to the mentioned regions and the factors of models (4.48), (4.49), (4.53). Their values are presented in Table 4.6. It should be pointed out that, in many cases, already at this stage of studies, the inverse correlation between the volumes of export, on one hand, and the foreign market price and exchange rate of the hryvnia for hard currency, on the other hand, was observed for products of ferrous metallurgy. A direct correlation with the producer's price and a weak statistical dependence on macroeconomic indices of importing countries took place (the value of the coefficients of correlation did not exceed 0.2). This contradicts the assumptions as to the influence of the mentioned factors used for construction of the models of oligopoly pricing and complicates their application for these commodity items.

As distinct to products of ferrous metallurgy, data of correlation analysis for the commodity flow 7601.a are coordinated with the structure of the mentioned models. A direct correlation of the volumes of export with the price in the foreign market and macroeconomic indicators of the importing country is observed here, as well as the inverse correlation with the producer's price.

After correlation analysis of the factors having effect on the volume of export from Ukraine to the mentioned regions, the parameters of the regression models considered above were identified. These parameters were determined by the least-squares method: with the help of a statistical package for the linear (linearized) model and using specially developed software for the nonlinear one. For solution of the optimization problem arising in parameter identification, the gradient method with variable metrics ($r$-algorithm) (see [54]) was used in this software.

Except for values of the parameters, the coefficient of determination $R2$ was calculated there. It shows the share of change of the resulting index (volume of export), which can be explained using the regression model proposed. The results

**Table 4.6** Values of coefficients of correlation between volume of exports and factors taken into account

| | Commodity flow, region | | | | | | |
|---|---|---|---|---|---|---|---|
| Factor | 7202.a | 7202.b | 7207.a | 7208.a | 7208.b | 7208.c | 7601.a |
| Price in foreign market | 0.2648 | −0.2453 | 0.6434 | −0.5192 | −0.550 | −0.03 | 0.2566 |
| Producer's price | 0. 5695 | 0.5073 | 0. 5432 | 0.53876 | 0.5551 | 0.5684 | −0.6225 |
| Exchange rate for hard currency | 0.346 | 0.369 | −0.4542 | 0.41797 | 0.4585 | −0.3956 | 0.4874 |
| Gross national product of the importer | −0.124 | −0.003 | −0.126 | 0.548 | 0.181 | −0.1375 | 0.5864 |
| Net investments of the importer | 0.216 | −0.1735 | 0.360 | −0.4412 | 0.064 | −0.4431 | 0.5213 |
| Average remuneration of labor | 0.6546 | −0.4667 | 0.1845 | 0.2773 | 0.179 | 0.2003 | 0.3325 |
| Rate of inflation of importer | 0.428 | 0.6430 | 0.3598 | 0.6612 | −0.475 | 0.5709 | −0.0409 |
| Rates of import duties | −0.227 | −0.355 | −0.380 | −0.292 | −0.3089 | −0.4416 | −0.6342 |
| Overdue credit arrears of all enterprises of the branch | 0.7997 | 0.747 | 0.827 | 0.8814 | 0.871 | 0.7488 | No data |

**Table 4.7** Average approximation error (%) and coefficients for regression models

| | Model of oligopoly pricing | | Model with regard for non-payments | |
|---|---|---|---|---|
| Commodity flow, region | Approximation error | Coefficient *R2* | Approximation error | Coefficient *R2* |
| 7202.a | 29.7 | 0.345 | 5.9 | 0.849 |
| 7202.b | 37.1 | 0.224 | 8.7 | 0.869 |
| 7207.a | 32.4 | 0.316 | 9.5 | 0.569 |
| 7208.a | 35.2 | 0.323 | 9.2 | 0.757 |
| 7208.b | 36.1 | 0.291 | 10.1 | 0.718 |
| 7208.c | 34.8 | 0.249 | 8.7 | 0.655 |

obtained are presented in Table 4.7. In this case, among the two possible models, the linearized and nonlinear, the model providing the least relative error of approximation (deviation of predicted values of export from actual ones) was selected for the considered period of observations.

The studies conducted yielded the following results.

The accuracy of the models of oligopoly pricing for commodity flows 7202.a, 7202.b, 7207.a, 7208.a, 7208.b, and 7208.c is not satisfactory. An exception is the linearized model of export for the commodity flow 7601.a, where the error of approximation was 21.6 %. Thus, the laws governing the dynamics of export of ferrous metals from Ukraine differed from those assumed in theoretical studies. For acceptable accuracy of prediction of these items, it was necessary to consider models taking into account factors differing from those traditionally considered.

Ferrous metallurgy belongs to raw-material intensive and power-consuming industries. An increase in the price for power resources, on one hand, and the impossibility (because of the absence of demand with increased price corresponding to the level of cost of products) of a further increase in price for the products made, on the other hand, forced the manufacturer to search for additional sources to cover its material expenditures. Such sources may be the circulating assets of enterprises-suppliers of raw material, which are appropriated by processing enterprises in the form of non-payments (overdue credit arrears) for the used raw material (energy carriers, electric power, etc.). Not paying for a part of its expenditures, the manufacturer actually decreased artificially the product prime cost by the value of the debt. This allowed him to obtain an acceptable "shadow" income with a smaller price for the products, and, in export–import operations, to achieve sufficient competitiveness by using pricing methods. Mechanisms of formation and development of crisis of nonpayments under this scenario were considered in [42] in greater detail. The similar situation was also analyzed in Sect. 1.4

Thus, the volumes of nonpayments became an important indicator of production activity. This hypothesis is confirmed by the large values (from 0.7 to 0.89) of the coefficient of correlation between the volumes of export and total overdue credit arrears of enterprises of ferrous metallurgy (Table 4.6). In view of these facts, the following linear regression model was proposed for prediction of volumes of export:

$$Z = a_1 H + a_2 P + a_3 Q + a_4 \Delta + a_5. \tag{4.54}$$

Here, $Z$ is the volume of export (of the given commodity item to a definite region), $H$ is the level of total overdue credit arrears of ferrous metallurgy enterprises, $P$ is the price for the products they make, in the countries where these products are exported, $Q$ is the exchange rate of the national currency for hard currency, $\Delta$ is the cumulative change of tariff rates; $a_1, \dots, a_5$ are coefficients determined from available statistical data. As in the models constructed before, in view of the above-stated considerations, all the price and cost indices were considered in the relative form. The coefficients $a_1, \dots, a_5$ were determined by the least-squares method with the help of "*EViews*" statistical software system. Along with their values, the same statistical indices as for construction of the linearized model were calculated. Data as to the accuracy of model (4.54), which takes into account the volume of nonpayments, are presented in Table 4.7. It is necessary to pay attention to the fact that negative values of the coefficient $a_3$ were obtained for the commodity flows considered. This means that inflation of the hryvnia should decrease, under otherwise equal conditions, the volumes of export of ferrous metals, which can be explained by the dependence of this branch on raw material import and its power consumption. An increase of the price (in national currency) of power resources with high power consumption of production leads to a larger increase in production cost as compared to the decrease of export prices.

It should be noted that the accuracy of the constructed linear models with regard for nonpayments was quite acceptable (the relative error of approximation was from 5.9 to 10.1 % and did not exceed 10 % for the majority of items, the value of the coefficient of determination was from 0.569 to 0.869 and was no less than 0.7 for the majority of items).

The calculations carried out with the help of an interbranch optimization model using real data made it possible to formulate a number of recommendations as to the development of the export potential of Ukrainian ferrous metallurgy.

It is necessary also to point out that export processes in the given branch cannot be adequately described using the well-known theoretical models, for example, the model of oligopoly pricing, whence we may conclude that specific mechanisms exist in the indicated sphere.

## *Concluding Remarks*

Application of econometric models had allowed us to determine the factors having a significant effect on the volumes of export of ferrous metallurgy products from Ukraine. First of all, this was the level of total overdue credit arrears, whose increase indicates development in this segment of the market of noncivilized, shadow forms of commodity-money relations based on "soft" budget constraints. It was natural that the imperfection of branch technologies influenced somehow this level; however, volumes of export of other forms of power-intensive products, for example, aluminium to Hungary, depended much more on factors typical of civilized markets. Thus, the structural changes were necessary for ferrous metallurgy; creation of a competitive environment and increase in financial discipline must play a definite role during these changes. When they would be realized, the use of oligopoly pricing will make it possible to increase the accuracy of prediction and to take into account more fully the influence of changes in standards of tariff-customs policy and other factors influencing the volumes and structure of export from Ukraine. However, another way of the problem solution so-called experiment in metallurgy was chosen in Ukraine at the end of the 1990s.

According to it, tax arrears to the state and local budgets of enterprises of ferrous metallurgy were written off and these enterprises were allowed not to pay several taxes including profit tax. The "experiment" means the legalization of the practice of "soft" budget constraints. Together with appropriate situation at the world market of metals it stimulated a sharp growth of volumes of metallurgical export in the first decade of the twenty-first century; the profits of enterprises also increased, but they were not transformed into investment to implementation of new technologies, modernization of equipment and other innovative project as it was assumed in economic background of this policy. Metallurgical firms used them profits to lobby state bodies and to realize their political projects. By fact, they spent their outcomes generated in the previous period to acquire the possibility not to pay taxes (and other "soft" privileges) in future. As a result, a low technological level together with rent-seeking behavior of top managers lead to catastrophic decline in metallurgy in the beginning of global financial crisis in late autumn of 2008.

## 4.6 Formalization of Nontransitive Preferences in Problems of Analysis of Foreign Economic Activity

In this section we consider the particular problem of decision-making in foreign trade based on preference relations. The last ones are often used in multicriteria optimization as models of nonformal pairwise comparison procedure. Experts and decision-makers implement this procedure, so the question concerning its transitiveness is a subject of the intensive discussions (see, for instance, [5, 19, 26]). On one hand, many approaches to development of numerical decision-making procedures are based on a transitiveness assumption and even formalization of a term of "the best decision" became a problem without it. On the other hand, it is difficult to check this assumption in real-data problems. In many cases, especially in collective decision-making, the assumption is obviously invalid. Considering all this, we can say that development of approaches to nontransitive decision-making is extremely important.

The essential results to relax transitivity of preferences have been obtained by Bouysson and Pirlot [5], Tsoukias [58].

The algorithms suggested in this Section can be considered as the only one possible way to the problem solution based on the alternative numerical description of preferences. The decision-making problem is transformed into the nondifferentiable optimization problem using the mentioned description. This approach is known; it has been developed, for instance, in [19, 20]. However, we will pay attention to some numerical algorithms for its implementation. The main goal of the Section is to demonstrate how decision-making problems in the nontransitive case can be reduced to the solution of problems for maximization of some scalar-value function. The considered economic model has mainly the illustrative goal and can be implemented to any similar problem if necessary.

Improvement of domestic products belongs to the important directions of export potential development. First, approximation competitiveness of products can be described by a price of production, its quality parameters and consumers' expectations. The last ones depend on advertising or are formed spontaneously.

A producer can change quality parameters and consumers' expectations by additional expenditures, which influence on a production price and must satisfy financial constraints. As a result, the problem to define expenditures value and their structure will arise, so they will provide the highest production competitiveness. This problem plays an important role in marketing analysis. Different methods and approaches, both formal and nonformalized, are used to solve it. Optimization modeling is the typical formal approach. Trying to apply it to the mentioned problem, the difficulties appear relative to formalization of competitiveness. The last one is determined by preferences of consumers, so binary relations can describe it mathematically. Such relations are specified at the set of $\{x, y, p\}$, where $x$ is a production quality parameter vector, $p$ is a production price and $y$ are the consumers' expectations. Let us denote mentioned relation by $R$. If production with quality parameters $x^{(1)}$, price $p^{(1)}$ and expectations $y^{(1)}$ is not less competitive than the production with $x^{(2)}$, $p^{(2)}$ and $y^{(2)}$, the relation $z^{(1)} R z^{(2)}$ holds, where $z^{(1)} = \left(x^{(1)}, y^{(1)}, p^{(1)}\right)$, $z^{(2)} = \left(x^{(2)}, y^{(2)}, p^{(2)}\right)$. Ordinary expert estimates, information obtained from consumers' interview and statistical information concerning the value of production to be sold can be sources of information about this relation. For instance, we can assume that a product $A$ is not less competitive than the similar product $B$ if an amount of selling for product $A$ is not less than for $B$.

The problem of searching for an additional expenditure structure $h$ has the following formulation.

Let the dependences $x(h)$, $y(h)$ and $p(h)$ between competitiveness parameters $(x, y, p)$ and $h$ be given and the set $H$ of admissible $h$ values be specified. It is necessary to obtain such $h^* \in H$ that for every $h \in H$

$$\left(x(h^*), y(h^*) p(h^*)\right) R \left(x(h), y(h), p(h)\right)$$

holds.

It is the preference optimization problem (see [41]) denoted as

$$\left(x(h), y(h), p(h)\right) \underset{R}{\longrightarrow} \text{pref}, \quad h \in H, \tag{4.55}$$

where $H$ is the set of admissible $h$ values. This set can be specified by financial (budgetary) constraints of producers, equations of mutual dependence between $x(h)$, $y(h)$, $p(h)$, etc.

The numerical methods for the solution of problems similar to (4.55) are considered in several papers [40, 41] under the assumption that the relation $R$ is complete, reflexive, transitive, continuous and convex [58].

In some papers (for instance, [33]), the algorithms of construction of utility function $U(z)$ were elaborated for the same assumptions. This function satisfies the inequality $U\left(z^{(1)}\right) \geq U(z^{(2)})$, if $z^{(1)} R z^{(2)}$ holds.

Unfortunately, the application of this approach to the problem (4.55) faces serious conceptual difficulties relative to refusal of assumption about transitiveness of relation $R$. This relation is either the model of collective preferences of majority of consumers (in a case of its statistical determination) or the model of individual preferences created at a family level by "averaging" of interests of several different persons (in a case of its determination based on "typical" consumer's behavior). In both cases, the result of aggregation of individual preferences using the "majority voting" principle (or another similar method) arises. Such aggregated preferences are not usually transitive even for the case of transitiveness of all individual preferences (see [2]). It should be noted that even the transitiveness of individual preference relation is often discussed question because a person doing pairwise comparison cannot remember all his previous answers. Therefore, we cannot assume the transitiveness of the relation $R$ in (4.55) and cannot apply the approach by Koshlai and Mikhalevich [33] to the problem solution. Any utility function $U(z)$ cannot satisfy the inequalities $U\left(z^{(1)}\right) \geq U(z^{(2)})$, $U(z^{(2)}) > U(z^{(3)})$ and $U(z^{(3)}) \geq U\left(z^{(1)}\right)$, which follow from nontransitive relations between solutions $z^{(1)}$, $z^{(2)}$ and $z^{(3)}$. Even the formalization of the conception of "the best solution" is problematic for such a case. In the example given above, the solution $z^{(3)}$ is directly better than $z^{(1)}$, but it is worse than $z^{(1)}$, in comparison through $z^{(2)}$.

The proposed approach is based on the preference indicator instead of a utility function.

This indicator is a function $F\left(z^{(1)}, z^{(2)}\right)$ that $F\left(z^{(1)}, z^{(2)}\right) \geq 0$ holds, if $z^{(1)} R z^{(2)}$, and $F\left(z^{(1)}, z^{(2)}\right) > 0$ holds, if $z^{(1)}$ is straightly preferable than $z^{(2)}$. The preference indicator can be determined as $F\left(z^{(1)}, z^{(2)}\right) = U\left(z^{(1)}\right) - U\left(z^{(2)}\right)$ for the case of existence of utility function $U(z)$. Really, the class of preferences, described by proposed technique is really wider, including nontransitive preference relations. For instance, consider the function

$$F(a,b) = ab\big(|a| - |b|\big),$$

where $a$ and $b$ are some real numbers. The inequalities

$$F\left(a^{(1)}, a^{(2)}\right) = 6 > 0, \quad F\left(a^{(2)}, a^{(3)}\right) = 2 > 0, \quad F\left(a^{(3)}, a^{(1)}\right) = 6 > 0,$$

hold for the solutions $a^{(1)} = 3$, $a^{(2)} = 1$, $a^{(3)} = -2$.

This indicator describes the strong nontransitive preference relation $R_1$ specified at $\left\{a^{(1)}, a^{(2)}, a^{(3)}\right\}$ as following:

$$a^{(1)} R_1 a^{(2)}, \quad a^{(2)} R_1 a^{(3)}, \quad a^{(3)} R_1 a^{(1)}.$$

The conditions of existence of the continuous preference indicator and such its properties as convexity, addictiveness and monotonicity have already been investigated in [20, 21].

Problem (4.55) can be formulated in terms of the preference indicator as the problem of such $h^* \in H$ search that, for every $h \in H$,

$$F\left(z(h^*), z(h)\right) \geq 0 \tag{4.56}$$

holds, $z(h^*) = \Big(x(h^*), y(h^*), p(h^*)\Big)$, $z(h) = \Big(x(h), y(h), p(h)\Big)$.

Condition (4.56) is equivalent to

$$\varphi(h^*) = \min_{h \in H} F\left(z(h^*), z(h)\right) \geq 0, \quad h^* \in H.$$

We will find the maximal point of the function $\varphi(h^*)$ determined on the set $H$ to solve the problem (4.56). If a function value at this point is nonnegative, the point will be a solution to the inequality, in the other case the problem (4.56) has no solutions, but the discrepancy of the inequality will be minimal at the mentioned point and it can be considered as the pseudosolution in the sense of Karmanov [30]. Thus, model (4.55) can be transformed into minimax optimization problem.

In any case, when the number of elements in $H$ is infinite it is impossible to obtain the exact solution of the problem

$$F\left(z(h^*), z(h)\right) \xrightarrow[h \in H]{} \min,$$

considered for the fixed $h^*$. It creates difficulties in calculation of subgradient of the function $\varphi(h^*)$ (see [55]).

The following numerical algorithm is proposed by authors to overcome the mentioned difficulties for the case of compact the set $H$.

The sequence $\{h^s\}$ of approximations to maximal point of $\varphi(h^*)$ function is constructing in a following way.

Let $h^0 = \tilde{h}^0$ be an arbitrary element from $H$. Describe the procedure of construction of $h^{s+1}, \tilde{h}^{s+1}$, when $h^s, \tilde{h}^s$ are known ($s = 0, 1, \ldots$).

1. Make a random choice of $\overline{h}^s$ so that for every $\varepsilon > 0$

$$P\left\{F\left(z(h^s), z(\overline{h}^s)\right) - \min_{h \in H} F\left(z(h^s), z(h)\right) \leq \varepsilon\right\} \geq \delta(\varepsilon) > 0.$$

   For instance, in the case of continuous preference indicator the value of $\overline{h}^s$ can be the $s$-th observation at the random variable, uniformly distributed at the set $H$.

2. Determine $\tilde{h}^{s+1}$ as

$$\tilde{h}^{s+1} = \begin{cases} \overline{h}^s, \text{ if } \; F\left(z(h^s), z(\overline{h}^s)\right) < F\left(z(h^s), z(\tilde{h}^s)\right), \\ \tilde{h}^s \text{ in other case.} \end{cases}$$

3. Calculate the subgradient $\nabla F\left(z(h^s), z(\overline{h}^s)\right)$ of the function

$$f(h) = F\left(z^{(1)}(h), z^{(2)}\right)$$

   at the point $z^{(1)}(h) = z(h^s), z^{(2)} = z(\tilde{h}^{s+1})$.

4. Determine $h^{s+1}$ as

$$h^{s+1} = \pi_H(h^s + \rho_s \nabla F\left(z(h^s), z(\hat{h}^{s+1})\right),$$

where $\pi_H(\cdot)$ is a projector on the set $H$, the choice of the step multiplier $\rho_s$ satisfies the conditions

$$\rho_s \geq 0, \quad \sum_{s=0}^{\infty} \rho_s = \infty, \quad \sum_{s=0}^{\infty} \rho_s^2 < \infty.$$

Using the results from [15], it is possible to prove that

$$E\left(F\left(z(h^s), z(\tilde{h}^{s+1})\right) / h^0, \ldots, h^s\right) = \nabla\varphi\left(h^s\right) + b^s,$$

where $\nabla\varphi(h^s)$ is the subgradient of the function $\varphi(h^*)$ at the point $h^* = h^s$, $\|b^s\| \underset{s\to\infty}{\longrightarrow} 0$. The convergence of $\{h^s\}$ to maximal point of the function $\varphi(h^*)$ follows from the last equality and the results of Birge and Louveaux [4], Ermol'ev and Wets [15].

It should be noted that the other criteria based on preference indicator could be applied for a choice of the best solution. For instance, the averaged value of the preference indicator

$$\psi_1(h^*) = \int\limits_{h\in H} F\left(Z(h^*), Z(h)\right)\alpha(h)dh$$

or the power of the solution set for which $h^*$ is "the best"

$$\psi_2(h^*) = \int\limits_{h\in\{t:F(Z(h^*),Z(t))\geq 0, t\in H\}} \alpha(h)dh$$

also can be considered as a such criteria. After maximizing $\psi_1(h^*)$ and $\psi_2(h^*)$, any positive-valued function $\alpha(h)$ that satisfies the condition

$$\int\limits_{h\in H} \alpha(h)dh = 1$$

can be used for $\psi_1(h^*)$ and $\psi_2(h^*)$ determination.

For instance, this function can be defined as the following:

$$\alpha(h) = \begin{cases} \frac{1}{N}, & \text{if} \quad h \in H, \\ 0, & \text{in other case,} \end{cases}$$

when $N$ is a cardinal number of the set $H$.

Therefore, the decision-making problem for the case with nontransitive preferences has become similar to the multicriteria optimization problem.

Difficulties connected with calculations of the values of $\psi_1(h^*)$ will appear, which can be considered as

$$\psi_1(h^*) = E\Big(F\big(z(h), z(\theta)\big)\Big),$$

where $\theta$ is a random variable with density $\alpha(h)$. The following numerical algorithm can be applied for the search of maximal point of such function.

Let $h^0$ be an arbitrary element from $H$. Every next approximation $h^{s+1}$ is calculated from known $h^s$ as the following

$$h^{s+1} = \pi_H\Big(h^s + \rho_s \nabla F\big(z(h^s), z(\theta^s)\big)\Big), \quad s = 0, 1, \ldots,$$

where $\theta^s$ is the $s$-th independent observation at the random variable $\theta$, $\pi_H(\cdot)$ is the projector on the set $H$, the choice of $\rho_s$ satisfies the same condition as for the previous algorithm. The convergence of $\{h^s\}$ to maximal point of $\psi_1(h^*)$ function is proved on the basis of the results [15].

The similar approach can be applied to maximize the function $\psi_2(h^*)$.

Let now consider the minimax problem that means to analyze the structure of the obtained solution. Let us define arbitrary closed subset $\widehat{H}$ from $H$.

The solution $\overline{h}^* \in \widehat{H}$ will be called as dominated under $\tilde{h}^* \in \widehat{H}$, if

$$F\big(z(\overline{h}^*), z(h)\big) \geq F\big(z(\tilde{h}^*), z(h)\big)$$

holds for every $h \in H$ and, at least for one $\widehat{h} \in H$, the strong inequality

$$F\big(z(\overline{h}^*), z(\widehat{h})\big) > F\big(z(\tilde{h}^*), z(\widehat{h})\big)$$

holds. For instance, $\overline{h}^*$ will be dominated under $\tilde{h}^*$, if both alternatives $z(\overline{h}^*)$ and $z(\tilde{h}^*)$ are better than other feasible solutions, but $z(\overline{h}^*)$ is more competitive than $z(\tilde{h}^*)$. For this case, $\widehat{h} = \overline{h}^*$, $F\big(z(\overline{h}^*), z(\overline{h}^*)\big) = 0$ since $R$ is the reflexive relation, but $F\big(z(\tilde{h}^*), z(\overline{h}^*)\big)$ is negative.

The solution $h^*$ will be called *Pareto optimal* on the set $\widehat{H}$, if there is no solution from this set dominated under it. The problem of construction of the set of all Pareto optimal solutions on $\widehat{H}$ can be solved by the following way.

Let us consider the function

$$G\big(z(h^*), z(h)\big) = F\big(z(h^*), z(h)\big) - \min_{h^* \in H} F\big(z(h^*), z(h)\big) + \varepsilon,$$

where $\varepsilon > 0$ is sufficiently small.

This is positive-valued monotonous transformation of preference indicator. It can be substituted by any other transformation saved the mentioned properties.

Assume that every component of $z(h)$ is the continuous and constrained on $H$ function as well as the function $F(x, y)$ is also continuous and constrained. Then, the function $G\left(z(h^*), z(h)\right)$ is also continuous and constrained on $H \times H$ positive-value function.

Consider now the following optimization problem

$$\psi(h^*) = \min_{h \in H}\left(\alpha(h) G\left(z(h^*), z(h)\right)\right) \xrightarrow[h^* \in \widehat{H}]{} \max, \tag{4.57}$$

where $0 < \alpha(h) \leq 1$ is some continuous function determined on the set $H$. The following statement is true.

**Theorem 4.3.** *Every Pareto optimal solution on the set $\widehat{H}$ can be constructed as a solution to optimization problem* (4.57) *for some function $\alpha(h)$.*

*Proof.* Let the solution $\overline{h}^*$ is arbitrary Pareto optimal on the set $\widehat{H}$. (It should be noted that the solution of the problem (4.56) belongs to the mentioned set which is not empty.)

Assume that

$$\overline{\alpha}(h) = \left(G\left(z(\overline{h}^*), z(h)\right)\right)^{-1} k,$$

where

$$k = \min_{h \in H}\left(G\left(z(\overline{h}^*), z(h)\right)\right), \quad 0 < \overline{\alpha}(h) \leq 1.$$

Let us consider the problem (4.57) for $\alpha(h) = \overline{\alpha}(h)$ and suppose that $\overline{h}^*$ is not the optimal solution to this problem. It means the existence of such $\tilde{h}^* \in \widehat{H}$ that $\psi(\tilde{h}^*) > \psi(\overline{h}^*)$ holds.

The last inequality is equivalent to

$$\min_{h \in H}\left(\frac{G\left(z(\tilde{h}^*), z(h)\right)}{G\left(z(\overline{h}^*), z(h)\right)}\right) > 1, \tag{4.58}$$

because of $\psi(\overline{h}^*) = k$ and $k$ is independent from $h$.

The inequality

$$G\left(z(\tilde{h}^*), z(h)\right) > G\left(z(\overline{h}^*), z(h)\right)$$

follows from (4.58) for every $h \in H$. It means that the solution $\tilde{h}^* \in \widehat{H}$ is dominated under $\overline{h}^*$. We obtain the contradiction with Pareto optimality for $\overline{h}^*$ on $\widehat{H}$. Theorem is proved.

The above-considered numerical algorithms can be applied to the solution of the problem (4.57).

The proposed approach needs the construction of the function $F\left(z^{(1)}, z^{(2)}\right)$. Ordinary regression methods (see [33]) can be used for this purpose.

First let us determine the analytical form of the preference indicator.

If the preference relation $R$ is complete, reflexive, and continuous, the analytical form of the indicator can be introduced by

$$F\left(z^{(1)}, z^{(2)}, \alpha, \beta\right) = Q\left(z^{(1)}, z^{(2)}, \alpha\right)\left(V\left(z^{(1)}, \beta\right) - V\left(z^{(2)}, \beta\right)\right),$$

where $Q\left(z^{(1)}, z^{(2)}, \alpha\right)$ and $V(z, \beta)$ are some continuous functions depending on unknown parameters $\alpha$ and $\beta$. If the relation $R$ is antisymmetric, i.e., from $z^{(1)}$ better $z^{(2)}$ always implies that $z^{(2)}$ is worse than $z^{(1)}$, the function $Q\left(z^{(1)}, z^{(2)}, \alpha\right)$ must satisfy the condition $Q\left(z^{(1)}, z^{(2)}, \alpha\right) = Q\left(z^{(2)}, z^{(1)}, \alpha\right)$ for every $z^{(1)}, z^{(2)}$. Therefore, the function $Q\left(z^{(1)}, z^{(2)}, \alpha\right)$ has, for instance, quadratic form with the symmetrical matrix $\alpha$, and $V(z, \beta)$ can be linear function with coefficients $\beta$.

The information about competitiveness of the proposed goods can be used for identification of values of the mentioned parameters.

Let us consider the consumer goods with parameters $z^{(1)}, z^{(2)}, \ldots, z^{(N)}$ and preferences $z^{(i)} R z^{(j)}$ that are known for some pairs of goods $(i, j)$. The set of all such pairs is denoted by $I$. Then, the parameters $\alpha$ and $\beta$ can be found as a solution to the inequality system

$$Q\left(z^{(i)}, z^{(j)}, \alpha\right) \cdot \left(V\left(z^{(i)}, \beta\right) - V\left(z^{(j)}, \beta\right)\right) \geq 0$$

for $(i, j) \in I$.

The numerical algorithms for a solution of such system based on the nonsmooth optimization methods from Shor [54] were elaborated in the same way as it was done for the methods used to solve the inequality (4.56).

## Conclusions

This approach has been applied to solve some real-data problems. It can be combined with other decision-aiding preference models in decision-support systems.

The proposed approach assumes the development of the decision-making algorithms based on nonsmooth optimization methods.

For instance, an arbitrary decision from generalized Pareto optimal set can be obtained as a solution of minimax optimization problem using preference indicator. This set in a whole can be constructed as a set of solutions of nonsmooth parametric programming problem.

The properties of elements of the mentioned set allow us to apply them as a background of decisions in real-data problems. The nondifferentiable optimization methods can be also used to construct the preference indicator based on ordinary expert estimates.

## 4.7 Models and Decision Support Tools for Transboundary Water Supply Problems

Hydroecological situation in many regions of the world has become particularly acute during the last decades of the twentieth century due to rapid drying of the fresh water lakes and rivers, deterioration of water quality resulting from industrial water pollution and intensive extraction of ground water. The centralized water supply in many towns of the mentioned regions is restricted during last three decades because of fresh water deficit. In addition, the water quality is unacceptable. Therefore, the ecological, economic, social, and other aspects of such situation are attracting with increasing frequency the attention of local policy-makers and scientists. The development of specialized tools including mathematical models, calculation algorithms, and specialized DSS is necessary to support their interaction.

One of the obvious reasons for the absence of a fresh water is the uncontrolled regional and transboundary use of the sources of the rivers basins to be crossed by state and regional borders. By economic theory, to provide the payment from water users will protect these sources. Therefore naturally raises the question of setting scientifically valid charges for water. This problem can be solved in the frame of specialized model-driven system; pricing models create its core. This system is created for a high management level (i.e., the special government or international agency).

The solution of pricing problem is complex for several reasons. First, in the absence of property rights in water it is difficult to develop the "pure" market relations with buying and selling of water resources, and market price formation mechanisms thus cannot be expected to provide the right signals. Second, under the specific regional conditions with its wide spread of irrigated agriculture, rapidly increased water demand in recreation area, the utility of water for its suppliers is much higher than the direct material costs associated with water delivery. The cost approach to pricing is therefore also inapplicable. Lack of information about the utility functions of water users, which often belong to different countries, combined with inaccurate and incomplete information about water consumption constitute an additional set of difficulties. The models for distributed water resource management using only indirect indicators of utility are proposed for such situation in present Section. The values of these indicators can be determined at a high management level using a multiregional ecological monitoring system.

In the frame of this system the procedure to determining of water price can be represented by the following model.

Consider $N$ users (firms, states, regional and local authorities, etc.) of a common water resource. First, assume that each user consumes the part of water without affecting the quality of the remainder. Let $F$ be the available water resource, $x_i$ water consumption of user $i$, $v_i(x_i)$ the profit of user $i$ derived from limit of water use in quantity $x_i$, $\overline{x}_i$ the upper limit on water consumption of this user. Then the profit maximization problem for all consumers is stated in the form

$$\sum_{i=1}^{N} v_i(x_i) \to \max, \tag{4.59}$$

$$\sum_{i=1}^{N} x_i \leq F, \tag{4.60}$$

$$0 \leq x_i \leq \overline{x}_i, \quad i = \overline{1, N}. \tag{4.61}$$

It follows from the results by Nikaido [48] that the solution of this problem is in the core of nonzero-sum games whose players are the $N$ consumers with the payoff functions $v_i(x_i)$.

Formally, the price of water may be considered as the shadow price (the value of dual variable which corresponds to constraint (4.7.2)) of the optimal water-use program. But the problem (4.59)–(4.61) cannot be solved by a single coordinating center (i.e., at a high management level), because water users are not interested in revealing their utility functions $v_i(x_i)$. The main difficulty is how to find the shadow prices without information about the user's utility functions. Decentralized approach is applied for these purposes.

Moreover, the concept of utility is usually difficult to understand for users and questions that they should answer do not hit their knowledge, understanding of the problem, or style of decision making. A good decision support technique should provide users who can holistically maximize their unstated utility functions, by allowing them to specify of desirable consequences of right decisions. These aspects also must be taken into account when decision support algorithm is constructed.

The following algorithm is proposed for the problem (4.59)–(4.61) solution.

Suppose that the coordinating center establishes a certain price for water. The users communicate to the center their demand for water at the given price. Depending on the relationship between the aggregate demand for water and the available resources, the coordinating center adjusts the price and notifies the new value to the users. The entire procedure is then repeated. To formalize the procedure, let write the Lagrange function for problem (4.59)–(4.61):

$$L(x,u) = \sum_{i=1}^{N} v_i(x_i) + u\left(F - \sum_{i=1}^{N} x_i\right) = \sum_{i=1}^{N} v_i(x_i) - u\sum_{i=1}^{N} x_i + uF. \tag{4.62}$$

If the optimal solution $u^*$ of the dual of problem (4.59)–(4.61) is known, we can consider the equivalent formulation (assuming concave utility functions)

$$L(x, u^*) \to \max, \tag{4.63}$$

$$0 \le x_i \le \overline{x}_i, \quad i = \overline{1, N}. \tag{4.64}$$

Using (4.62), we obtain from (4.63), (4.64) $N$ independent problems

$$v_i(x_i) - u^* x_i \to \max, \tag{4.65}$$

$$0 \le x_i \le \overline{x}_i, \quad i = \overline{1, N}, \tag{4.66}$$

which can be solved by users at a low management level.

At a given water price $u^*$, the solution of the optimal water use problem (4.59)–(4.61) is decentralized. The same also holds for an arbitrary price $u$ substituted instead of $u^*$ in (4.65), (4.66), but in this case the resulting solution is not an optimal solution of problem (4.59)–(4.61).

In reality, the exact optimal solution of the dual problem is unknown. However, we can construct an adaptive procedure to find $u^*$ using the current imbalance between the aggregate water demand $\sum_{i=1}^{N} x_i$ and the available resource $F$, i.e., the difference $\sum_{i=1}^{N} x_i - F$. The proposed adaptive procedure has the following form. Take an arbitrary number as the initial approximation $u^0$. Suppose that $u^s$, the price of water at time $s$, $s = 0, 1, \ldots$, has been determined. Given this price, each water user determines the own demand $x_i^s$ by solving the problem

$$v_i(x_i) - u^s x_i \to \max, \tag{4.67}$$

$$0 \le x_i \le \overline{x}_i, \quad i = \overline{1, N}. \tag{4.68}$$

The new price $u^{s+1}$ is set by changing $u^s$ in proportion to the difference between the aggregate demand $\sum_{i=1}^{N} x_i^s$ and the available supply $F$ subject to nonnegativity of $u^{s+1}$:

$$u^{s+1} = \max\left(0, u^s + \rho_s\left(\sum_{i=1}^{N} x_i^s - F\right)\right). \tag{4.69}$$

Note that $\sum_{i=1}^{N} x_i^s - F$ in (4.69) is the generalized gradient of the objective function $\varphi(u) = \max_{0 \le x_i \le \overline{x}_i} L(x, u)$, $i = \overline{1, N}$. (Under our assumptions this function is nondifferentiable for $u = u^s$, when at least one of the problems (4.67), (4.68) has a nonunique optimal solution). The procedure (4.69) itself is a generalized gradient descent method (see [54]) $\rho_s$ therefore can be chosen under the standard conditions for this method:

$$\rho_s \geq 0, \rho_s \xrightarrow[s \to \infty]{} 0, \quad \sum_{s=0}^{\infty} \rho_s = \infty, \quad \sum_{s=0}^{\infty} \rho_s^2 < \infty.$$

Note that the coordinating center cannot compute the value of the function $\varphi(u)$ given the available information. The optimal price $u^*$ therefore cannot be determined by numerical methods that use this function value.

The condition $\sum\limits_{s=0}^{\infty} \rho_s^2 < \infty$ guarantees convergence of sequences to one of the optimal solutions $u^*$ of the dual of problem (4.59)–(4.61). Without this condition we can only assert that any convergent subsequence from $\{x^s\}$ converges to $\{u^*\}$ (see [54]).

As we have noted previously, $u^*$ may be interpreted as the optimal price of water and $x_i^s$ as the demand for water of user $i$ at the price $u^s$. Relationship (4.69) is therefore a difference analogue of the Walras equation that describes the dynamics of prices under specific market conditions and a procedure for the formation of an optimal (i.e., equilibrium) price.

The proposed algorithm simulates a procedure of market price formation for a "common" resource—water. If $x_i$ in model (4.67), (4.68) is an $m$-dimensional nonnegative vector, and not a scalar (it means that $i$ is the number of water users' group (state or region) and every group includes $m$ independent water users), then problem (4.67), (4.68) is replaced with the following problem:

$$v_i(x_i) - u^s \sum_{j=1}^{m} x_{ij} \to \max, \tag{4.70}$$

$$x_i \in X_i, \quad i = \overline{1, N}, \tag{4.71}$$

where $x_i = (x_{i1}, \ldots, x_{im}), i = \overline{1, N}$, $X_i$ is the feasible set for the group $i$.

Note that in reality the users pollute water and its quality deteriorates due to effluents and discharges. We therefore need a model that allows for these factors. In this model, the objective function (4.59) and the constraints (4.60), (4.61) are augmented with the constraints

$$\sum_{i=1}^{N} q_i^k(x_i) \leq Q_k, k = \overline{1, K}, \tag{4.72}$$

where $q_i^k(x_i)$ is the dependence between the volume of water use and the increase of the concentration of the $k$-th pollutant in water, $Q_k$ is the maximum admissible concentration of the $k$-th pollutant in water, $k$ is the number of pollutants introduced in the model. In this case, alongside the price $u$ of water we should also impose a tax $w_k$ per unit of pollutant $k$ discharged into the water. The adaptive process forming the water price and the pollution tax takes the following form.

Take the arbitrary values for $u_0$ and $w_k^0$, $k = \overline{1, K}$. We will describe a procedure that constructs $u^{s+1}$ and $w_k^{s+1}$ given $u^s$ and $w_k^s$ $(s = 0, 1, \ldots, k = \overline{1, K})$:

1. the optimal structure of water use $x_i^s$ is determined independently by each user from the condition of profit maximization:

$$v_i(x_i) - u^s x_i - \sum_{k=1}^{K} w_k^s q_i^k(x_i) \to \max,$$

$$0 \le x_i \le \overline{x}_i, \quad i = \overline{1, N};$$

2. the new water price $u^{s+1}$ is determined from the condition

$$u^{s+1} = \max\left(0, u^s + \rho_s\left(\sum_{i=1}^{N} x_i^s - F\right)\right); \tag{4.73}$$

3. the new pollution charges $w_k^{s+1}$ are calculated from the formula

$$w_s^{s+1} = \max\left(0, w_k^s + \rho_s\left(\sum_{i=1}^{N} q_i^k(x_i) - Q_k\right)\right), \quad k = \overline{1, K}. \tag{4.74}$$

The convergence conditions of this algorithm are similar to the previous case. We additionally assume convexity of the functions $q_i^k(x_i)$ and existence of an interior point in the set defined by the constraints (4.72).

The procedure to determine the water prices and the pollution tax can be considered as the model of negotiations between coordinating center (further it is called as "agency") and water users in the frame of model-driven DSS. This system is based on a tight local network of work stations and computerized users' working places, created by an "agent," with a gateway to external networks. The negotiations are organized in a following way. The "agent" announces through DSS communications the current water prices $u^s$ and the pollution charges $w_k^s$. Then the users, independently of one another, determine the optimal water use levels $x_i^s$ and report them to the "agent." They also compute the expected levels of water pollution $q_i^k(x_i^s)$ using the ecological models received from the "agency." The described above pricing model located in the "agent" aggregates the incoming information and calculates the differences $\sum_{i=1}^{N} x_i^s - F$ and $\sum_{i=1}^{N} q_i^k(x_i) - Q_k$, $k = \overline{1, K}$.

Then this model computes from (4.73), (4.74) the new water price $u^{s+1}$ and the new environmental tax rates $w_k^{s+1}$. Note that these negotiations are easily organized on a computer network linking the "agency" with the users. The "agency" initially has only information about the aggregated quantities $F$ and $Q_k$. Information about the functions $v_i(x)$ and $q_i^k(x)$ is available only to the users. The values of $x_i^s$ and $q_i^k(x_i^s)$ are received from the users directly at the network central computer (working station), which computes the aggregated quantities $\sum_{i=1}^{N} x_i^s$ and $\sum_{i=1}^{N} q_i^k(x_i)$. Protected communications between computers prevent undesirable disclosure of information, which is confidential for each user. Compared with the previous price formation procedure based on monitoring data, these negotiations are also advantageous to the users because they avoid the negative effect on their economies of the high nonoptimal prices set in the intermediate stages of the price-formation procedure and allow them to realize the holistic approach to DSS development. This approach is treating users as a learning individuals whose way of reaching conclusions depends on the extent of activity during the decision making process.

Considered approach was verified for several model and real-data examples. One of them deals with situation in the Aral region at the end of the twentieth century (see [16]). Uncontrolled water supply in basins of the Syr Daria and Amu Daria rivers lead to the large-scaled ecological catastrophe, when the area of the Aral Sea decreased more then twice and the concentration of salt in its water increased by an order of magnitude. Calculations made for data from Chimby region in the delta of Amudaria demonstrated that introduction of the pay for water could prevent this catastrophe or, at list, essentially restrict its consequences. These calculations demonstrated that for the mentioned region the equilibrium price suggests that water use will decrease by as much as 30 % without, however, significant economic losses. Of course, some changes in the structure of agriculture will be result of the payment introduction. The production of cotton and rise (cultures with the largest water consumption) will fall close to zero. However, the growth of olive and livestock production could compensate this negative effect. The idea of creation of water-trade agency was intensively discussed in the Middle Asia countries during the last years of the twentieth century, but it was rejected due, mainly, by non-economic reasons.

Let return now to description of tools for negotiation support. The above-mentioned DSS is developed as open menu-driven software, available for the wide range of PC hardware and requiring no special computer training from the users. The system does not control the users. Rather, it lets the users' expertise control in decision-making process. Its other important features–modularity and extensibility–facilitate quick adaptation to special requirements.

The protocol of negotiations is automatically recorded under the "agency" control. Besides recording the actual sequence of propositions and other users' actions, some of the information discussed can be put aside with a view to drafting the final documents.

The current research dealing with this system concentrates on building efficient and robust optimization algorithms which can be applied instead of the procedure (4.73), (4.74), improving the user-computer interaction and network implementation of decision-support procedure. Such procedure can be both centralized and decentralized, as it is described below.

Procedure of bilateral trade of water use levels is proposed for the case when the "agency" cannot be created due to non-economic reasons. For instance, it can be applied for water distribution between users from different states with different legislation systems. Such situation was typical in the Aral Sea basin. Further, we consider such pricing procedure for the simple case, i.e., for the model (4.59)–(4.61), but it can be generalized for more complicate problem solution. The following assumptions have been made for these purposes.

Let us assume that all users' utility functions $v_i(x_i)$ are positive, concave, differentiable and the values of $\overline{x}_i$ are so large that constrains $x_i \leq \overline{x}_i$ can not be taken into account for every $i = \overline{1,N}$. In the bilateral trading mechanism, user with relatively low marginal utility from water consumption searches for a trading partner with relatively high marginal utility to sell the part of his limit for water use. The values of marginal utility become equal for both users after a purchase. This multistep process can go on as long as there are two or more users with different marginal utilities. Prices may and normally will differ between the sequential trades that are made, but under stationary conditions (see [17]), such a trading system converges to an equilibrium where total profit from water consumption are maximized, with marginal utility of all users equal to previously determined water price.

Now we will give the formal description of such procedure. Let $x^0 = (x_1^0, \ldots, x_N^0)$ be a vector of limits of water use that satisfies (4.60), (4.61), but may not be a maximum point of function (4.59). Let us describe the procedure for the construction of the next approximation $x^{k+1} = (x_1^{k+1}, \ldots, x_n^{k+1})$, if the value of $x^k = (x_1^k, \ldots, x_n^k)$ is known, $k = 0, 1, 2, \ldots$.

1. Consider any two users, $i_k$ and $j_k$ of the step $k$, such that $v'_{i_k}(x^k_{i_k}) \neq v'_{j_k}(x^k_{j_k})$. Without loss of generality, we can assume that user $j_k$ has the higher marginal utility, i.e., $v'_{j_k}(x^k_{j_k}) > v'_{i_k}(x^k_{i_k})$.
2. Find the value $\Delta_k$ such that

$$v_{j_k}\left(x^k_{j_k} + \Delta_k\right) + v_{i_k}\left(x^k_{i_k} - \Delta_k\right) = \max_{x^k_{i_k} \geq \Delta \geq 0} \left(v_{j_k}\left(x^k_{j_k} + \Delta\right) + v_{i_k}\left(x^k_{i_k} - \Delta\right)\right).$$

According to our assumptions for small enough $\Delta > 0$

$$v_{j_k}\left(x^k_{j_k} + \Delta\right) + v_{i_k}\left(x^k_{i_k} - \Delta\right) - v_{j_k}\left(x^k_{j_k}\right) - v_{i_k}\left(x^k_{i_k}\right) = \Delta\left(v'_{j_k}\left(x^k_{j_k}\right) - v'_{i_k}\left(x^k_{i_k}\right)\right) + o(\Delta) > 0$$

holds, therefore $\Delta_k > 0$ and $v'_{j_k}(x^k_{j_k} + \Delta_k) = v'_{i_k}(x^k_{i_k} - \Delta_k)$.

3. Determine

$$x_{j_k}^{k+1} = x_{j_k}^k + \Delta_k,\ x_{i_k}^{k+1} = x_{i_k}^k - \Delta_k,\ x_j^{k+1} = x_j^k,\ j = \overline{1,N},\ j \neq j_k,\ j \neq i_k.$$

Every $r$, $0 < r < r^k = v_{j_k}(x_{j_k}^k + \Delta_k) - v_{j_k}(x_{j_k}^k)$ can be a price of exchange of water use limits between the users $i_k$ and $j_k$. The convergence of the sequence $\{x^k\}$ to optimal solution $x^*$ of the problem (4.59)–(4.61) is proved similar to the procedure considered in [17]. It shows that $\frac{r^k}{\Delta_k} \xrightarrow[k\to\infty]{} u^*$, the optimal water price is determined by the algorithm (4.69).

It should be noted that the proposed approach can not be applied to the solution of the problem (4.59)–(4.61), (4.72). Examples of disconvergence of the sequence $\{x^k\}$ are presented in [17] for the similar problem dealing with markets for tradable emission.

## *Concluding Remarks*

Two main types of procedures determining a price for water use can be applied in specialized model-driven decision-support tools.

First type procedures are similar to a Walrasian auction. In the auction, there comes first stage in which the auctioneers determine their demands for water at given prices. This stage repeated for different prices until supply–demand equilibrium is achieved. As a result, the equilibrium price is found using the methods of nonsmooth optimization. Only after that the actual transactions are made in the second stage.

This approach can be realized in the frame of model-driven DSS based on computer network, which communicates all main water users. Creation of special "agency" is necessary for this realization. The proposed procedures are rather universal, but they are also complicated.

Bilateral trade is the second type of procedures. They converge to market equilibrium and maximize total utility under additional assumptions about users' utility functions and problem constrains. They are simple for realization, but not as universal as the procedures of the first type. These procedures do not need a creation of "agency" for water trade.

Both proposed approaches have its advantages and disadvantages, therefore their rational combination in the frame of distributed DSS can be applied for water resource management dealing with transboundary river basins.

## References


1. Alesina, A., Roubini, N., Cohen, G.: Political Cycles and the Macroeconomy. MIT, Cambridge (1997)
2. Arrow, K.: Social Choice and Individual Values. Wiley, New York (1951)
3. Aslund, A.: Why has Ukraine returned to economic growth? IER Working Paper No. 15 (2002)
4. Birge, J.R., Louveaux, F.: Introduction to Stochastic Programming. Springer, Berlin (1997)
5. Bouysson, D., Pirlot, H.: Conjoint Measurement Without Additivity and Transitivity. In: Advances in Decision Analysis, pp. 13–29. Kluwer Academic, Dordrecht (1999)
6. Bunday, B.D.: Basic Linear Programming. Edward Arnold, Bradford (1984)
7. Burakovskii, I.: International Trade and Economic Development of Countries with a Transition Economy. Theoretical and Methodological Analysis of Correlation Mechanism. Nichlava, Kiev (1998, in Ukrainian)
8. Cagan, P.: The monetary dynamics of hyperinflation. In: Friedman, M. (ed) Studies in the Quantity Theory of Money, pp. 25–117. Univ. Chicago Press, Chicago (1956)
9. Samchenko, V.V. (ed.): The Economy of Ukraine in 1993: Statistical Yearbook/Ministry of Statistics of Ukraine, 494 pp. Tekhnika, Kyiv (1994)
10. Samchenko, V.V. (ed.): The Economy of Ukraine in 1994: Statistical Yearbook/Ministry of Statistics of Ukraine, 519 pp. Tekhnika, Kyiv (1994)
11. CRU Steel Index. http://www.cruindices.com/
12. Diaz-Alejandro, C.: Good-bye financial repression, hello financial crash. J. Dev. Econ. **19**(1–2), 1–24 (1985)
13. Dolan, E.G., Lindsey, D.E.: Macroeconomics, 6th ed. Dryden Press, Chicago (1991)
14. Dombrovsky, M.: Is the economic growth in Ukraine sustainable? In: Cramon-Taubadel, S., Akimova, I. (eds) Fostering Sustainable Growth in Ukraine. Physica-Verlag, Heidelberg (2001)
15. Ermol'ev, Yu.M., Wets, R.J.: Numerical Technique for Stochastic Optimization Problems. Springer, Berlin (1988)
16. Ermol'ev, Yu.M., Mikhalevich, M.V., Uteuliev, N.U.: Economic Modeling of International Water Use (The Case of the Aral Sea Basin). International Institute of Applied Systems Analysis, Memorandum RR-95-5 (1995)
17. Ermol'ev, Yu.M., Michalevich, M., Nentjes, A.: Markets for tradeable emission and ambient permits: A dynamic approach. Environ. Resource Econ. **15**(1), 39–56 (2000)
18. Fischer, S., Sahay, R., Vegh, C.: Stabilization and growth in transition economies: the early experience. J. Econ. Perspect. **10**(2), 45–66 (1996)
19. Fishburn, P.C.: Nontransitive measurable utility. J. Math. Psychol. **26**, 31–67 (1982)
20. Fishburn, P.C.: Nontransitive preferences in decision theory. J. Risk Uncertain. **4**, 113–134 (1991)
21. Fishburn, P.C.: Preference structures and their numerical representation. Theor. Comput. Sci. **217**(2), 359–383 (1999)
22. Burakovsky, I.V.: International trade and economic development of countries in transition economics. Sc.D. Thesis overview, T. Shevchenko National University, Kyiv, 35 pp. (1998, in Ukrainian)
23. Friedman, M.: Money and the stock market. J. Polit. Econ. **96**(2), 221–245 (1988)
24. Gavrilenkov, E.: Achievements or missed opportunities: Factors of economic growth in Russia. What lessons are relevant to Ukraine? In: Cramon-Taubadel, S., Akimova, I. (eds) Fostering Sustainable Growth in Ukraine. Physica-Verlag, Heidelberg (2002)
25. Hetemaki, M, Kaski, E.L.: KESSU IV: An Econometric Model of the Finish Economy. Ministry of Finance, Helsinki (1992)
26. Huber, O.: Nontransitive multidimensional preferences. Theor. Decis. **10**, 147–165 (1979)
27. Hawrylyshyn, B.: Quarterly Predictions, #15, April 2001. The International Centre for Policy Studies, Kyiv, 13 pp. (2001)
28. Johnson, J.D., DiNardo, J.: Econometric Methods. McGraw-Hill, New York (1997)
29. Kall, P., Wallace, S.W.: Stochastic Programming. Wiley, Chichester (1994)

30. Karmanov, V.G.: Mathematical Programming. Nauka, Moscow (1980, in Russian)
31. Kolodko, G.W.: The World Economy and Great Post-Communist Change. Nova Science Publishers, New York (2006)
32. Kornai, J.: From Socialism to Capitalism: What Is Meant by the Change of System. Social Market Foundation, London (1998)
33. Koshlai, L.B., Mikhalevich, M.V.: Decision-support multifunctional system "Ukrainian Budget". In: Tangian, A., Gruber, J. (eds) A Construction and Applying Objective Functions, Lecture Notes in Economics and Mathematical Systems, vol. 510, pp. 349–366. Springer, Berlin (2001)
34. Krugman, P.: Toward a Counter-Revolution in Development Theory. In: Proceedings of the World Bank's Annual Conference on Development Economics, pp. 15–38, April 30–May 1, 1992, Washington, DC (1992)
35. Krugman, P.: Competitiveness: a dangerous obsession. Foreign Affairs **73**(2), 1–17 (1994)
36. Kuznetsov, Ye.: Multiply Inflation/Output Equilibria in the Noisy Economy. Research Memorandum RM-21, CEPAL, Santiago and ICRET, Moscow (1993)
37. Leontief, W.: Essays in Economics: Theories, Theorizing, Facts and Policies. Transaction, New Brunswick, Oxford (1985)
38. Mankovska, N., Guicci, R., Betliy, O.: The Influence of the Real Effective Exchange Rate on the Trade Balance in Ukraine. Ukrainian Economic Trends, UEPLAC Quarterly Issue (March 2002)
39. Mikhalevich, M.V.: Generalized stochastic method of centers. Cybernetics **16**(2), 292–296 (1980)
40. Mikhalevich, M.V.: Stochastic Approaches to Interactive Multicriteria Optimization Problems. WP-86-10. IIASA, Laxenburg (1986)
41. Mikhalevich, M.V., Koshlai, L.B.: Some Questions of Stability Theory of Stochastic Economical Models. Lecture Notes in Mathematics, vol. 982, pp. 123–136 (1983)
42. Mikhalevich, M., Koshlai, L.: On modeling of inflationary and postinflationary processes in Ukraine in 1990–1996. In: Kulikowski, N.Z., Owsinski, J. (eds) Economic Transformation and Integration: Problems, Arguments, Proposals, pp. 261–270, Systema Research Institute, Warsaw (1998)
43. Mikhalevich, V.S., Mikhalevich, M.V.: Dynamic pricing macromodels for a transition economy. Cybern. Syst. Anal. **31**(3), 409–420 (1995)
44. Mikhalevich, M.., Podolev, I.: modeling of selected aspects of the state's impact on pricing in a transitional economy. Working Paper WP-95-12. IIASA (1995)
45. Mikhalevich, M.V., Sergienko, I.V., Koshlai, L.B., Khmil, R.V.: Allowing for external factors in production-financial models of transition economy. Cybern. Syst. Anal. **32**(4), 511–518 (1996)
46. Mirzoakhmedov, F., Mikhalevich, M.V.: Applied Aspects of Stochastic Programming. Maorif, Dushanbe (1989, in Russian)
47. Movchan, V., Glucci, R.: Economic growth in Ukraine: do reforms matter? In: Factors of Economic Growth in Ukraine and Neighboring Countries, pp. 35–41, Altpress, Kiev (2004)
48. Nikaido, H.: Convex Structures and Economic Theory. Academic, New York (1967)
49. Pugachev, V.F., Pitelin, A.K.: Inflation in a technologically backward monopolized economy. Economika i mathematicheskye metody **31**(1), 30–35 (1995, in Russian)
50. Mikhalevich, M.V., Sergienko, I.V., Koshlai, L.B.: Simulation of foreign trade activity under transition economy conditions. Cybern. Syst. Anal. **37**(4), 515–532 (2001)
51. Stiglitz, J.E.: Globalization and Its Discontents. W.W. Norton & Company, New York (2002). ISBN 9780393051247
52. Sachs, J.D., Larrain, F.B.: Macroeconomics in the Global Economy. Prentice Hall International, Upper Saddle River (1993)
53. Sargent, T.I., Wallace, N.: Rational expectations and the dynamics of hyperinflation. Int. Econ. Rev. **14**, 328–350 (1973)
54. Shor, N.Z.: Minimization Methods for Non-differentiable Functions. Springer, New York (1985)

55. Shor, N.Z.: Nondifferentiable Optimization and Polynomial Problems. Kluwer Academic, Boston (1998)
56. Stiglitz, J.E.: Globalization and Its Discontents. W.W. Norton & Company, New York (2002)
57. Stiglitz, J., Weiss, A.: Credit rationing in markets with imperfect information. Am. Econ. Rev. **71**(3), 393–410 (1981)
58. Tsoukias, A., Ozturk, M., Vincke, P.: Preference Modelling. Springer, Berlin (2005)
59. Vasylenko, Y.: Consequences of Devaluation for Efficiency of Exports. Visnyk NBU (February 2001)
60. Zhang, W.B.: Economic Dynamics - Growth and Development. Springer, Berlin (1990)
61. Zhyliaev, I., Movchan, V.: Non-monetary settlements of the budget. In: Szyrmer, J. (ed) The Barter Economy: Non-Monetary Transactions in Ukraine's Budget Sector, pp. 23–32, Harvard University Ukraine Project (2000)

Printed in the United States
By Bookmasters